# CISM International Centre for Mechanical Sciences

## Courses and Lectures

## Volume 607

For more than 40 years the book series edited by CISM, "International Centre for Mechanical Sciences: Courses and Lectures", has presented groundbreaking developments in mechanics and computational engineering methods. It covers such fields as solid and fluid mechanics, mechanics of materials, micro- and nanomechanics, biomechanics, and mechatronics. The papers are written by international authorities in the field. The books are at graduate level but may include some introductory material.

Stéphane Vincent · Jean-Luc Estivalèzes ·
Ruben Scardovelli

# Small Scale Modeling and Simulation of Incompressible Turbulent Multi-Phase Flow

Springer

Stéphane Vincent
MSME
Université Gustave Eiffel
Marne-La-Vallée, France

Jean-Luc Estivalèzes
ONERA/DMPE
Université de Toulouse
Toulouse, France

Ruben Scardovelli
Dipartimento di Ingegneria Industriale
University of Bologna
Bologna, Italy

ISSN 0254-1971 ISSN 2309-3706 (electronic)
CISM International Centre for Mechanical Sciences
ISBN 978-3-031-09263-3 ISBN 978-3-031-09265-7 (eBook)
https://doi.org/10.1007/978-3-031-09265-7



This Springer imprint is published by the registered company Springer Nature Switzerland AG
The registered company address is: Gewerbestrasse 11, 6330 Cham, Switzerland

# Acknowledgements


## Collaborators

This work is a synthesis of more than 20 years of research. Most of the material of this course comes from works of our former Ph.D. students and postdocs. We would like to warmly thank them: Davide Zuzio, Pierre Trontin, Fredéric Couderc, Arthur Sartou, Bastien DiPierro, Jorge Brändlde de Motta, Amine Chadil, Céline Caruyer and Ali Ozel. We thank all contributors and collaborators, and in particular Jean-Paul Caltagirone, for their numerous and fruitful discussions. Jean-Paul Caltagirone was at the origin of the development of many ideas and collaborations summarized here.

## Technical Aspects

We are grateful for access to the computational facilities of the French CINES (National Computing Center for Higher Education), IDRIS (Institut du Développement et des Ressources en Informatique Scientifique, CNRS) CCRT (National computing center of CEA) under project number x20XX2b6115 and A00X2b06115 and CALMIP (CALcul en MIdi-Pyrénées) under project P0633. The Agence Nationale de la Recherche (ANR) through the MODEMI and Player projects is also associated with this work. The PARAMESH software used in this work was developed at the NASA Goddard Space Flight Center and Drexel University under NASA's HPCC and ESTO/CT projects and under grant NNG04GP79G from the NASA/AISR project.

## Financial Support

The present study has been granted by the Foundation STAE-RTRA-research program COFFECI. The authors are grateful to the STAE-RTRA Foundation for having funded this study and especially the postdoc of Dr. B. DiPierro.

# Contents

# Chapter 1
# Introduction and Motivations

Two-phase flows occur in many academic and applied fluid mechanics problems such as boiling crisis in a nuclear plant, chemical reactors, material coating by plasma projection, bubbles in pipes, wave breaking, oil extraction in porous media, powder and fluidized beds processes, fuel injection in engines to cite a few. Exhaustive presentation of various unsteady or turbulent multi-phase flows is given for example in [1, 2]. As soon as turbulence and interface interact in a non linear way with macroscopic interfacial deformations inducing ligaments, coalescence or rupture, the experimental characterization of these flows is difficult due to the heterogenous character of the multi-phase medium. Modeling and numerical simulation thus represent an interesting way to study the physical processes that control these flows.

The present book was written following a CISM course devoted to the modeling and numerical simulation of multi-phase flows with fictitious domain approaches (2005). The objective of the course was to bring an exhaustive presentation, discussion and validation of models and numerical methods for the simulation of multiphase problems for which the interfacial scale is resolved, i.e. the size of the mesh cell is smaller than the size of the interfacial structures. This book has also to be related to the special issue of Acta Mechanica published in 2019 [3].

The one-fluid model is the basis of the representation of moving and deformable interfaces, even if Ghost Fluid techniques are also undertaken. The typical scales of interfaces being assumed larger than the numerical resolution scale, i.e. the local size of the Eulerian flow mesh, in a kind of Direct Numerical Simulation of interfacial structures, while the turbulent flow characteristics can be modeled by means of DNS (in the sence) or Large Eddy Simulation (LES). Different interface tracking techniques are presented and compared as well as the management of capillary forces and incompressibility and solid constraints. In particular, penalty methods are extensively used for the numerical treatment of boundary conditions, pressure-velocity coupling or immersed obstacle representation. Various validations and applications are discussed in order to stress the capability of the model and numerical methods.

 
S. Vincent et al., *Small Scale Modeling and Simulation of Incompressible Turbulent Multi-Phase Flow*, CISM International Centre for Mechanical Sciences 607,
https://doi.org/10.1007/978-3-031-09265-7_1

In the present manuscript, much attention is paid on resolved scale fluid-fluid interfaces [4–20], with also informations on fluid-solid interactions and resolved scale particles, which can also be investigated with similar approaches [21–31]. Different aspects of modeling and simulation of multi-phase flows will be presented in a common concept for all types of multi-phase flows (free surfaces, liquid-liquid, particulate flows) that is called the fictitious domain approach. Attention will be paid to models, interface tracking, validations and applications.

## 1.1 Governing Equations for DNS of Multiphase Flows

The first consideration to be done concerns the scale and the nature of the physical phenomenon to be studied. In the case of fluid observation, the ratio between the mean free path and the characteristic length, known as the *Knudsen* number, determines whether a continuum ($Kn \leqslant 0.1$) or a discrete model ($Kn > 0.1$) of the fluid itself is possible. In the fist case the physical quantities are defined for a small control volume, and obtained by taking the mean of all the molecules contained in the volume. On the other case, it is not possible to neglect the strong variation of physical properties of the molecules and atoms; if they are rarefied enough, then a particle based approach must be employed. The fluid model used in this work is always a continuum model; the term fluid may without distinction refer to a liquid or a gas.

### 1.1.1 Mass Conservation

Under the continuum hypothesis, the Navier-Stokes equations describe the properties of the fluids as functions of space and time $(\mathbf{x}, t)$, which is from an Eulerian point of view. The first equation describes the temporal evolution of the fluid mass, or the density $\rho$ integrated at each instant of time in the control volume $\Omega$ included in the surface $\partial\Omega$. The conservation of the matter implies that, in absence of chemical reactions, each variation in time of the mass is balanced by the net flux crossing the surface

$$\frac{d}{dt}\int_{\Omega} \rho \, dV = -\int_{\partial\Omega} \rho . \boldsymbol{u} \, dS \tag{1.1}$$

where $V$ and $S$ are the unit of volume and surface, $\boldsymbol{n}$ is the normal vector oriented outside the surface. The Gauss-Ostrogradsky theorem affirms that the outward flux of a vector field through a closed surface is equal to the volume integral of the divergence on the region inside the surface. It allows the transformation of the right hand term into a volume integral:

$$\int_{\Omega} \left( \frac{\partial \rho}{\partial t} + \nabla.(\rho \boldsymbol{u}) \right) dV = 0 \tag{1.2}$$

Under the hypothesis of a continuous function, this can be reduced to the local equation of mass conservation:

$$\frac{\partial \rho}{\partial t} + \nabla.(\rho \boldsymbol{u}) = 0 \tag{1.3}$$

### 1.1.2 Momentum Conservation

In a similar way, the momentum $(\rho \boldsymbol{u})$ of the fluid particle of volume $\Omega$ and surface $\partial\Omega$ follows the Newton second law, which states that the net force on a particle is equal to the time rate of change of its linear momentum:

$$\frac{d}{dt} \int_{\Omega} \rho \boldsymbol{u} \, dV = \int_{\Omega} \rho \boldsymbol{f} \, dV + \int_{\partial\Omega} \boldsymbol{T}.\boldsymbol{n} \, dS \tag{1.4}$$

The tensor $\boldsymbol{T}$ contains the surface forces, while the vectorial function $f$ contains the volume forces, like gravity. The temporal derivative is a Lagrangian derivative, as it is written in a reference system which is placed on the volume particle and follows it. In order to return to an Eulerian description, the total derivative definition is applied:

$$\frac{d\boldsymbol{u}}{dt} = \frac{\partial \boldsymbol{u}}{\partial t} + (\boldsymbol{u}.\nabla)\boldsymbol{u} \tag{1.5}$$

By using the Gauss-Ostrogradsky theorem again, the Eq. 1.4 becomes

$$\rho \left( \frac{\partial \boldsymbol{u}}{\partial t} + (\boldsymbol{u}.\nabla)\boldsymbol{u} \right) = \nabla.\boldsymbol{T} + \boldsymbol{f} \tag{1.6}$$

### 1.1.3 Fluid Assumptions

Some assumptions can be made on the general form of the Navier-Stokes equations in order to simplify them. At first, given the typical velocities that appear in the liquid sheet disintegration are largely inferior to the speed of sound, both fluids can be considered as incompressible. The volumetric mass can be considered as constant in space and time, with a step discontinuity localized between the fluids. The time derivative of Eq. 1.3 becomes equal to zero, and the equation itself degenerate to the zero divergence condition

$$\nabla.(\rho \boldsymbol{u}) = 0 \tag{1.7}$$

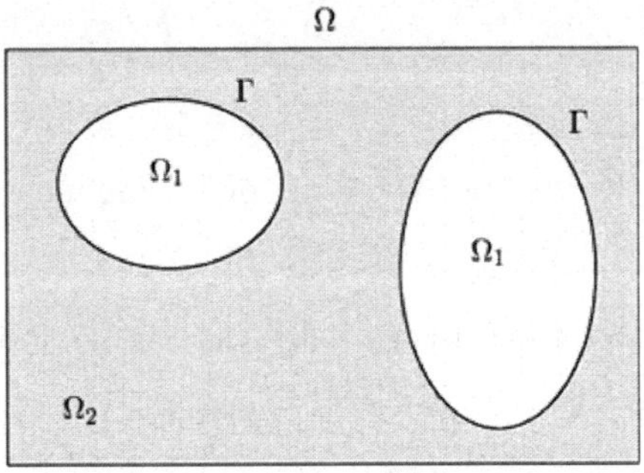


**Fig. 1.1** Representation of two fluids $\Omega_1$ and $\Omega_2$ and the interface which separates them, $\Omega = \Omega_1 \cup \Omega_2 \cup \Gamma$

The description of the surface forces is contained in the $\boldsymbol{T}$ tensor. It can be divided in two parts, the first including the isotropic normal stresses corresponding to the thermodynamic pressure, and the anisotropic stresses caused by the fluid viscosity:

$$\boldsymbol{T} = -p\mathbf{I} + \mathbf{D} \tag{1.8}$$

If the Newtonian fluid model is used together with the incompressibility assumption, the $\mathbf{D}$ tensor elements $e_{ij}$ depend on the instantaneous deformation values:

$$\boldsymbol{D}_{ij} = 2\mu e_{ij} = \mu \left( \frac{\partial u_i}{\partial x_j} + \frac{\partial u_j}{\partial x_i} \right) \tag{1.9}$$

## 1.2 Interface and Jump Conditions

Once the equations have been written for the for the two fluids, the two phase aspects must be dealt with. The computational domain is therefore divided in two regions (Fig. 1.1), which are separated by an interface. The interface as it is considered in this work is a separation zone of negligible thickness[1], where the function representing the physical quantities are discontinuous (or their derivatives are), so that the fluid properties jump instantaneously in space. This is reflected in the model by the imposition of jump conditions across an infinitely thin interface. For a space of dimension $n$, the interface is an object of dimension $n - 1$.

### *1.2.1 Surface Tension*

The cohesive Van Der Waals forces between liquid molecules are responsible for the force known as surface tension. The molecules inside the fluid share their cohesive force with all the surrounding neighbouring molecules, resulting subject to a zero

[1] Other approaches may consider it of finite, even if very small, thickness.

net total force. But molecules at the surface do not have other like molecules on all sides of them, and consequently they receive a net force which pulls them inwards. The inter molecular attractive forces on the surface are enhanced to achieve force balance: this enhancement is called surface tension. In energetic terms, the surface tension balance the energy loss due to the missing molecular bonds on the surface of the fluid; the minimization of this surface energy is the reason for the spherical shape of a droplet in absence of gravity, as it is the shape which minimizes the free surface. One of the consequences of this phenomenon is the generation of an overpressure inside the concave side of the surface. This difference of pressure is predicted by the Laplace's law

$$[p] = \sigma\left(\frac{1}{R_1} + \frac{1}{R_2}\right) = \sigma\kappa \tag{1.10}$$

where $R_{1,2}$ are the main curvature radius, $\kappa$ the associated curvature and $\sigma$ the surface tension coefficient, which depends on the difference of the molecular bond energy and $[p] = p_1 - p_2$ the pressure jump. A bubble or a droplet are always in an overpressure situation in comparison to the external environment.

### 1.2.2 Viscosity

The Laplace's law does not fully describe the jump conditions when the interface is moving: under the hypothesis of Newtonian fluids the viscous forces act on the interface, depending on the velocity derivatives. The tensor $\boldsymbol{T}$ contains these surface forces, their jumps being defined by

$$\begin{cases} \boldsymbol{n}.[\boldsymbol{T}].\boldsymbol{n} = \sigma\kappa \\ \boldsymbol{t}.[\boldsymbol{T}].\boldsymbol{n} = 0 \end{cases} \tag{1.11}$$

where $[\boldsymbol{T}] = \boldsymbol{T}_1 - \boldsymbol{T}_2$ represents the jump between the two fluids. The viscous stresses on the interface impose the continuity of the tangential velocity, while the conservation of mass imposes the continuity of the normal component, so that

$$[\boldsymbol{u}] = 0 \tag{1.12}$$

The balance of the tangential velocities comes from the balance of the stress force applied on the two sides of the interface, which are the product of the dynamic viscosity and the velocity gradient, which describes the jump in the first derivative of tangential velocity on the interface, as depicted in Fig. 1.2.

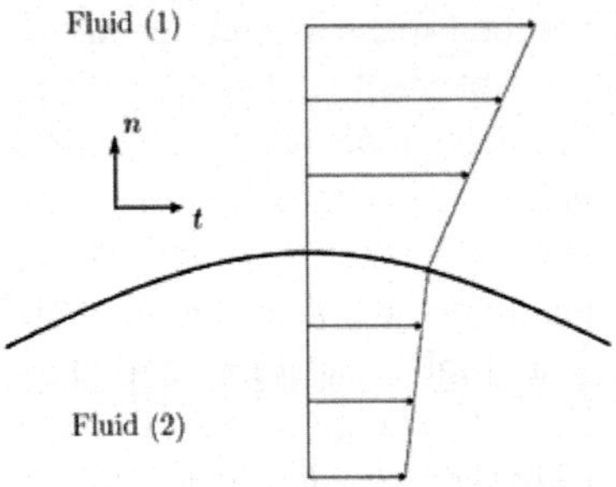


**Fig. 1.2** Representation of the tangential velocity profile, continuous on the interface but discontinuous in the first derivative

## 1.3 The Final Model

The final model consists into the incompressible Navier-Stokes equations solved in each of the two fluids. The computational domain is divided in the corresponding two regions separated by a sharp interface as in Fig. 1.1. In each of them the model is:

$$\begin{cases} \nabla \cdot \boldsymbol{u} = 0 \\ \dfrac{\partial \boldsymbol{u}}{\partial t} + (\nabla . \boldsymbol{u})\, \boldsymbol{u} = \dfrac{1}{\rho} (\nabla \cdot \mathbf{T} + \boldsymbol{f}) \end{cases} \tag{1.13}$$

With the following hypothesis:

$$\begin{cases} \mathbf{T} = -p\mathbf{I} + \mathbf{D} \\ \mathbf{D} = \mu \left( \nabla \boldsymbol{u} + (\nabla \boldsymbol{u})^T \right)) \end{cases} \tag{1.14}$$

where $\boldsymbol{u} = [u, v]^T$ represents the velocity vector field, $p$ the hydrodynamic pressure, $\rho$ the density, $\mu$ the dynamic viscosity of the fluid and $f$ the external forces like gravity. Across the interface, the jump conditions are imposed by capillarity and viscosity; the velocity respects a no-slip condition:

$$\begin{cases} [p] - \boldsymbol{n}.\left[\mu \left( \nabla \boldsymbol{u} + (\nabla \boldsymbol{u})^t \right)\right].\boldsymbol{n} = \sigma \kappa \\ \boldsymbol{t}.\left[\mu \left( \nabla \boldsymbol{u} + (\nabla \boldsymbol{u})^T \right)\right].\boldsymbol{n} = 0 \\ [\boldsymbol{u}] = 0 \end{cases} \tag{1.15}$$

Here the notation $[\cdot] = (\cdot)_1 - (\cdot)_2$ represents the jump across the interface, $\boldsymbol{n}$ the normal vector pointing from 1 to 2, $\boldsymbol{t}$ the tangential one. $\sigma$ is the surface tension, $\kappa$ the interface curvature.

# Chapter 2
# DNS of Resolved Scale Interfacial and Free Surface Flows with Fictitious Domains

The numerical simulation of multi-phase flows with resolved scale interfaces could be investigated following two different numerical strategies: unstructured or structured grids. This important choice is motivated by the representation of the complex interface shapes. On the one hand, the more natural solution seems to be the implementation of an unstructured body-fitted grid to simulate interface separating different phases present in the multi-phase flow. Building such a finite-volume or finite-element mesh in three-dimensions is not easy and requires automatic mesh generators as the interface moves and deforms according to time [32, 33]. The remeshing process at each calculation step is time consuming and can be very difficult to manage automatically in computer softwares when the shape of the interface is complex. On the other hand, it can be imagined to use a fixed structured grid to simulate the two-phase evolutions with coalescence and rupture of part of the interface. In this case, the mesh of the fluid domain is simple, i.e. a fixed Cartesian grid for example. The difficulty lies in the taking into account of the complex interface shape on a grid non conforming to the interface topology. This type of numerical problem belongs to the class of fictitious domains [34]. The modeling strategy developed hereafter is based on this approach.

Irregular Cartesian underlaying calculation grids are considered in the rest of the article for their easy programming and the possibility to refine the interaction zone between interface and walls for example. As presented in Fig. 2.1, the fictitious domain method consists in considering each different phase (for example air, water, solid particle, ...) as a fluid domain of specific rheological properties which is located by a phase function $C_i$, associated to each phase $i$. By definition, $C_i = 1$ in phase $i$ and 0 elsewhere. The interface between media or phases $i$ and $k$ is called $\Sigma_{ik}$.



S. Vincent et al., *Small Scale Modeling and Simulation of Incompressible Turbulent Multi-Phase Flow*, CISM International Centre for Mechanical Sciences 607,
https://doi.org/10.1007/978-3-031-09265-7_2

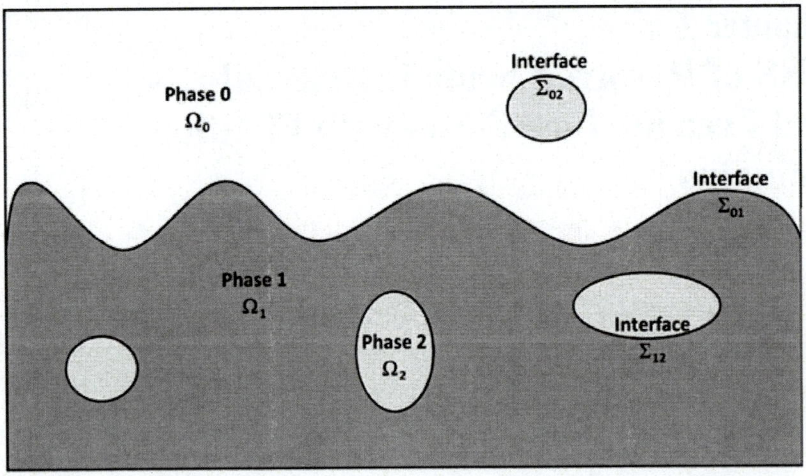


**Fig. 2.1** Definition sketch of the fictitious domain method for a multi-phase flow

## 2.1 One-Fluid Model

The modeling of incompressible multi-phase flows involving separated phases can be achieved by solving the incompressible Navier-Stokes equations with phase functions $C_i$. In the example provided in Fig. 2.1, only $i = 1, 2$ are considered as $C_0$ is obtained as $C_0 = 1 - C_1 - C_2$. As explained by Kataoka [35], the resulting model takes implicitly into account the jumps relations at the interface [18, 36] and the interface evolutions are described by an advection equation on function $C$:

$$\nabla \cdot \boldsymbol{u} = 0 \quad (2.1)$$

$$\rho(\frac{\partial \boldsymbol{u}}{\partial t} + (\boldsymbol{u} \cdot \nabla)\boldsymbol{u}) = -\nabla p + \rho \boldsymbol{g} + \nabla \cdot \big((\mu + \mu_t)(\nabla \boldsymbol{u} + \nabla^t \boldsymbol{u})\big) + \boldsymbol{F}_{st} \quad (2.2)$$

$$\frac{\partial C_i}{\partial t} + \boldsymbol{u} \cdot \nabla C_i = 0 \quad (2.3)$$

where $\boldsymbol{u}$ is the velocity, $p$ the pressure, $t$ the time, $\boldsymbol{g}$ the gravity vector, $\rho$ and $\mu$ respectively the density and the viscosity of the equivalent fluid. These are obtained in the one fluid model as functions of $C_i$ volume fractions. It reads for example for an arithmetic average [37]

$$\rho = C_2\rho_2 + (1 - C_2)C_1\rho_1 + (1 - C_2)(1 - C_1)\rho_0 \quad (2.4)$$

$$\mu = C_2\mu_2 + (1 - C_2)C_1\mu_1 + (1 - C_2)(1 - C_1)\mu_0 \quad (2.5)$$

Other density and viscosity averages could be utilized. Comparisons will be given in Sect. 2.5.2.

Concerning the turbulence modeling (see presentation of Lakehal [19, 20]), a deterministic Large Eddy Simulation (LES) approach can be used to take into account the under-resolved sub-grid scale turbulence structures by means of a mixed scale model [38] for example. This model is obtained as a combination of the Smagorinsky [39] and Turbulent Kinetic Energy (TKE) [40] models. It lies on a dissipative representation, through the turbulent viscosity $\mu_t$, of the small scale turbulent structures. It is formulated as

$$\mu_T = \rho C_S^{2\alpha} C_{TKE}^{1-\alpha} \overline{\Delta}^{1+\alpha} \left(2 S_{ij} S_{ij}\right)^{\alpha/2} \left(\frac{1}{2}\overline{\boldsymbol{u}'\boldsymbol{u}'}\right)^{\frac{1-\alpha}{2}} \tag{2.6}$$

where $\alpha$ is generally chosen equal to 0.5, $C_S$ is the Smagorinsky constant chosen in the range 0.15–0.2, $C_{TKE}$ is equal to 0.2 and the fluctuation of the sub-grid scales $\overline{\boldsymbol{u}'} = \overline{\boldsymbol{u}} - \widetilde{\overline{\boldsymbol{u}}}$ is the difference between the filtered velocity due to LES and the test filtered LES velocity. It results from the application of a specific test filter $\tilde{.}$ of compact support larger than the LES filter, applied to the LES filtered velocity $\overline{\boldsymbol{u}}$. The mixed scale model used in this work has been developed for single-phase flows. It is well-known that specific multi-phase LES terms should be modeled in addition to the classical inertial sub-grid terms [14, 15, 19, 41–43]. However, we consider that in the present course, the turbulent interactions (if the multi-phase flow is turbulent) between the two phases are negligible at the sub-grid scale. The framework of this course is the Direct Numerical Simulation (DNS) of multi-phase flows. Detailed analysis and discussions on LES of multi-phase flows will be given by D. Lakehal.

The surface tension forces are taken into account thanks to a volume force $F_{st} = \sigma \kappa \boldsymbol{n}_i \delta_i$. The surface tension coefficient $\sigma$ is assumed constant. The local curvature of the interface is $\kappa$ whereas the normal to the interface is $\boldsymbol{n}_i$ and $\delta_i$ is a Dirac function indicating interface. In order to close the one-fluid model, Brackbill and co-workers [44] have proposed to use the Continuum Surface Force (CSF) model to formulate the curvature and the normal to the interface as functions of the indicator functions $C_i$. The surface tension force so obtained reads

$$\boldsymbol{F}_{st} = \sigma \nabla \cdot \left(\frac{\nabla C_i}{\|\nabla C_i\|}\right) \nabla C_i \tag{2.7}$$

with $\kappa = \nabla \cdot \left(\frac{\nabla C_i}{\|\nabla C_i\|}\right)$ and $\boldsymbol{n}_i \delta_i = \nabla C_i$. Another possible approach to deal with capillary forces in the one-fluid model is to use directly the jump equations at the interface with a Ghost Fluid method [45]. Details on formulation and approximation of capillary forces will be given by J.-L. Estivalezes and R. Scardovelli.

The one-fluid model is almost identical to the classical incompressible Navier-Stokes equations, except that the local properties of the equivalent fluid ($\rho$ and $\mu$) depends on $C_i$, the interface location requires the solving of an additional equation and a specific volume force is added at the interface to account for capillary effects.

## 2.2 General Discretization and Solvers

The whole numerical discretizations and solvers are based on finite volumes [46], discretized on MAC staggered grids [47]. The time discretizations can be of explicit or implicit type depending on the code and numerical methods (projection or augmented Lagrangian). Details are given bellow in the following subsections.

### 2.2.1 *Pressure-Velocity Coupling and Solvers*

In this section the numerical method used to solve the Navier-Stokes momentum equations (2.9) is presented. At first the attention is focused on the resolution of a single phase flow; the treatment of the two phase jump conditions will be added later on. The unknown of the equation system are the pressure and velocity field $(p, \boldsymbol{u})$, in each phase. Some assumptions on the equations have already been done in the previous chapter: in particular, the assumption of constant density (2.8) imposes the choice of the numerical method among the incompressible solvers. If some advantages come from this formulation, such as a less restrictive time step, some difficulties arise as well. For example, the boundary conditions are not easy to define. Numerically, an elliptic equation must be solved, adding an expensive iterative solver to the algorithm; this problem becomes more and more difficult to solve as the composite mesh is used and the two phase aspect is considered.

$$\nabla \cdot \boldsymbol{u} = 0 \tag{2.8}$$

$$\frac{\partial \boldsymbol{u}}{\partial t} + (\nabla \boldsymbol{u})\, \boldsymbol{u} = \frac{1}{\rho} (\nabla \cdot \mathbf{T} + f) \tag{2.9}$$

Boundary conditions are needed to close the system; their description is given later in this chapter.

There are, in literature, different methods to solve this system of equation. Among these are the fundamental pressure correction methods (SIMPLE, SIMPLER), the artificial compressibility methods, the penalization methods. However, one of the most widely spread algorithms are the projection or predictor-corrector methods, developed by [48] and [49]. This procedure needs two resolution of equations plus a correction, when the variables are arranged in a staggered way. For this reason the variables in the code are so located, the pressure lies on the cell center, the $u$ component of the velocity vector on the right cell face and the $v$ component on the upper one (Fig. 2.2).

The projection method comes from the Helmholtz-Hodge theorem, which states that any velocity field $\boldsymbol{u}$ defined on a simply connected domain can be uniquely decomposed into a divergence free (solenoidal) part $\boldsymbol{u}_{\text{sol}}$ and an irrotational part $\boldsymbol{u}_{\text{irrot}}$. Given the mathematical property $\nabla \times \nabla \Psi = 0$, the velocity vector can be rewritten as the solenoidal part plus the gradient of a scalar function:

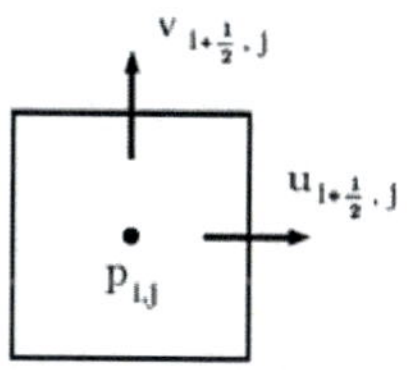


**Fig. 2.2** Position of the unknowns on the staggered mesh

$$\boldsymbol{u} = \boldsymbol{u}_{\text{sol}} + \boldsymbol{u}_{\text{irrot}} = \boldsymbol{u}_{\text{sol}} + \nabla\Psi \tag{2.10}$$

If the divergence of equation (2.10) is taken, the results is that, since $\nabla.\boldsymbol{u}_{\text{sol}} = 0$, the divergence of the velocity is equal to the Laplacian of the potential:

$$\nabla.\boldsymbol{u} = \Delta\Psi \tag{2.11}$$

This is the origin of the Poisson equation associated with this method, where the pressure is related to the potential $\Psi$ through a factor $\Delta t$.

**Projection method**

The Chorin projection method, as it was at first developed, de-couples velocity and pression by using first a totally explicit discretization for the effects of advection and diffusion in a predictor step, and then enforcing the compliance with a velocity divergence constraint obtained in a subsequent elliptic equation solve. The algorithm to advance the velocities from time $t$ to time $t + \Delta t$ can be summarized as follows:

- Prediction step: resolution of the momentum equation without pressure term

$$\boldsymbol{u}^* = \boldsymbol{u}^n + \Delta t^n \left( -\left(\boldsymbol{u}^n.\nabla\right)\boldsymbol{u}^n + \frac{\mu}{\rho}\nabla.\left(\nabla\boldsymbol{u}^n + (\nabla\boldsymbol{u}^n)^T\right)\right) \tag{2.12}$$

- Pressure computation: resolution of a Poisson equation with the divergence of the predicted velocities as right hand side

$$\nabla.\left(\frac{1}{\rho}\nabla p^{n+1}\right) = \frac{\nabla.\boldsymbol{u}^*}{\Delta t^n} \tag{2.13}$$

- Correction step: correction of the intermediate velocities by the pressure gradient

$$\boldsymbol{u}^{n+1} = \boldsymbol{u}^* - \frac{\Delta t^n}{\rho}\nabla p^{n+1} \tag{2.14}$$

This results in a non linear parabolic advection equation for the momentum without pressure terms (non incremental form) and an elliptic equation for the pressure to solve. The third step is a correction of the velocity vector $\boldsymbol{u}^*$ with the pressure gradient to make it a zero divergence field. The temporal convergence order of this original algorithm is, following the work of [50, 51], an $O(\Delta t)$ for both velocity and pressure, when using Dirichlet boundary conditions for $\boldsymbol{u}^n$ and $\boldsymbol{u}^*$ and homogeneous Neumann for the second. If the order of convergence can be raised for the velocity with a more accurate $\partial \boldsymbol{u}/\partial t$ discretization, the imposition of the numerical boundary layer from the $\boldsymbol{n}\nabla.p^{n+1}$ condition does not allow the same for the pressure. It has been found in the same works that the order of convergence can be raised to two by two means, the first being to incorporate in the prediction step the gradient of pressure $\nabla p^n$ (incremental form), the second an implicit treatment of the viscous terms $\boldsymbol{T}^n$ (which should in this case be called $\boldsymbol{T}^*$). However, in order to avoid the construction of the linear system associated to this implicit solution, an explicit form is maintained.

Other type of qpproaches for handling incompressible or compressible two-phase flows with be undertaken in the next sections. They will be associated to coupled solving of all velocity unknowns with time spliting, *i.e.* augmented Lagrangian like methods [52–57].

**Numerical resolution**

For sake of clarity, we will deal with 2D Navier-Stokes equations, considering that 3D extensions are quite straightforward.

**The first approximation**

The discretization on the staggered mesh of the Eqs. (2.12), (2.13), (2.14) becomes respectively:

- Prediction step

$$
\begin{aligned}
u^*_{i+\frac{1}{2},j} &= u^n_{i+\frac{1}{2},j} - \Delta t^n \left( u^n \frac{\partial u^n}{\partial x} + v^n \frac{\partial u^n}{\partial y} \right)_{i+\frac{1}{2},j} + \Delta t^n \frac{\mu}{\rho} (\nabla.\nabla u^n)_{i+\frac{1}{2},j} \\
v^*_{i,j+\frac{1}{2}} &= v^n_{i,j+\frac{1}{2}} - \Delta t^n \left( u^n \frac{\partial v^n}{\partial x} + v^n \frac{\partial v^n}{\partial y} \right)_{i,j+\frac{1}{2}} + \Delta t^n \frac{\mu}{\rho} (\nabla.\nabla v^n)_{i,j+\frac{1}{2}}
\end{aligned}
\tag{2.15}
$$

- Pressure step

$$
\begin{aligned}
&\frac{\partial}{\partial x}\left(\frac{1}{\rho}\frac{\partial p^{n+1}}{\partial x}\right)_{i,j} + \frac{\partial}{\partial y}\left(\frac{1}{\rho}\frac{\partial p^{n+1}}{\partial y}\right)_{i,j} = \\
&\frac{1}{\Delta t^n}\left( \frac{u^*_{i+\frac{1}{2},j} - u^*_{i-\frac{1}{2},j}}{\Delta x} + \frac{v^*_{i,j+\frac{1}{2}} - v^*_{i,j-\frac{1}{2}}}{\Delta y} \right)
\end{aligned}
\tag{2.16}
$$

- Correction step

$$
\begin{aligned}
u^{n+1}_{i+\frac{1}{2},j} &= u^*_{i+\frac{1}{2},j} - \frac{\Delta t^n}{\rho_{i+\frac{1}{2},j}} \left( \frac{p^{n+1}_{i+1,j} - p^{n+1}_{i,j}}{\Delta x} \right) \\
v^{n+1}_{i,j+\frac{1}{2}} &= v^*_{i,j+\frac{1}{2}} - \frac{\Delta t^n}{\rho_{i,j+\frac{1}{2}}} \left( \frac{p^{n+1}_{i,j+1} - p^{n+1}_{i,j}}{\Delta y} \right)
\end{aligned}
\tag{2.17}
$$

- The pressure equation
  A standard five points, second order centered discretization is used for the Laplacian operator in Eq. (2.16):

$$\frac{\partial}{\partial x}\left(\frac{1}{\rho}\frac{\partial p^{n+1}}{\partial x}\right)_{i,j}+\frac{\partial}{\partial y}\left(\frac{1}{\rho}\frac{\partial p^{n+1}}{\partial y}\right)_{i,j}=$$
$$\frac{1}{\rho_{i+\frac{1}{2},j}}\left(\frac{p^{n+1}_{i+1,j}-p^{n+1}_{i,j}}{\Delta x^2}\right)-\frac{1}{\rho_{i-\frac{1}{2},j}}\left(\frac{p^{n+1}_{i,j}-p^{n+1}_{i-1,j}}{\Delta x^2}\right)$$
$$\frac{1}{\rho_{i,j+\frac{1}{2}}}\left(\frac{p^{n+1}_{i,j+1}-p^{n+1}_{i,j}}{\Delta y^2}\right)-\frac{1}{\rho_{i,j-\frac{1}{2}}}\left(\frac{p^{n+1}_{i,j}-p^{n+1}_{i,j-1}}{\Delta y^2}\right) \tag{2.18}$$

  Under the hypothesis $\Delta x = \Delta y$, this discretization leads to a linear system with a symmetrical matrix of five diagonals in 2D. The resolution of this equation will be detailed in the forthcoming chapter , where all the problems that arises with the use of the adaptive mesh are discussed.
- The temporal derivative
  There are multiple choices for the temporal resolution of a partial derivative equation. The advection/diffusion equation is clearly simpler than the full Navier-Stokes system. Among the different methods, the Runge-Kutta and the linear multi-step methods are the most spread. The more performing high order Runge-Kutta methods have a severe drawback, however: they require as many applications of the methods as the desired order of accuracy. This would mean, for the incompressible solver, multiple solving of the pressure equation, which, in turn, costs on average more than three quarter of the total computational time (up to more than 95% in the most difficult cases). This is not a problem for the multi-step methods, where the results at the previous time steps, instead of computed intermediate ones, are used to improve the accuracy of the solution.
  The Adams-Bashford scheme belong to this category. It numerically solves the problem

$$\frac{\partial \boldsymbol{u}}{\partial t}=P(\boldsymbol{u}) \tag{2.19}$$

  where $P(\boldsymbol{u})$ is the projection operator applied to the velocity $\boldsymbol{u}$. The time integration gives:

$$\boldsymbol{u}^{n+1}-\boldsymbol{u}^{n}=\Delta t^{n}\int_{t^n}^{t^{n+1}} P(\boldsymbol{u}(t))\,dt \tag{2.20}$$

  One approximation of $P(\boldsymbol{u})$ between the two instants of time can be obtained by a Lagrange polynomial fit:

$$L(t)=P(\boldsymbol{u}^{n-1})\left(\frac{t-t^{n}}{t^{n-1}-t^{n}}\right)+P(\boldsymbol{u}^{n})\left(\frac{t-t^{n-1}}{t^{n}-t^{n-1}}\right) \tag{2.21}$$

  so that the integral (2.20) can be approximated as

$$
\begin{aligned}
\int_{t^n}^{t^{n+1}} L(t)\,dt &\approx \frac{\Delta t^n}{2}\left(L(t^{n+1}) + L(t^n)\right) \\
&\approx \frac{\Delta t^n}{2}\left(P(\boldsymbol{u}^{n-1})\left(\frac{t^{n+1}-t^n}{t^{n-1}-t^n}\right) + P(\boldsymbol{u}^n)\left(\frac{t^{n+1}-t^{n-1}}{t^n-t^{n-1}}\right) + P(\boldsymbol{u}^n)\right) \\
&\approx \frac{\Delta t^n}{2}(P(\boldsymbol{u}^n)\left(\frac{\Delta t^n + 2\Delta t^{n-1}}{\Delta t^{n-1}}\right) - P(\boldsymbol{u}^{n-1})\left(\frac{\Delta t^n}{\Delta t^{n-1}}\right)
\end{aligned}
\tag{2.22}
$$

which allows the writing of the final form of the Adams-Bashford scheme described in [58]:

$$
\boldsymbol{u}^{n+1} = \boldsymbol{u}^n + \frac{\Delta t^n}{2}\left(P(\boldsymbol{u}^n)\left(\frac{\Delta t^n + 2\Delta t^{n-1}}{\Delta t^{n-1}}\right) - P(\boldsymbol{u}^{n-1})\left(\frac{\Delta t^n}{\Delta t^{n-1}}\right)\right) \tag{2.23}
$$

One last approximation, $\Delta t^n \approx \Delta t^{n-1}$, leads the simplified form

$$
\boldsymbol{u}^{n+1} = \boldsymbol{u}^n + \frac{\Delta t^n}{2}\left(3\,P(\boldsymbol{u}^n) - P(\boldsymbol{u}^{n-1})\right) \tag{2.24}
$$

- The advection terms
  The spatial derivatives of the velocity of the non linear advective term $\boldsymbol{u}.\nabla\boldsymbol{u}$ is discretized using the same fifth order WENO5 scheme from [59] that we used for level-set advection. The details about this discretization are given in Chap. 3, where the linear advection equation solver is presented. For the velocities it is however necessary to interpolate the $u$ and $v$ components because of the staggered mesh and the location of the advected variables.

$$
\begin{aligned}
&u_{i,j+1/2} = 1/4\left(u_{i+1/2,j} + u_{i-1/2,j} + u_{i+1/2,j+1} + u_{i-1/2,j+1}\right) \\
&v_{i+1/2,j} = 1/4\left(v_{i,j+1/2} + v_{i,j-1/2} + v_{i+1,j+1/2} + v_{i+1,j-1/2}\right) \\
&u_{i,j} = 1/2\left(u_{i-1/2,j} + u_{i+1/2,j}\right) \quad v_{i,j} = 1/2\left(v_{i,j+1/2} + v_{i,j-1/2}\right)
\end{aligned}
\tag{2.25}
$$

- The viscous terms
  The discretization of the viscous terms is totally explicit second order centered scheme

$$
\begin{aligned}
&(\nabla.\nabla u^n)_{i+\frac{1}{2},j} = \\
&\frac{u^n_{i+\frac{3}{2},j} - 2u^n_{i+\frac{1}{2},j} + u^n_{i-\frac{1}{2},j}}{\delta x^2} + \frac{u^n_{i+\frac{1}{2},j+1} - 2u^n_{i+\frac{1}{2},j} + u^n_{i+\frac{1}{2},j-1}}{\delta y^2} \\
&(\nabla.\nabla v^n)_{i,j+\frac{1}{2}} = \\
&\frac{v^n_{i+1,j+\frac{1}{2}} - 2v^n_{i,j+\frac{1}{2}} + v^n_{i-1,j\frac{1}{2}}}{\delta x^2} + \frac{v^n_{i,j+\frac{3}{2},j+1} - 2v^n_{i,j+\frac{1}{2}} + v^n_{i,j-\frac{1}{2}}}{\delta y^2}
\end{aligned}
\tag{2.26}
$$

**Time step**

In order to ensure the temporal stability of the explicit projection step, an adaptative time step $\Delta t^n$ in respect of the convective, viscous, gravitational and capillary effects can be defined as in [60]:

$$\begin{aligned} C_{\text{conv}} &= \max\left(\frac{\|u\|_\infty}{\Delta x}, \frac{\|v\|_\infty}{\Delta y}\right) \\ C_{\text{visc}} &= \max\left(\frac{\mu_l}{\rho_l}, \frac{\mu_g}{\rho_g}\right)\left(\frac{2}{\Delta x^2} + \frac{2}{\Delta y^2}\right) \\ C_{\text{grav}} &= \sqrt{\max\left(\text{abs}\left(\frac{g_x}{\Delta x}\right), \text{abs}\left(\frac{g_y}{\Delta y}\right)\right)} \\ C_{\text{capl}} &= \frac{\sigma \|\kappa\|_\infty}{\min\left(\rho_g, \rho_l\right) \min\left(\Delta x, \Delta y\right)^2} \end{aligned} \tag{2.27}$$

that, combined, give

$$\Delta t^n \left(C_{\text{conv}} + C_{\text{visc}} + \sqrt{(C_{\text{conv}} + C_{\text{visc}})^2 + 4(C_{\text{grav}} + C_{\text{capl}})}\right) \leq 2\,\text{CFL} \tag{2.28}$$

where the parameter CFL is typically less or equal to 0.5, depending on the simulation parameters.

## 2.2.2 *Jump Conditions*

Once the single phase Navier-Stokes has been set up, the next step is to realize the coupling between the two fluids. The jump conditions given by the physical model are here taken from Sect. 1.1:

$$[p] - \boldsymbol{n}.\left[\mu\left(\nabla \boldsymbol{u} + (\nabla \boldsymbol{u})^T\right)\right].\boldsymbol{n} = \sigma\kappa \tag{2.29}$$

$$\boldsymbol{t}.\left[\mu\left(\nabla \boldsymbol{u} + (\nabla \boldsymbol{u})^T\right)\right].\boldsymbol{n} = 0 \tag{2.30}$$

$$[\boldsymbol{u}] = 0 \tag{2.31}$$

The jump conditions are treated by the Ghost Fluid method, as detailed in Sect. 1.1 compute the jumps for the new velocity field $\boldsymbol{u}^{n+1}$. The objective is to get a good coupling between the two fluids for the search of a final zero divergence velocity field.

**Interface geometrical properties**

The information given by the Level Set is the distance of the grid point to the interface in each cell. The imposition of the jump conditions require two more informations, which are relatively easy to derive from a Level Set formulation, as, differently from methods based on VOF, no reconstruction of the interface is needed. These are the normal (and consequently the tangential) vector and the curvature, that take part in the computation of pressure jump in the Laplace law. The normal vector and the curvature can be derived from the derivatives of $\phi$:

$$\boldsymbol{n} = \frac{\nabla \phi}{\|\nabla \phi\|} \tag{2.32}$$

$$\kappa = \nabla \cdot \left( \frac{\nabla \phi}{\|\nabla \phi\|} \right) \tag{2.33}$$

The divergence of gradient of the Level Set function is evaluated as $\nabla.(\nabla \phi) = 2\phi_x \phi_y \phi_{xy} - \phi_x^2 \phi_{yy} - \phi_{xx} \phi_y^2$. The first and second derivatives can be disretized with a centered second order scheme as follows:

$$\begin{aligned} \phi_x &= \frac{\phi_{i+1,j} - \phi_{i-1,j}}{2\Delta x} \qquad \phi_{xx} = \frac{\phi_{i+1,j} + \phi_{i-1,j} - 2\phi_{i,j}}{\Delta x^2} \\ \phi_y &= \frac{\phi_{i,j+1} - \phi_{i,j-1}}{2\Delta y} \qquad \phi_{yy} = \frac{\phi_{i,j+1} + \phi_{i,j-1} - 2\phi_{i,j}}{\Delta y^2} \\ \phi_{xy} &= \frac{\phi_{i+1,j+1} + \phi_{i-1,j-1} - \phi_{i-1,j+1} - \phi_{i+1,j-1}}{4\Delta x \Delta y} \end{aligned} \tag{2.34}$$

while the denominator of the Eqs. (2.32) and (2.33) can be expressed as

$$\|\nabla \phi\| = \left(\phi_x^2 + \phi_y^2\right)^{3/2} \tag{2.35}$$

If this quantity is zero, the derivatives are computed with a first order de-centered scheme.

**Evaluation of jumps**

- The pressure jump
  The Ghost Fluid allows an explicit treatment of jump conditions across an interface. In order to use this method the discontinuities of the fluid properties and unknowns as well as their first derivatives must be explicitly evaluated in each direction. Given a two dimensional case where the vector $\boldsymbol{u}$ contains the components $(u, v)$ and the normal vector the components $(n_x, n_y)$, the diffusion term of Eq. (2.29) can be written as

$$\boldsymbol{n}.\left(\nabla \boldsymbol{u} + (\nabla \boldsymbol{u})^T\right) = \begin{pmatrix} \nabla u.\boldsymbol{n} \\ \nabla v.\boldsymbol{n} \end{pmatrix} + \begin{pmatrix} n_y \frac{\partial v}{\partial x} - n_y \frac{\partial u}{\partial y} \\ n_x \frac{\partial u}{\partial y} - n_x \frac{\partial v}{\partial x} \end{pmatrix} \tag{2.36}$$

  and then

$$\boldsymbol{n}.\left(\nabla \boldsymbol{u} + (\nabla \boldsymbol{u})^T\right).\boldsymbol{n} = \begin{pmatrix} \nabla u.\boldsymbol{n} \\ \nabla v.\boldsymbol{n} \end{pmatrix}.\boldsymbol{n} \tag{2.37}$$

Then the normal jump in equation (2.29) can be expressed as

$$[p] - \left[ 2\mu \begin{pmatrix} \nabla u.\boldsymbol{n} \\ \nabla v.\boldsymbol{n} \end{pmatrix} \right] .\boldsymbol{n} = \sigma \kappa \tag{2.38}$$

[60] develops the equation which, with the incompressibility of the fluids, becomes

$$\left[ \begin{pmatrix} \nabla u.\boldsymbol{n} \\ \nabla v.\boldsymbol{n} \end{pmatrix} .\boldsymbol{n} \right] = 0 \tag{2.39}$$

The pressure jump can be written as

$$[p] - 2\,[\mu] \begin{pmatrix} \nabla u.\boldsymbol{n} \\ \nabla v.\boldsymbol{n} \end{pmatrix} .\boldsymbol{n} = \sigma \kappa \tag{2.40}$$

which is easy to implement on the code once the normal direction computed: the discontinuity is now relative to $p$ and $\mu$ alone.

- The viscous term jump
  In the (2.30) the tangential equation allows the computation of the velocity tangential derivatives across the interface. The passage from a Cartesian base formulation to a local base one has been avoided by [60], who give the following Cartesian based system of equations

$$\begin{aligned} \begin{pmatrix} [\mu u_x] & [\mu u_y] \\ [\mu v_x] & [\mu v_y] \end{pmatrix} &= [\mu] \begin{pmatrix} \nabla u \\ \nabla v \end{pmatrix} \begin{pmatrix} 0 \\ \boldsymbol{t} \end{pmatrix}^T \begin{pmatrix} 0 \\ \boldsymbol{t} \end{pmatrix} \\ &+ [\mu]\, \boldsymbol{n}^T \boldsymbol{n} \begin{pmatrix} \nabla u \\ \nabla v \end{pmatrix} \boldsymbol{n}^T \boldsymbol{n} \\ &- [\mu] \begin{pmatrix} 0 \\ \boldsymbol{t} \end{pmatrix}^T \begin{pmatrix} 0 \\ \boldsymbol{t} \end{pmatrix} \begin{pmatrix} \nabla u \\ \nabla v \end{pmatrix}^T \boldsymbol{n}^T \boldsymbol{n} \end{aligned} \tag{2.41}$$

  This equation allows the computation of the jump on the velocity derivatives from the computations of the Ghost Fluid forcing terms. The Navier Stokes equations can now be solved in each fluid independently, as it is coupled to the other by the imposition of the jump conditions which act as boundary conditions for the interface border.

### 2.2.3 Boundary Conditions

In this section the ensemble of the boundary conditions for the pressure and velocity used for the simulations are presented. They are necessary to close the equation system, and they must be carefully chosen in order to maintain the convergence rate of the solver. Once Dirichlet boundary conditions are imposed in the projected and

Fig. 2.3 Variables and ghost values on the right boundary

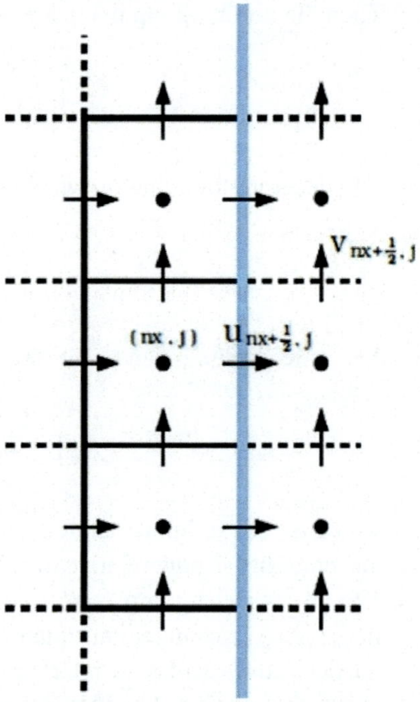


corrected velocity $\boldsymbol{u}^*$ and $\boldsymbol{u}^{n+1}$, the pressure conditions are imposed by the correction equation (2.14) as homogeneous Neumann in the normal direction:

$$\boldsymbol{n}.\nabla p^{n+1} = \frac{\rho}{\Delta t^n}(\boldsymbol{n}.\boldsymbol{u}^* - \boldsymbol{n}.\boldsymbol{u}^{n+1}) = 0 \tag{2.42}$$

and the convergence rate of the solver is formally set. For the others conditions it is more difficult to establish, in particular for the inflow and outflow conditions.

In the solver guardcells are allocated to store the boundary values. The number of allocated cells depends upon the spatial extension of the numerical stencil. The number is imposed to 3 by the WENO spatial discretization scheme of the velocity. Only one cell is needed by the Poisson equation discretization. The image 2.3 shows the localization of boundary values on the right boundary. This is taken as an example to show the different kind of conditions; the other three boundaries are treated in the same way with the opportune index change.

**Periodic conditions**

These are the simplest boundary conditions, as they should represent an infinite domain which repeats indefinitely in space. They are obtained by replacing the boundary cells values with the first values on the opposite face of the computed vector:

$$\begin{aligned} u_{nx+1/2,j} &= u_{1/2,j} \qquad & v_{nx+1/2,j} &= v_{1/2,j} \\ u_{nx+3/2,j} &= u_{3/2,j} \qquad & v_{nx+3/2,j} &= v_{3/2,j} \qquad & p_{nx+1,j} &= p_{1,j} \\ u_{nx+5/2,j} &= u_{5/2,j} \qquad & v_{nx+5/2,j} &= v_{5/2,j} \end{aligned} \tag{2.43}$$

These conditions are useful for convergence tests, because they are intrinsically "exact". They are also useful for temporal simulations where a fixed wavelength of a phenomenon is followed in time.

**Slip conditions**

Slip conditions are used when the physical boundary layer on a wall is negligible if compared to the boundary cell size. In this way the tangential velocity is not zero, but the Navier-Stokes solution value. The normal velocity is equal to zero; as the tangential velocity on the staggered mesh is not positioned on the real boundary but shifted of a half cell, symmetry conditions are used for this component:

$$\begin{aligned} u_{nx+1/2,j} &= 0 \qquad v_{nx+1/2,j} = v_{nx-1/2,j} \\ u_{nx+3/2,j} &= 0 \qquad v_{nx+3/2,j} = v_{nx-3/2,j} \\ u_{nx+5/2,j} &= 0 \qquad v_{nx+5/2,j} = v_{nx-5/2,j} \end{aligned} \tag{2.44}$$

The condition is applied to both $\boldsymbol{u}^*$ and $\boldsymbol{u}^{n+1}$, so the pressure has to respect the Eq. (2.42) and its boundary condition is an homogeneous Neumann.

**No slip conditions**

No slip conditions impose the respect of zero tangential velocity on the walls, as for the normal component. It is used when the physical boundary layer is important and has to be numerically taken in account. As for the previous case, the staggered mesh does not allow a direct imposition of tangential velocity on the physical boundary. The value in this location can be imposed by an interpolation of the velocity nodes on both sides. Moreover, the wall may move with its own speed. The usual method to adapt the no slip conditions to a wall moving at velocity $V$ is to mirror them across the boundary:

$$\begin{aligned} u_{nx+1/2,j} &= 0 \qquad v_{nx+1/2,j} = 2V - v_{nx-1/2,j} \\ u_{nx+3/2,j} &= 0 \qquad v_{nx+3/2,j} = 2V - v_{nx-3/2,j} \\ u_{nx+5/2,j} &= 0 \qquad v_{nx+5/2,j} = 2V - v_{nx-5/2,j} \end{aligned} \tag{2.45}$$

As in the slip conditions, it is adapted to both $\boldsymbol{u}^*$ and $\boldsymbol{u}^{n+1}$, and the pressure respects a zero gradient condition again.

**Inflow conditions**

An inflow condition imposes a mass flow entering the domain with a defined velocity profile $u_{inj}(y)$, perpendicular to the boundary. If the flow is turbulent, some models have to be used, but none of them is employed in this work. Theoretically the pressure should respect a non zero gradient depending on the diffusive term

$$\boldsymbol{n}.\nabla p^{n+1} = \mu \frac{\partial u_{inj}(y)}{\partial y} \tag{2.46}$$

If the same condition is applied to the projected velocity too, the pressure will follow a zero gradient condition. However, different conditions can be applied to the projected and corrected velocities in order to better define the pressure gradient

$$\frac{\rho}{\Delta t^n}(\boldsymbol{n}.\boldsymbol{u}^* - \boldsymbol{n}.\boldsymbol{u}^{n+1}) = \frac{\rho}{\Delta t^n}(\boldsymbol{n}.\boldsymbol{u}^* - u_{inj}) = \boldsymbol{n}.\nabla p^{n+1} \tag{2.47}$$

which gives the new boundary conditions

$$\begin{aligned} \boldsymbol{n}.\boldsymbol{u}^{n+1} &= u_{inj}(y) \\ \boldsymbol{n}.\boldsymbol{u}^* &= u_{inj}(y) + \frac{\Delta t^n}{\rho}\mu\frac{\partial u_{inj}(y)}{\partial y} \\ \boldsymbol{n}.\nabla p^{n+1} &= \mu\frac{\partial u_{inj}(y)}{\partial y} \end{aligned} \tag{2.48}$$

However, into the projection method the right hand side of the pressure equation make the term $\nabla.\boldsymbol{u}^*$ and $\boldsymbol{n}.\nabla p^{n+1}$ to elide themselves, so that the actual conditions revert to

$$\begin{aligned} \boldsymbol{n}.\boldsymbol{u}^{n+1} &= u_{inj}(y) \quad \boldsymbol{n}.\nabla.(\boldsymbol{t}.\boldsymbol{u}^{n+1}) = 0 \\ \boldsymbol{n}.\boldsymbol{u}^* &= u_{inj}(y) \quad \boldsymbol{n}.\nabla.(\boldsymbol{t}.\boldsymbol{u}^*) = 0 \\ \boldsymbol{n}.\nabla p^{n+1} &= 0 \end{aligned} \tag{2.49}$$

**Outflow conditions**

The outflow conditions should represent an "interruption" on the computational domain as it were a window for the observation of the flow. This is a numerical problem, because there are no informations available from the outside of the domain, so that no discretization stencil can be built, even at first order. In [58] two main axes of research for this problem are described. The first involves the resolution of a linear advection equation with the estimated advection speed on the boundary. Knowing the injected mass flow, this velocity can be evaluated to match the total mass flow. This method is reported, however, to generate a numerical aspiration towards the boundaries. Moreover, it is very difficult to adapt it to the two phase flow simulations. In the second approach, the mass conservation is imposed for $\boldsymbol{u}^{n+1}$ on all the last cells belonging to the computational domain, so that all the mass entering the boundary cells is forced to flow outside. The ensemble of the conditions becomes then

$$\begin{aligned} \nabla.\boldsymbol{u}^{n+1} &= 0 \qquad \boldsymbol{n}.\nabla.(\boldsymbol{t}.\boldsymbol{u}^{n+1}) = 0 \\ \boldsymbol{n}.\boldsymbol{u}^* &= 0 \quad \boldsymbol{n}.\nabla.(\boldsymbol{t}.\boldsymbol{u}^*) = 0 \\ p^{n+1} &= 0 \end{aligned} \tag{2.50}$$

The convergence of the so defined boundary conditions has been demonstrated by Guermond [50] for a rotational projection method; these condition are retained for the sheet simulations for the last chapter.

### 2.2.4 Poisson Pressure Solver

In the second step of the Chorin projection method the pressure field is computed by solving the elliptic equation (2.13) in order to project the velocity into a divergence free field. This equation is linear, but the coefficients, containing the densities of the two fluids, are only constants inside each fluid. They are discontinuous on the interface, the jump proportional to the density ratio. A strong variation on some of the discretization matrix coefficients, particulary if one of the fluid is a liquid and the other a gas, makes the conditioning number to rise, resulting in a linear system more difficult to solve. Among the different techniques for solving a Poisson problem, the natural choice falls on iterative solvers, where an initial guess of the solution is more and more approached to the exact one, and can be stopped when the approximation is good enough (the convergence of the algorithm depends on the characteristics of the matrix, which are satisfied for the given problem). Several algorithm belong to this family: the classical iterative methods: Jacobi, Gauss-Seidel, Krylov subspace methods, multigid. For dealing with such systems which can be ill-conditionned, a conjugate gradient method preconditioned by a geometric multigrid method has been chosen [61, 62], this will ensure robustness and performance the linear system solver. Moreover, the parallelization of the algorithm will be quite simple. Here is detailed the algorithm of the MGCG method:

---

**Algorithm 1** The MGCG algorithm for the $Ax = b$ linear system.

---

Given $x_0,\ \epsilon$

**while** $(\|r_i\|_2/\|b\|_2 > \epsilon)$ **do**

$z_{i-1} = V_{\text{cycle}}(A, r_{i-1}\|)$ one multigrid V cyle
$\rho_{i-1} = (r_{i-1} \cdot z_{i-1})$

**if** $i = 1$ **then**

$p_i = z_0$

**else**

$\beta_{i-1} = \rho_{i-1}/\rho_{i-2}$
$p_i = z_{i-1} + \beta_{i-1} p_{i-1}$

**end if**

$q_i = Ap_i$
$\alpha_i = \rho_{i-1}/(p_i \cdot q_i)$
$x_i = x_{i-1} + \alpha_i p_i$
$r_i = r_{i-1} - \alpha q_i$

**end while**

---

where multigrid algorithm is given by the recursive algorithm.

**Algorithm 2** Recursive V-cycle multigrid algorithm for $Ax = f$.

```
Procedure MG(l)
if (l > 0) then
  if (l = l_max) then
    e_lmax = x
  else
    e_l = 0
  end if
  e_l ← RELAX(A_l, f_l)
  r_l = f_l − A_l e_l
  f_{l−1} ← Restrict (r_l)
  MG(l − 1)
  δe_l ← Prolong (e_{l−1})
  e_l = e_l + δe_l
  e_l ← RELAX(A_l, f_l)
else
  e_l = 0
  e_0 ← RELAX(A_0, f_0)
end if
```

In the multigrid procedure, restriction and prolongation are used to transfer data from fine grid to coarse one (respectively from coarse one to fine grid). In our implementation, as we use finite volume discretization, the restriction and prologation operators are repectively given as follow:

$$u_{i_c,j_c} = \frac{1}{4}\left(u_{i_f,j_f} + u_{i_f+1,j_f} + u_{i_f,j_f+1} + u_{i_f+1,j_f+1}\right) \tag{2.51}$$

and for the prolongation

$$\begin{aligned} u_{i_f,j_f} &= u_{i_c,j_c} \\ u_{i_f+1,j_f} &= u_{i_c,j_c} \\ u_{i_f,j_f+1} &= u_{i_c,j_c} \\ u_{i_f+1,j_f+1} &= u_{i_c,j_c} \end{aligned} \tag{2.52}$$

Figure 2.4 shows the prolongation and restriction in the context of cell-centered finite volume approach Fig. 2.4a, b.

The matrix of different levels of grid $A_l$ is calculated by rediscretization of the problem at the corresponding level $l$.

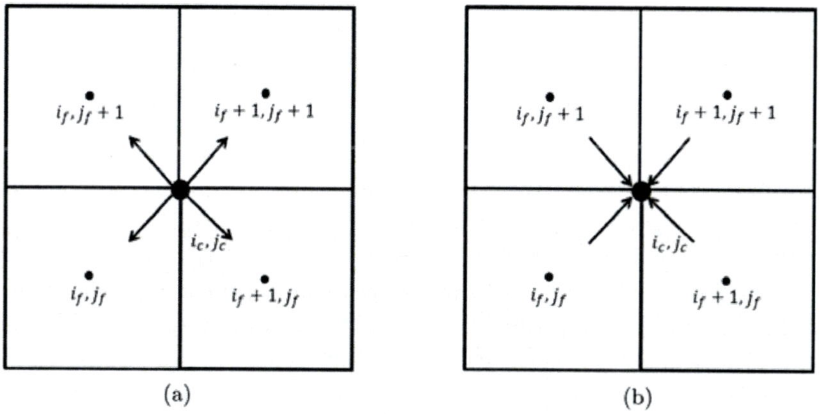


**Fig. 2.4** **a** Prologation from the coarse to fine grid, **b** restriction from the fine to coarse grid

## 2.3 Methods for Handling Interfaces

At molecular length scale, an interface is a transition region where some molecules of the first fluid are mixed with molecules of the second fluid. Consequently, the thickness of this region is equal to few mean free molecular path. Depending on this dimension, the interface can be considered thick when the properties change gradually, or punctual, meaning a step change in the fluid properties. The first case can be described by the Cahn-Hilliard equation [63] as a transition zone. The other approach, used in this work, considers the interface as a discontinuity zone. The exact localization of the interface ensures the mass conservation in time when the fluids evolve and cause its topological deformation. Several numerical methods have been developed to localize and track this separation and to follow the topological changes, such as break-up or coalescence. Two main families can be distinguished.

The first contains the methods where the interface is defined on a second computational grid of which the nodes are evolved in a Lagrangian way. They are referred to as the *front tracking* methods, because they perform an explicit track of the interface by advection of the nodes on which it is defined. They allow a very accurate resolution of the properties jumps; they are, however, expensive and somehow complex to realize, especially when extended to three dimensions. To this class belongs the work of [47, 64], where a free surface flow is simulated by a Lagrangian advection of particles moving with an interpolated velocity coming from the fixed Navier-Stokes grid. This is the very first method to simulate two-phase interfacial flows. Later works, such as [65], allow a full two-phase flow simulation by limiting the markers to the interface itself instead of a whole phase (*Markers and Cells methods*). Its weaknesses are mainly the *reseeding* problem, arising when the markers are accumulated in certain regions due to the shape of the velocity field, and the necessity of a coalescence/break-up model. The method has been also coupled with the discontinuous Galerkin solver in [66]. Some evolution of this philosophy involve some

Euler-Lagrange hybrid treatment of the mesh (ALE, *arbtrary Lagrangian-Euler*), such as in [67–69], where the mesh evolves with a different velocity from the mean flow. The last option is to treat in a fully Lagrangian way the flow and the interface with the so called SPH (*smoothed particle hydrodynamics*), where the fluid model consists of distinct particles with a set of physical properties.

The second class includes the methods called *front capturing*, which use a fixed Eulerian Navier-Stokes grid with a scalar field defined on to implicitly represent the interface. The values relative to the interface such as the velocity are interpolated from the Eulerian field. The two main currents of this methodology are the Level Set and the Volume of Fluid (VOF). The first, developed mainly by [70–72], consists in considering the zero contour of a signed distance function as interface. Its main advantages are the simplicity in its concept and numerical implementation, and the easiness in retrieving the topological properties as the normal vector and the curvature; its drawback is the lack of explicit conservation of mass as the simulation advances. The second one contains the methods based on the transport of a volume fraction (the *colour function*), varying from 0 to 1, which describes the percentage of presence of one fluid inside the cell. Conversely to the Level Set methods, the mass is exactly conserved in time, but the reconstruction of the interface is difficult. The VOF method can be found in the works of [17, 73], and other variants, mainly in the interface reconstruction, in [74–77].

Some details on these methods are given in this chapter; the most part will be dedicated to the description and the numerical resolution of the solution which has been retained in this work, the Level Set.

## 2.3.1 *Interface Tracking Methods*

### Mass trackers methods

In the precursor work of [47] one of the two phase is described by a set of particles without mass (*markers and cells*), as illustrated in Fig. 2.5a, superimposed to a fixed grid on which the velocity field is defined. The markers presence or absence define the interface; their position in time is solution of the following linear advection equation

$$\frac{d\boldsymbol{x}_i}{dt} = \boldsymbol{u}_i \tag{2.53}$$

which means that markers are following the fluid particles. This method was the first one in the meddle sixties to tacke two-phase interfacial flows. The method shows its full capabilities in [78] where it is able to perform a simulation of a Rayleigh-Taylor instability. One of its remarkable advantages is the exact conservation of mass. However several drawback hampers it. First of all, it is limited to the free surface flows, where only one fluid moves, the other being considered only to provide boundary conditions. The interface is not precisely defined, so that the computation of the topological characteristics as normal or curvature is complex. Moreover, the mark-

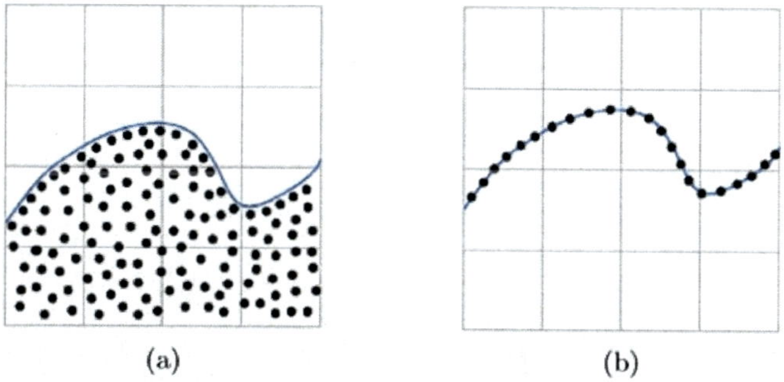

Fig. 2.5 Representation of different markers interface tracking methods. **a** Mass trackers, **b** interface trackers

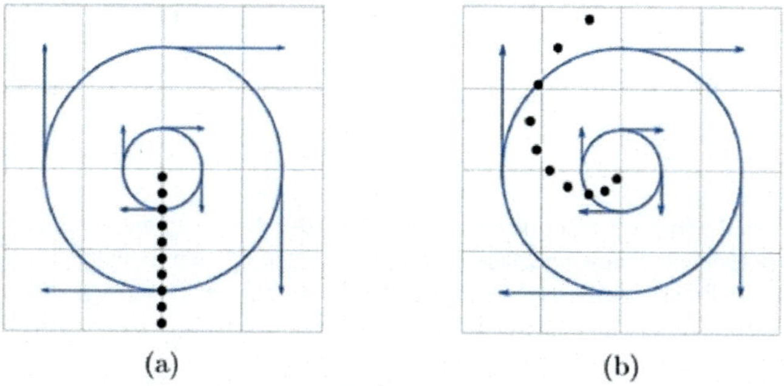

Fig. 2.6 Representation of the advection of massless markers by a whirling velocity field. The markers can be seen concentrating towards the center of the whirl, while becoming more rarefied towards the periphery. **a** Initial condition, **b** after some time

ers have to be always concentrated on the proximity of the interface when important deformations occur, in order to avoid an overload of computational resources, and the flow can induce excessive concentration or rarefaction of the markers.

### Front-Tracking (explicit interface)

One simple alternative to the mass trackers is the use of markers to define the interface alone instead of one whole phase. The list of massless markers is again evolved by a Lagrangian advection, and the interface is each time step redefined by its new position, as shown in Fig. 2.5b. The Markers are connected by polynomial or *spline* interpolation of the desired order to get more or less accuracy (and oscillations).

This method, introduced by [64, 79], has the obvious advantage over the previous one of allowing a real two fluid simulations, and to have the interface always exactly defined by the interpolation of the markers. The Lagrangian advection of the zero mass particles is still subject to the effects of the underlying flow, and they tend to concentrate on "calm" flow regions and conversely to rarefy on strongly inertial flow ones, as can be seen represented in Fig. 2.6.

An important issue concerning this method applied to the interfacial flow mechanics is the *coalescence* process, by which two or more droplets or particles merge

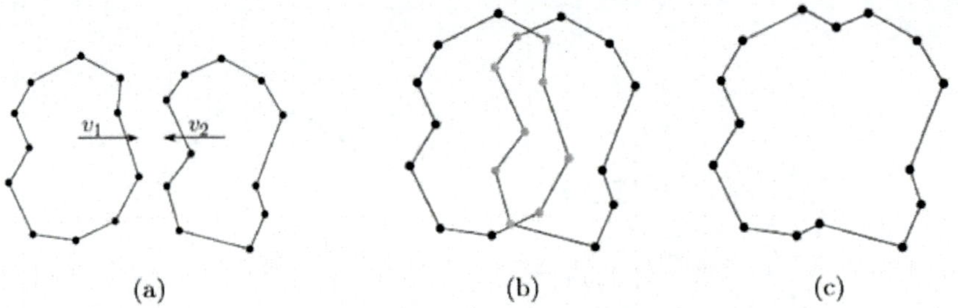


**Fig. 2.7** Coalescence of two structures with Lagrangian method. **a** Initial situation, two distinct droplets approach, **b** overlapping of the droplets, the grey points are inside the new interface. **c** Final interface: the internal markers have been removed and the interface reconstructed

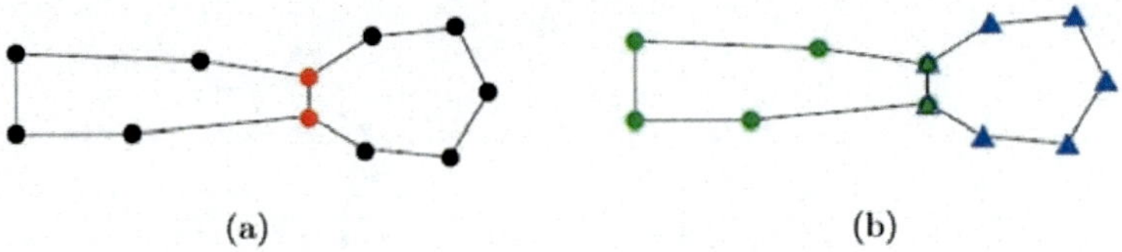


**Fig. 2.8** Artificial break-up criterion for Lagrangian advected ligament. **a** When two markers not consecutive are too near, a new interface is created between the two points. **b** The left (circles) and right (triangles) define now two different items newly created

during contact to form a single final structure. If two markers-traced interfaces representing the starting droplets (Fig. 2.7a) are brought from their mean velocity to overlap, if no action is taken some of the markers end up inside the interface of the merging droplets Fig. (2.7b). They have to be numerically removed because they do not represent an interface any more. The new resulting interface representing the final "big" drop keeps all the old markers outside the coalescence zone, and possibly adds markers on this latter to rebuild the merging point. Another criterion could be a minimum distance of the markers where the two old interfaces are broken and a new bond is created between the droplets.

The opposite problem arises as well: the markers defined interface do not have any natural break-up criterion. As for the coalescence, a minimum distance of two markers inside a thin ligament can be used as the limit for the imposition of the division in two of the ligament, as can be represented in Fig. 2.8.

The numerical algorithms which should deal with those problems are difficult and numerically "heavy", mainly when extended in three dimensions. Despite the strong surjery needed to manage the front topology, this method has been rediscover and improved a lot in the nineties par [65] for incompressible two-phases interfacial flows. Between the middle eighties to nineties, Glimm et al. [80, 81] have developed the FronTier code based on front-tracking surface markers but for compressible two-phase interfacial flows. Some advanced methods in which there is no explicit connection between markers have been developed by [82] by employing an "index function" of which the 0.5 contour should represent the interface. But apart from this enhancement, the markers based interface suffers from the same problems as the above mentioned methods.

### 2.3.2 *Front-Capturing (Implicit Interface)*

#### Volume of Fluid

This is the first method based on a discontinuous field Eulerian approach. It was originally applied to multiphase flow by [83]. In this method a scalar value representing the volumetric percentage of one fluid is given on each cell. This function, called *the color function* $\chi$ can vary from zero to one, being the two extremes the presence of only one of the two fluids in the cell:

$$\begin{cases} \chi\,(\mathbf{x},t) = 1, & \text{if fluid 1} \\ \chi\,(\mathbf{x},t) = 0, & \text{if fluid 2} \end{cases} \tag{2.54}$$

The method consists essentially in two steps. The first is the advection of the color function; the second is the reconstruction of the interface from the colour function. The colour function is the solution of the linear advection equation

$$\frac{\partial \chi}{\partial t} + \boldsymbol{u}.\nabla \chi = 0 \tag{2.55}$$

$\boldsymbol{u}$ being the Eulerian velocity field. The interface is represented by the discontinuity of $\chi$. The discrete correspondent volume fraction $C_{ij}$ defined on the $(i, j)$ cell of volume $v_{ij}$ is the integral value of the volumetric fraction of the one of the two fluids (example in Fig. 2.9a):

$$C_{ij} = \frac{1}{\Omega_{ij}} \int\limits_{\Omega_{ij}} \chi(\mathbf{x})\, dv \tag{2.56}$$

The value $C_{ij}$ is the solution of the conservation law in the sense of a finite volume discretization; the transport equation can be written in conservative and non conservative form given the $\nabla.\boldsymbol{u} = 0$ condition:

$$\frac{\partial C_{ij}}{\partial t} + \boldsymbol{u}.\nabla C_{ij} = 0\,, \text{ non conservative form} \tag{2.57}$$

$$\frac{\partial C_{ij}}{\partial t} + \nabla.(\boldsymbol{u} C_{ij}) = 0\,, \text{ conservative form} \tag{2.58}$$

Equations (2.57) and (2.58) are hyperbolic conservation laws, and their numerical solution is well documented in literature. The conservative resolution of such equation should guarantee the exact conservation of the mass. However, this equation is

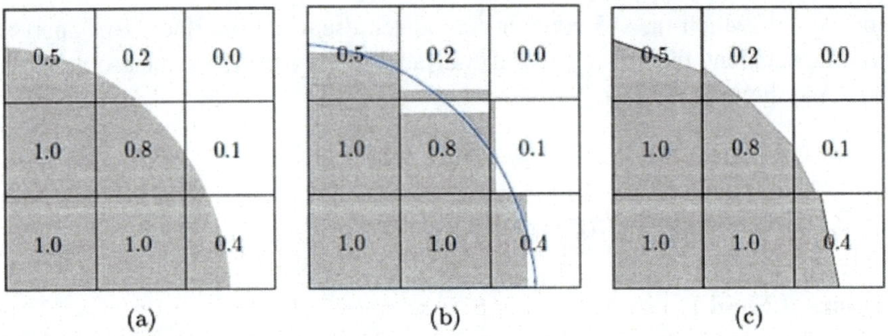


**Fig. 2.9** VOF reconstruction of the interface. **a** Interface position, **b** SLIC constant piecewise reconstruction, **c** PLIC linear piecewise reconstruction

not usually directly solved because of the discontinuous nature of the discrete function $C$: the numerical diffusion would smooth the jumps as the simulation advances in time. Instead, the interface is reconstructed with the aim of building numerical fluxes from geometrical considerations, in order to ensure the local mass conservation. In this sense the differences amongst the VOF variants are mainly found into the interface reconstruction. The topological information are, on the other hand, needed to project the jump conditions in the local normal and tangential reference system. The quality of the overall method strongly depends on the accuracy of the reconstruction.

The simplest algorithm is the SLIC (*Simple Line Interface Calculation*), initially developed by [74]. It fills each cell with its value of $C_{ij}$ by a constant value reconstruction, its position on the cell decided by the local velocity vector (Fig. 2.9b). It is at best a first order method. A straightforward enhancement of this algorithm is the PLIC (*Piecewise Linear Interface Calculation*, Fig. 2.9c), which tries to build a linear piecewise reconstruction always conserving the integral quantity of mass, thus obtaining a normal to the interface from the discontinuous $C^1$ reconstruction.

The values of $C_{ij}$ from the surrounding $3 \times 3$ stencil is used to compute the parameters of the reconstruction. To this category belong the works of [84, 85], which use in different ways the stencil to obtain their geometry (the first computes the center of mass, the second the concentration gradient directly). A second order reconstruction has been developed by [86, 87], with respectively the LVIRA and ELVIRA methods, thanks to a minimization of a $L_2$ norm. Another approach has been presented by [88]: in their work, several candidates linear (planar in three dimensions) interpolations are selected by a least-square fit procedure. This method is reported to be second order when applies iteratively. As summary, the VOF allows a natural treatment of the interface and its topological changes, breakup and coalescence. The numerical transport of the volume of fluid allows an exact conservation of the mass. On the other side, the position of the interface is not known explicitly, and its reconstruction may prove difficult to realize, mainly in three dimensions; the imposition of jump conditions when solving the Navier-Stokes equations may be difficult as a consequence.

### Level-Set

The Level Set method was originally developed by [70] to model an interface moving with a velocity dependent on its curvature. Knowing the difficulty to compute the curvature in the VOF methods, they represented the interface as a contour of a smooth function (the ensemble of point $(x, y)$ for which the function $\phi(x, y) = \text{const}$) where this and other topology informations would be easier. Usually this function is chosen to be hyper regular ($\|\nabla\phi\| = 1$), so that it is equivalent in its definition to a distance function. The Level Set function is passively advected by the local velocity field, it is therefore the solution of the transport equation

$$\frac{\partial\phi}{\partial t} + \boldsymbol{u}.\nabla\phi = 0 \tag{2.59}$$

This formulation for two phase flow problem is mathematically simple and elegant. It is however bound to the redistance problem. This mathematical problem consists in the loss of the distance properties as the initial function evolves and changes its shape. The redistance algorithm is necessary to reimpose this property at each time step. Additionally to the discretization error, this algorithm may generate mass losses by artificial shift of the interface and additional numerical diffusion, as well as the round-off of sharp corners. For this reason The Level Set equation is usually discretized with high order schemes. For the code described in this work the Level Set formulation has been chosen because of its simple definition and its advantages in the interface topology quantification. For this reason a more detailed overview of this method is given in the next section.

### Hybrid methods

- Levels Set/VOF
  Given some opposite advantages and disadvantages of the described methods, some hybrid algorithms have been developed to obtain advantage of both the original ones. The first interesting case is the coupling between Level Set and VOF, a technique pioneered by [89] to simulate the dynamics of an axisymmetric rising bubble, with a remarkable application in the liquid jet primary atomization in [90]. The method is called CLSVOF (*Coupled Level Set Volume Of Fluid*). Its main purpose is to use the Level Set function to precisely compute the interface characteristics and to correct the mass conservations errors with a VOF colour function. The flow-field is computed with a Level-Set method, which is advantageous for computing interface normals and curvature and regularization of discontinuities. In order to conserve mass, the interface is advected by a VOF (PLIC) method. After each interface update the coupling between the Level-Set and VOF colour function is done: some small local (meaning on each cell) Level Set shifts are performed to recover any artificial mass alteration. A drawback might be that the elaborateness of the VOF methods is imported. The advantage of the method is that the mass-conservation properties are shown to be comparable with "pure" VOF methods. [91] proposed a variant where no VOF reconstruction is necessary, called MCLS (Mass Conserving Level Set). Differently from the CLSVOF, the

only present VOF element is the volume fraction. Under the hypothesis of piecewise linear interface, a simple relation between this quantity and $\phi$ is set: a good mass conservation is achieved with a simpler method than the PLIC reconstruction.

- Levels Set/Lagrangian
  The results of the interface tracking given by the LS/VOF coupling are remarkable, but, being Eulerian, these methods still suffer from the smoothing of the sharp corners. In [92] the Level Set interface is enhanced by a set of massless particles disposed in both sides, which are treated as seeding particles as in the Lagrangian methods. A particle crossing the interface is a detector of a local error, and consequentially the Level Set is corrected to match the information from the particles. Another possible application of the particles to the Level Set is the disposition of markers on the interface itself, as proposed in [93]. Still in his work the accuracy of the Lagrangian method is not fully exploited, as there is no coupling with the momentum equation. [94] developed a particle-Level Set method fully coupled to the momentum equation, obtaining impressive results in the interface tracking.

### 2.3.3 SPH Methods

Another alternative for the interfacial flow simulation is a mesh-less Lagrangian technique, where fluid interfaces are advected with very little numerical diffusion. It is called Smoothed Particle Hydrodynamics (SPH), and was originally developed to model astrophysical problems; however, it has since been extended to model a wide variety of problems in computational physics. The fully Lagrangian nature of SPH maintains sharp fluid interfaces without employing high-order advection schemes or explicit interface reconstruction, so it fits very well the two phase flow simulations. Several possible implementations of surface tension have been tested in [95], giving some first successful simulations. Some work has still to be done concerning high density or high viscosity ratios. Finally, its high computational cost can hamper the diffusion of this algorithm for this kind of problems.

## 2.4 Capillary Effects and Jump Conditions at Interface

In the Level Set formulation of the two phase problem, the space is divided into two sub-domains which represent two distinct fluids. In each of them, the couple velocity-pressure are solution of the Navier-Stokes equations. The resolution of such equations needs boundary conditions to close the system. The jump conditions allow the connection of the two solutions: each sub-domain is solved in an independent way by receiving its boundary conditions from the other one. The system solved in a "single fluid" way is then correctly solved by the respecting of the boundary conditions on the domain limits and the jump conditions across the interface. There are multiple possible approaches to the numerical implementation of the jump conditions. Two main categories can be distinguished.

The first contains the methods which consider the effects of the interface as source terms in the right hand side of the equations. In this family is the so called "Continuum Surface Force" (CSF) introduced by [96] and derived from the pioneer work of [97] on the "Immersed Boundary" method for biological flows. In these algorithms, as the interface does not coincide with the Eulerian mesh nodes, the interfacial surface force is transformed into a volume force in the region near the interface, using a delta function $f_v(x) \approx \int_\Gamma f_s(x)\delta^\epsilon(x - x_s)\, d\Gamma$ where $\delta^e$ is the Dirac function centered on the $x_s$ point. The results of the application of the CSF and IB methods model the discontinuities as slightly smoothed variations on some cells around the interface.

The second category includes the immersed interface methods ("IIM") from [98, 99], where the discontinuity is taken explicitly in the equation discretization. This method has its basis into a Taylor's series development of the discrete solution, obtained taking in account the position of the interface and inserting the jumps inside the stencils. It is a second order method, but it is somewhat difficult to implement, especially for the three dimensional simulations. Another drawback is the consequent generation of a Poisson's equation matrix no more symmetrical, hurting the performances of the iterative solver. It happens, however, that the matrix resulting from the discretization on the adaptive mesh has the same consequence.

A simpler and symmetrized version of the IIM has been originally developed by Fedkiw for discontinuities like shocks, detonation and deflagration in Euler equations [100] and viscous flows [101], the "Ghost Fluid". This method can be employed to model the Navier-Stokes equations without adding any volume source term, and is easy to implement. Its technique is basically to extend each fluid characteristic to the opposite side of the interface (generating the "ghost fluid" zone) so that standard stencils can be used, and to bring to the right side of the equation the contribution of the discontinuities. Better results in interfacial flow solutions are reported by [58] in comparison with the CSF solution. For this reason the GFM has been chosen on the bounce of his compared analysis.

### 2.4.1 Ghost Fluid

In this section a quick overview to the Ghost Fluid method applied to the pressure equation is given (2.29). We first provides the detail of the method for a one dimensionnal elliptic pde. Two complementary domains are given, $\Omega^+$ and $\Omega^-$. The problem to be solved is

$$\begin{cases} \dfrac{\partial}{\partial x}\left(\beta(x)\dfrac{\partial u(x)}{\partial x}\right) = f(x) \\ [u]_\Gamma = u^+ - u^- = a(x) \quad x \in \Gamma \\ \left[\beta(x)\dfrac{\partial u(x)}{\partial x}\right]_\Gamma = \left(\beta(x)\dfrac{\partial u(x)}{\partial x}\right)^+ - \left(\beta(x)\dfrac{\partial u(x)}{\partial x}\right)^- = b(x) \quad x \in \Gamma \end{cases} \tag{2.60}$$

This case is focused on $C^0$ and $C^1$ discontinuities, as described in Eq. (2.60), because both appear in the two phase solution, the $C^0$ in the pressure and the $C^1$ in the tangential velocity derivatives.

**Jump of the unknown at the interface**

In this configuration in the system (2.60) there are $a(x) \neq 0$ and $b(x) = 0$. This is the case described in [102], where a Poisson equation is solved in a complex domain described by the interface. The imposition of the jump conditions becomes the imposition of the right boundary conditions to the pressure equation. The standard discretization of this equation in the $x_i$ (the left point, the right one $x_{i+1}$ is treated symmetrically) cell is:

$$\frac{\beta_{i+\frac{1}{2}}\left(\dfrac{u_{i+1}-u_i}{\Delta x}\right) - \beta_{i-\frac{1}{2}}\left(\dfrac{u_i - u_{i-1}}{\Delta x}\right)}{\Delta x} = f_i \tag{2.61}$$

In order to evaluate the first and second derivatives, the jump values in $C^0$ and $C^1$ are introduced into the definitions of the derivatives themselves. These values allow the computation of a projected value of the quantity to the other side of the discontinuity, finally completing the discretization stencil. In practice the contribution of the interface is in the addition of a right hand side term that gives the discontinuous discrete solution.

As it can be seen in Fig. 2.10, the interface located in $x_I$ between $x_i$ and $x_{i+1}$ divides the two domains $\Omega^+$ and $\Omega^-$. The "ghost fluid" consists into the prolongation of the two domains on both sides of the interface, so the $x_i^+$ and $x_{i+1}^-$ points. Their values are

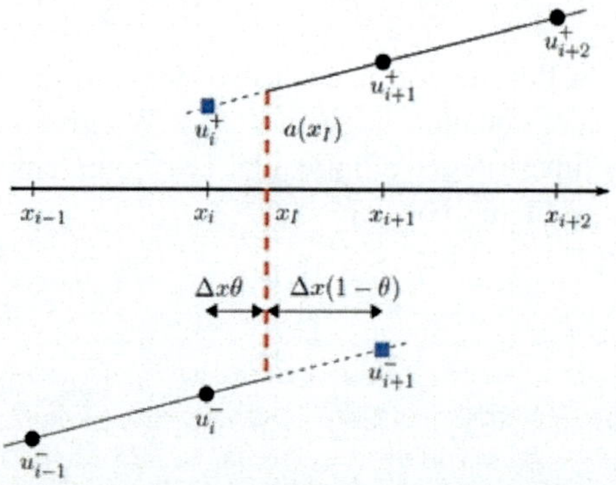


**Fig. 2.10** $C^0$ discontinuous solution $u$ and creation of the ghost points

$$\begin{cases} u_i^+ = u_i^- + a(x_I) \\ u_{i+1}^- = u_{i+1}^+ - a(x_I) \end{cases} \tag{2.62}$$

Then the Eq. (2.61) can be rewritten for the left side as

$$\frac{\beta_{i+\frac{1}{2}}\left(\frac{u_{i+1}^- - u_i^-}{\Delta x}\right) - \beta_{i-\frac{1}{2}}\left(\frac{u_i^- - u_{i-1}^-}{\Delta x}\right)}{\Delta x} = f_i \tag{2.63}$$

By substituting (2.62)

$$\frac{\hat{\beta}\left(\frac{u_{i+1} - a(x_I) - u_i}{\Delta x}\right) - \beta_{i-\frac{1}{2}}\left(\frac{u_i - u_{i-1}}{\Delta x}\right)}{\Delta x} = f_i \tag{2.64}$$

and

$$\frac{\hat{\beta}\left(\frac{u_{i+1} - u_i}{\Delta x}\right) - \beta_{i-\frac{1}{2}}\left(\frac{u_i - u_{i-1}}{\Delta x}\right)}{\Delta x} = f_i + \frac{a(x_I)}{\Delta x^2}\hat{\beta} \tag{2.65}$$

where the value $\hat{\beta}$ is a weighted value of the two surrounding $\beta$ on $x = x_I$

$$\hat{\beta} = \frac{\beta_{i+1}\beta_i}{\beta_{i+1}\theta + \beta_i(1-\theta)} \tag{2.66}$$

once obtained $\theta$ from the Level Set values

$$\theta = \frac{|\phi_i|}{|\phi_{i+1}| + |\phi_i|} \tag{2.67}$$

Equation (2.65) recalls exactly the original discretization of the Poisson equation (2.61) with an added forcing term and a modified coefficient $\beta$, revealing how the Ghost Fluid method allow the use of the single phase discretization for the two phase flows, without altering the symmetry of the discretization matrix (which will, however, be lost within the block based AMR). The value of the jump $a(x_I)$ can be obtained, as for the value $\hat{\beta}$, from a weighting of the values of the interface

$$a(x_I) = \frac{a_i|\phi_{i+1}| + a_{i+1}|\phi_i|}{|\phi_{i+1}| + |\phi_i|} \tag{2.68}$$

**Jump of the normal derivative of the unknown at the interface**

This time the solution $u$ of the system (2.60) is continuous through the interface, but not its first derivative, as shown in Fig. 2.11.

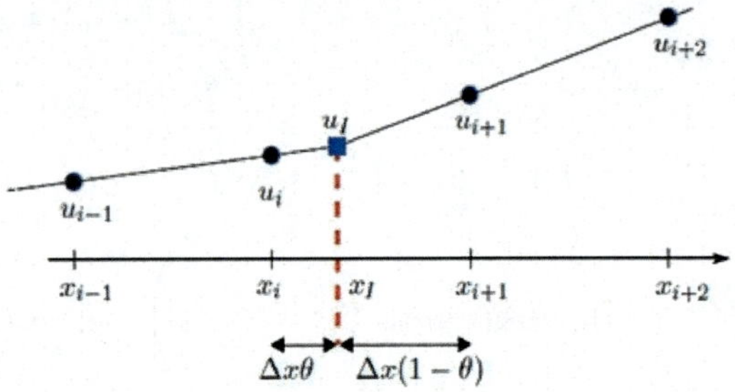


**Fig. 2.11** $C^1$ discontinuous solution $u$, identification of the $u_I$ point

This concretely happens in the viscous terms $\frac{\mu}{\rho}(\nabla.\boldsymbol{u} + \nabla.\boldsymbol{u}^T)$. The jump condition for the Eq. (2.60) can be written as

$$\beta_{i+1}\left(\frac{u_{i+1}-u_I}{(1-\theta)\Delta x}\right) - \beta_i\left(\frac{u_I - u_{i-1}}{\theta\Delta x}\right) = b(x_I) \tag{2.69}$$

and it can be solved for $u_I$

$$u_I = \frac{\beta_{i+1}u_{i+1}\theta + \beta_i u_i(1-\theta) - b(x_I)\theta(1-\theta)\Delta x}{\beta_{i+1}\theta + \beta_{i-\frac{1}{2}}(1-\theta)} \tag{2.70}$$

so that approximate derivatives can be written on both sides of the interface.

$$\begin{cases} \beta_{i+1}\left(\dfrac{u_{i+1}-u_I}{(1-\theta)\Delta x}\right) = \hat{\beta}\left(\dfrac{u_{i+1}-u_i}{\Delta x}\right) - \dfrac{\hat{\beta}b(x_i)\theta}{\beta_i} \\ \beta_i\left(\dfrac{u_I - u_i}{\theta\Delta x}\right) = \hat{\beta}\left(\dfrac{u_{i+1}-u_i}{\Delta x}\right) - \dfrac{\hat{\beta}b(x_i)(1-\theta)}{\beta_{i+1}} \end{cases} \tag{2.71}$$

As for the previous case, the discretization of the Eq. (2.60) for the point $x_i$ can be reverted to the original form with an addition to the right hand side:

$$\frac{\hat{\beta}\left(\frac{u_{i+1}-u_i}{\Delta x}\right) - \beta_{i-\frac{1}{2}}\left(\frac{u_i - u_{i-1}}{\Delta x}\right)}{\Delta x} = f_i + \frac{\hat{\beta}b(x_i)(1-\theta)}{\beta_{i+1}} \tag{2.72}$$

**The final form**

The final form of the Poisson equation can finally be represented as

$$\frac{\hat{\beta}_{i+\frac{1}{2}}\left(\frac{u_{i+1}-u_i}{\Delta x}\right) - \hat{\beta}_{i-\frac{1}{2}}\left(\frac{u_i - u_{i-1}}{\Delta x}\right)}{\Delta x} = f_i + F_i^L + F_i^R \tag{2.73}$$

where $F_i^L$ $\hat{\beta}_{i-\frac{1}{2}}$ are calculated as follows

- if $\phi_{i-1} \cdot \phi_i > 0$

$$\hat{\beta}_{i-\frac{1}{2}} = \beta_{i-\frac{1}{2}} \quad F_i^L = 0 \tag{2.74}$$

- if $\phi_{i-1} \cdot \phi_i \leq 0$

$$\theta = \frac{|\phi_{i-1}|}{|\phi_i| + |\phi_{i-1}|}$$

$$a_\Gamma = \frac{a_i|\phi_{i-1}| + a_{i-1}|\phi_i|}{|\phi_i| + |\phi_{i-1}|}$$

$$b_\Gamma = \frac{b_i|\phi_{i-1}| + b_{i-1}|\phi_i|}{|\phi_i| + |\phi_{i-1}|}$$

$$\text{if} \quad \phi_{i-1} < 0 \quad \text{and} \quad \phi_i \geq 0$$

$$\begin{cases} \hat{\beta}_{i-\frac{1}{2}} = \dfrac{\beta_i \beta_{i-1} \left(|\phi_i| + |\phi_{i-1}|\right)}{\beta_i|\phi_{i-1}| + \beta_{i-1}|\phi_i|} \\[2ex] F_i^L = -\dfrac{\hat{\beta}_{i-\frac{1}{2}} a_\Gamma}{\Delta x^2} + \dfrac{\hat{\beta}_{i-\frac{1}{2}} b_\Gamma \theta}{\beta_{i-\frac{1}{2}} \Delta x} \end{cases} \tag{2.75}$$

$$\text{if} \quad \phi_{i-1} \geq 0 \quad \text{and} \quad \phi_i < 0$$

$$\begin{cases} \hat{\beta}_{i-\frac{1}{2}} = \dfrac{\beta_i \beta_{i-1} \left(|\phi_i| + |\phi_{i-1}|\right)}{\beta_i|\phi_{i-1}| + \beta_{i-1}|\phi_i|} \\[2ex] F_i^L = \dfrac{\hat{\beta}_{i-\frac{1}{2}} a_\Gamma}{\Delta x^2} - \dfrac{\hat{\beta}_{i-\frac{1}{2}} b_\Gamma \theta}{\beta_{i-\frac{1}{2}} \Delta x} \end{cases}$$

and $F_i^R$, $\hat{\beta}_{i+\frac{1}{2}}$ by

- if $\phi_i \cdot \phi_{i+1} > 0$

$$\hat{\beta}_{i+\frac{1}{2}} = \beta_{i+\frac{1}{2}} \quad F_i^R = 0 \tag{2.76}$$

- if $\phi_i \cdot \phi_{i+1} \leq 0$

$$\theta = \frac{|\phi_i|}{|\phi_i| + |\phi_{i+1}|}$$

$$a_\Gamma = \frac{a_i|\phi_{i+1}| + a_{i+1}|\phi_i|}{|\phi_i| + |\phi_{i+1}|}$$

$$b_\Gamma = \frac{b_i|\phi_{i+1}| + b_{i+1}|\phi_i|}{|\phi_i| + |\phi_{i+1}|}$$

$$\text{if} \quad \phi_i \geq 0 \quad \text{and} \quad \phi_{i+1} < 0$$
$$\begin{cases} \hat{\beta}_{i+\frac{1}{2}} = \dfrac{\beta_i \beta_{i+1} \left(|\phi_i| + |\phi_{i+1}|\right)}{\beta_{i+1}|\phi_i| + \beta_i|\phi_{i+1}|} \\ \\ F_i^R = -\dfrac{\hat{\beta}_{i+\frac{1}{2}} a_\Gamma}{\Delta x^2} + \dfrac{\hat{\beta}_{i+\frac{1}{2}} b_\Gamma \theta}{\beta_{i+\frac{1}{2}} \Delta x} \end{cases} \tag{2.77}$$

$$\text{if} \quad \phi_i \leq 0 \quad \text{and} \quad \phi_{i+1} > 0$$
$$\begin{cases} \hat{\beta}_{i+\frac{1}{2}} = \dfrac{\beta_{i+1} \beta_i \left(|\phi_i| + |\phi_{i+1}|\right)}{\beta_{i+1}|\phi_i| + \beta_i|\phi_{i+1}|} \\ \\ F_i^R = \dfrac{\hat{\beta}_{i+\frac{1}{2}} a_\Gamma}{\Delta x^2} - \dfrac{\hat{\beta}_{i+\frac{1}{2}} b_\Gamma \theta}{\beta_{i+\frac{1}{2}} \Delta x} \end{cases}$$

**Extension in two dimensions**

We will provides in this section the formula for the pression equation jump conditions in two dimensions, extension in three dimensions is straighforward. The Poisson like equation writes:

$$\begin{gathered} \frac{\hat{\beta}_{i+\frac{1}{2},j}\left(u_{i+1,j} - u_{i,j}\right)}{\Delta x^2} - \frac{\hat{\beta}_{i-\frac{1}{2},j}\left(u_{i,j} - u_{i-1,j}\right)}{\Delta x^2} \\ + \\ \frac{\hat{\beta}_{i,j+\frac{1}{2}}\left(u_{i,j+1} - u_{i,j}\right)}{\Delta y^2} - \frac{\hat{\beta}_{i,j-\frac{1}{2}}\left(u_{i,j} - u_{i,j-1}\right)}{\Delta y^2} \\ = \\ f_{i,j} + F_{i,j}^L + F_{i,j}^R + F_{i,j}^B + F_{i,j}^T \end{gathered} \tag{2.78}$$

where $\hat{\beta}_{i-\frac{1}{2},j}$, $F^L_{i,j}$, are given by

- if $\phi_{i-1,j} \cdot \phi_{i,j} > 0$

$$\hat{\beta}_{i-\frac{1}{2},j} = \beta_{i-\frac{1}{2},j} \quad F^L_{i,j} = 0 \tag{2.79}$$

- if $\phi_{i-1,j} \cdot \phi_{i,j} \leq 0$

$$\theta = \frac{|\phi_{i-1,j}|}{|\phi_{i,j}| + |\phi_{i-1,j}|}$$

$$a_\Gamma = \frac{a_{i,j}|\phi_{i-1,j}| + a_{i-1,j}|\phi_{i,j}|}{|\phi_{i,j}| + |\phi_{i-1,j}|}$$

$$b_\Gamma = \frac{b_{i,j}n_{x i,j}|\phi_{i-1,j}| + b_{i-1,j}n_{x i-1,j}|\phi_{i,j}|}{|\phi_{i,j}| + |\phi_{i-1,j}|}$$

$$\text{if} \quad \phi_{i-1,j} < 0 \quad \text{and} \quad \phi_{i,j} \geq 0$$

$$\begin{cases} \hat{\beta}_{i-\frac{1}{2},j} = \dfrac{\beta_{i,j}\beta_{i-1,j}\left(|\phi_{i,j}| + |\phi_{i-1,j}|\right)}{\beta_{i,j}|\phi_{i-1,j}| + \beta_{i-1,j}|\phi_{i,j}|} \\ \\ F^L_{i,j} = -\dfrac{\hat{\beta}_{i-\frac{1}{2},j}a_\Gamma}{\Delta x^2} + \dfrac{\hat{\beta}_{i-\frac{1}{2},j}b_\Gamma\theta}{\beta_{i-\frac{1}{2},j}\Delta x} \end{cases} \tag{2.80}$$

$$\text{if} \quad \phi_{i-1,j} \geq 0 \quad \text{and} \quad \phi_{i,j} < 0$$

$$\begin{cases} \hat{\beta}_{i-\frac{1}{2},j} = \dfrac{\beta_{i,j}\beta_{i-1,j}\left(|\phi_{i,j}| + |\phi_{i-1,j}|\right)}{\beta_{i,j}|\phi_{i-1,j}| + \beta_{i-1,j}|\phi_{i,j}|} \\ \\ F^L_{i,j} = \dfrac{\hat{\beta}_{i-\frac{1}{2},j}a_\Gamma}{\Delta x^2} - \dfrac{\hat{\beta}_{i-\frac{1}{2},j}b_\Gamma\theta}{\beta_{i-\frac{1}{2},j}\Delta x} \end{cases}$$

and $\hat{\beta}_{i,j-\frac{1}{2}}$, $F^B_{i,j}$ by:

- if $\phi_{i,j-1} \cdot \phi_{i,j} > 0$

$$\hat{\beta}_{i,j-\frac{1}{2}} = \beta_{i,j-\frac{1}{2}} \quad F^B_{i,j} = 0 \tag{2.81}$$

- if $\phi_{i,j-1} \cdot \phi_{i,j} \leq 0$

$$\theta = \frac{|\phi_{i,j-1}|}{|\phi_{i,j}| + |\phi_{i,j-1}|}$$

$$a_\Gamma = \frac{a_{i,j}|\phi_{i,j-1}| + a_{i,j-1}|\phi_{i,j}|}{|\phi_{i,j}| + |\phi_{i,j-1}|}$$

$$b_\Gamma = \frac{b_{i,j}n_{y_{i,j}}|\phi_{i,j-1}| + b_{i,j-1}n_{y_{i,j-1}}|\phi_{i,j}|}{|\phi_{i,j}| + |\phi_{i,j-1}|}$$

$$\text{if} \quad \phi_{i,j-1} < 0 \quad \text{and} \quad \phi_{i,j} \geq 0$$

$$\begin{cases} \hat{\beta}_{i,j-\frac{1}{2}} = \dfrac{\beta_{i,j}\beta_{i,j-1}\left(|\phi_{i,j}| + |\phi_{i,j-1}|\right)}{\beta_{i,j}|\phi_{i,j-1}| + \beta_{i,j-1}|\phi_{i,j}|} \\ \\ F^B_{i,j} = -\dfrac{\hat{\beta}_{i,j-\frac{1}{2}}a_\Gamma}{\Delta y^2} + \dfrac{\hat{\beta}_{i,j-\frac{1}{2}}b_\Gamma\theta}{\beta_{i,j-\frac{1}{2}}\Delta y} \end{cases} \tag{2.82}$$

$$\text{if} \quad \phi_{i,j-1} \geq 0 \quad \text{and} \quad \phi_{i,j} < 0$$

$$\begin{cases} \hat{\beta}_{i,j-\frac{1}{2}} = \dfrac{\beta_{i,j}\beta_{i,j-1}\left(|\phi_{i,j}| + |\phi_{i,j-1}|\right)}{\beta_{i,j}|\phi_{i,j-1}| + \beta_{i,j-1}|\phi_{i,j}|} \\ \\ F^L_{i,j} = \dfrac{\hat{\beta}_{i,j-\frac{1}{2}}a_\Gamma}{\Delta y^2} - \dfrac{\hat{\beta}_{i,j-\frac{1}{2}}b_\Gamma\theta}{\beta_{i,j-\frac{1}{2}}\Delta y} \end{cases}$$

Formula for $\hat{\beta}_{i+\frac{1}{2},j}$ $\hat{\beta}_{i,j+\frac{1}{2}}$, $F^R_{i,j}$ $F^T_{i,j}$ are given by symmetry. One of the main feature of ghost-fluid method for dealing with jump conditions of unknown of coefficient of elliptic Poission like equation, is that the matrix coming for the discretization with ghost-fluid method is always symmetric allowing use of fast linear solver based on multigrid or preconditonned conjugate gradient methods.

### 2.4.2 *Continuum Surface Force*

The surface tension force appearing in the one fluid model depends on the interfacial properties. This force can be modeled by a Continuum Surface Force (CSF) approach based on the work of Brackbill et al. [44]:

$$\boldsymbol{F}_{st}(C) = \sigma\kappa\boldsymbol{n}\delta_i = \sigma\nabla\cdot\left(\frac{\nabla C}{||\nabla C||}\right)\nabla C \tag{2.83}$$

The notation $\kappa$ denotes the local mean curvature of the interface, $\sigma$ the constant surface tension coefficient, $\boldsymbol{n}$ the normal to the interface and $\delta_i$ the Dirac function indicating the interface.

Discretizing the surface tension force (2.83) requires using a smooth approximation of the volume fraction $C$. This can be achived in different ways, by using a level set function, by smoothing $C$ [103] or by building height functions according to $C$. For a complete presentation of the differents techniques and approaches, the reader can refer to [104].

## 2.5 Validations of Interface Tracking and Fictitious Domains

### 2.5.1 *Comparison of Interface Tracking Methods*

• Solid body rotation
In order to assess the capability of the different methods to adequately transport thin structures and sharp corners, the solid body rotation of a notched circle is simulated [105]. In this test case, Zalesak's slotted disk is rotated around the center of a $[0, 100\,\text{m}]^2$ square domain using a uniform vorticity field given by $\boldsymbol{u} = (u, v)$:

$$\begin{aligned} u(x, y) &= \frac{\pi}{314}(50 - y) \\ v(x, y) &= \frac{\pi}{314}(x - 50) \end{aligned} \tag{2.84}$$

A disk of radius $R = 15\,\text{m}$ and slot width $H = 5\,\text{m}$ is initially centered at (50 m, 75 m). The disk returns back to its initial position at $t = 628\,\text{s}$ with its shape perfectly maintained during the advection of the disk. In Fig. 2.12, results are presented after one revolution is completed.

A zoom of the down-left corner is presented too. With the Level Set method approximated with Weno schemes (LSET-WENO technique [106]), the notch is damped after one revolution, indicating a dissipative behavior of WENO schemes used for the spatial discretization. With the Front Tracking approach associated to spline shape elements and Level Set (LSET-PART) techniques [106], no dissipation of the notch can be seen. Indeed, a reconstruction step is used in this last method to build $\phi_i$ without any spatial discretization scheme (with this method, the Lagrangian elements locating $\Sigma_i$ are used to build the exact level set by tacking the Euclidian distance between an Eulerian mesh vertex and a Lagrangian surface element).

About interface shape, LSET-PART is more accurate than Piecewise Linear Interface Construction (VOF-PLIC) method, especially near the notch. In large curvature zones, the LSET-PART method is more accurate than the Level Contour Reconstruction Front Tracking (FTR-LCR) of Shin and Juric [82], for which corners are rounded off due to the projection of the markers (Eq. (3.127)).

To compare accuracies between the different techniques in a quantitative way, the $L1$ errors between the initial configuration (time $t_0$) and the shape after one revolution

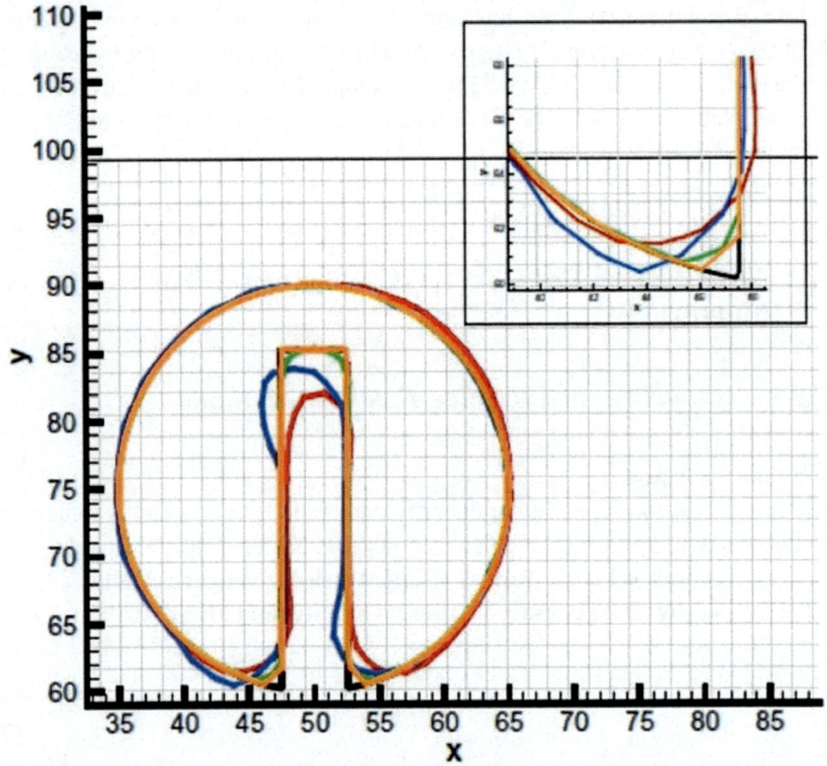


**Fig. 2.12** Comparison of interface tracking methods for the Zalesak advection test [106] after one revolution—exact solution (black), level set (red), VOF-PLIC (blue), front tracking level contour reconstruction (LCR) [82] (green) and front tracking level set particle (LSET-PART) [106] (orange). A $64^2$ grid is used

(time $t_1$) are performed in Fig. 2.13. The $L1$ error is defined by:

$$e_{L_1} = \frac{1}{N} \sum_{ij} \| C_{ij}^{t_1} - C_{ij}^{t_0} \| \tag{2.85}$$

where $N$ is the total number of points in the numerical domain. To perform such comparisons, a color function C is built from the level-set function $\phi_i$ for the methods LSET-WENO and LSET-PART according to a Heaviside function given by Sussman and Fatemi [107]:

$$C_i = \begin{cases} 1 & \text{if } \phi > \Delta x \\ 0 & \text{if } \phi < \Delta x \\ \frac{1}{2}\left(1 + \frac{\phi}{\Delta x} + \frac{1}{\pi}\sin\left(\pi \frac{\phi}{\Delta x}\right)\right) & \text{otherwise} \end{cases} \tag{2.86}$$

| Method | Grid size | $L_1$ error | Convergence rate |
|---|---|---|---|
| *LSET-WENO* | 256 × 256 | $2.56\,10^{-4}$ | - |
| | 128 × 128 | $1.61\,10^{-3}$ | 2.65 |
| | 64 × 64 | $6.09\,10^{-3}$ | 1.92 |
| | 32 × 32 | $2.08\,10^{-2}$ | 1.77 |
| | 16 × 16 | $1.91\,10^{-2}$ | - |
| *VOF-PLIC* | 256 × 256 | $3.70\,10^{-3}$ | - |
| | 128 × 128 | $3.53\,10^{-3}$ | - |
| | 64 × 64 | $9.07\,10^{-3}$ | 1.36 |
| | 32 × 32 | $4.15\,10^{-2}$ | 2.19 |
| | 16 × 16 | $1.25\,10^{-1}$ | 1.59 |
| *FTR-LCR* | 256 × 256 | $6.49\,10^{-4}$ | - |
| | 128 × 128 | $1.28\,10^{-3}$ | 0.98 |
| | 64 × 64 | $2.48\,10^{-3}$ | 0.95 |
| | 32 × 32 | $3.85\,10^{-3}$ | 0.63 |
| | 16 × 16 | $4.39\,10^{-3}$ | 0.19 |
| *LSET-PART* | 256 × 256 | $9.74\,10^{-6}$ | - |
| | 128 × 128 | $1.37\,10^{-5}$ | 0.49 |
| | 64 × 64 | $1.01\,10^{-4}$ | 2.88 |
| | 32 × 32 | $2.65\,10^{-4}$ | 1.39 |
| | 16 × 16 | $2.47\,10^{-3}$ | 3.22 |

**Fig. 2.13** Errors of interface tracking methods for the Zalesak advection test [106] after one revolution

The LSET-PART technique is the most accurate technique compared to the other ones, especially in the case of coarse grids, where the use of markers allows local over-resolution. Nevertheless, the order of convergence seems to sink as the spatial resolution increases.

- Deformation of a circle by a single vortex

The single vortex test of Bell et al. [108] is now under consideration. The deformation of a circle by a single vortex has been considered to assess the ability of numerical methods to resolve thin filaments. In this test, a circle is deformed with a field defined by the stream function:

$$\Psi = \frac{1}{\pi} \sin^2(\pi x) \sin^2(\pi y) \tag{2.87}$$

Initially the circle has a radius of 0.15 m and is centered at (0.5 m, 0.75 m) in a $(0.1\text{ m})^2$ box. The simulation time is $t = 3$ s. In Fig. 2.14 results are presented for a $64^2$ grid. The head of the streamer is perfectly captured by LSET-PART. Moreover, no thickening is observed at the tail but successive droplets appear. Indeed when the thickness of the streamer becomes lower than the grid cell, the continuous body of the streamer degenerates into a set of droplets. Such droplets can be seen with VOF-PLIC but they do not respect the shape of the streamer. For LSET-WENO,

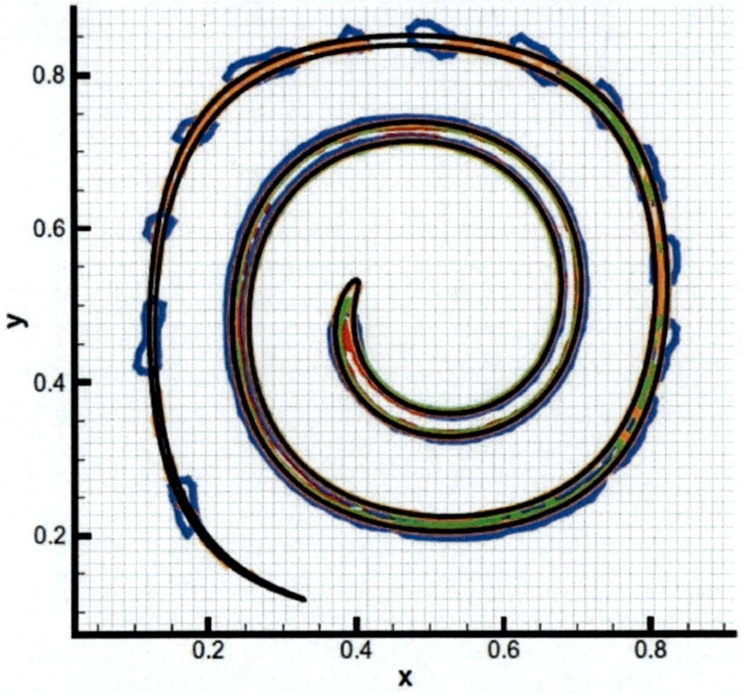


**Fig. 2.14** Comparison of interface tracking methods for the deformation of a circle [106] in a shearing velocity field—exact solution (black), level set (red), VOF-PLIC (blue), front tracking level contour reconstruction (LCR) [82] (green) and front tracking level set particle (LSET-PART) [106] (orange). A $64^2$ grid is used

VOF-PLIC and FTR-LCR methods, the streamer is not properly captured when its width is of the order of $\Delta x$. This can be seen not only where large local curvatures are observed (the head for instance) but also along the body of the streamer. About the FTR-LCR results, only the projected, i.e. Eulerian, shape of the interface is considered here, and the markers are located accurately on the reference solution.

Figure 2.15 shows the relative mass time evolution for the different methods. The relative mass is computed as follows by $\text{mass}_{\%} = \dfrac{\text{mass}(t)}{\text{mass}(t=0)} \times 100$ where $t$ is the time (in seconds). For the coarse grid ($64^2$), mass losses appear with the LSET-WENO method (up to 50%). Mass conservation is better for the VOF-PLIC method where losses are less than 5%. FTR-LCR and LSET-PART methods are very accurate (losses less than 2%). For fine grids ($128^3$ and $256^3$), mass losses are reduced for the LSET-WENO method. For the FTR-LCR and LSET-PART methods, mass losses are drastically reduced so that $\text{mass}_{\%} \approx 100\%$ at every time. After the 3 s, the streamer is run backward in time during 3 s up to its initial position.

The $L1$ errors between the shape after 6 s ($t_1$) and the initial one ($t_0$) are compared in Fig. 2.16. The error is given by (2.85). For LSET-WENO and LSET-PART methods, a colour function is built using Eq. (2.86). The LSET-PART method is the most accurate

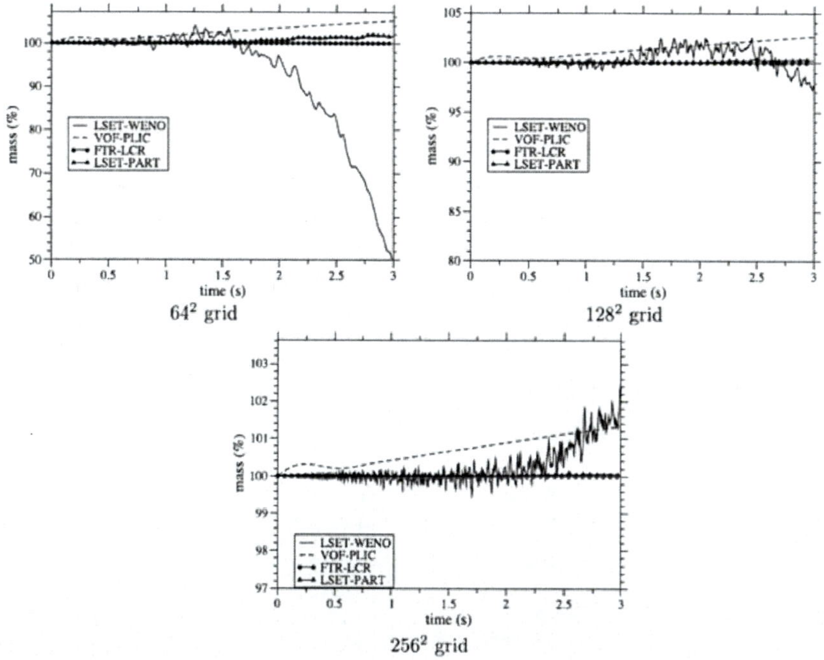


**Fig. 2.15** Comparison of interface tracking methods for the deformation of a circle [106] in a shearing velocity field—time evolution of relative mass error is presented. 100% corresponds to the initial ($t = 0$) mass. Three different grids are considered: $64^2$, $128^2$ and $256^2$

and the convergence rate is the highest. VOF-PLIC technique is more accurate than LSET-WENO for low resolutions (322 and 642) but the rate of convergence for VOF-PLIC is lower than for LSET-WENO and VOF-PLIC is not as accurate for fine grids (from 1282). FTR-LCR is more accurate than LSET-WENO and VOF-PLIC methods but the levels of accuracy remain of the same order for a given grid. The convergence rate is around one.

The most interesting results here concern coarse grids. Indeed, for example in the framework of direct numerical simulation (DNS) of atomization process, a maximum of ten grid points inside a droplet can be expected to be computed. Indeed, the use of more grid points to describe the droplets would lead to not affordable DNS in term of CPU time and memory. The LSET-PART method is very accurate even for coarse grids (typically 322) and will be of great interest in the context of DNS of atomization process for instance.

### 2.5.2 *Density and Viscosity Averages*

This section is concerned with the choice of interpolation functions for estimating, respectively, density $\rho$ and viscosity $\mu$ across the interface. One example of possible

| Method | Grid size | $L_1$ error | Convergence rate |
|---|---|---|---|
| *LSET-WENO* | 1024 × 1024 | $2.6510^{-4}$ | - |
| | 512 × 512 | $7.9210^{-4}$ | 1.58 |
| | 256 × 256 | $4.9310^{-3}$ | 2.64 |
| | 128 × 128 | $1.3410^{-2}$ | 1.44 |
| | 64 × 64 | $6.6410^{-2}$ | 2.31 |
| | 32 × 32 | $7.1110^{-2}$ | 0.1 |
| *VOF-PLIC* | 1024 × 1024 | $3.5710^{-3}$ | - |
| | 512 × 512 | $7.1510^{-3}$ | 1.00 |
| | 256 × 256 | $1.4410^{-2}$ | 1.01 |
| | 128 × 128 | $2.8410^{-2}$ | 0.98 |
| | 64 × 64 | $3.9210^{-2}$ | 0.47 |
| | 32 × 32 | $3.9810^{-2}$ | 0.02 |
| *FTR-LCR* | 1024 × 1024 | $1.1610^{-3}$ | - |
| | 512 × 512 | $5.3410^{-4}$ | - |
| | 256 × 256 | $1.1410^{-3}$ | 1.09 |
| | 128 × 128 | $2.5710^{-3}$ | 1.18 |
| | 64 × 64 | $6.5210^{-3}$ | 1.35 |
| | 32 × 32 | $1.5310^{-2}$ | 1.23 |
| *LSET-PART* | 1024 × 1024 | $3.3610^{-13}$ | - |
| | 512 × 512 | $3.6310^{-13}$ | 0.11 |
| | 256 × 256 | $1.2810^{-12}$ | 1.82 |
| | 128 × 128 | $3.2010^{-11}$ | 4.64 |
| | 64 × 64 | $1.0210^{-9}$ | 5.00 |
| | 32 × 32 | $3.2910^{-8}$ | 5.00 |

**Fig. 2.16** Errors of interface tracking methods for the deformation of a circle [106] in a shearing velocity field

choice for the one-fluid model was given in Eqs. (2.4 and 2.5). The way of estimating average density and viscosity at the interface neighborhood is somewhat empirical and needs rigorous testing, and particularly as characteristic gradients between gas and liquid are very high. Practically, the interpolation scheme does rule the respective weight of gas and liquid into interfacial cells.

As aforementioned, very few works have been found in literature in which averaging techniques are described and thoroughly assessed, even though they are used widely. One can mention three common schemes for evaluating $\rho$ and $\mu$ as functions of $(C_i)$. Let us consider a function f (the density or the viscosity) and its constant values in each phase, denoted as $f_i$, we can write:

- discontinuous averaging:

$$f(C_i) = f_i \text{ if } C_i \geq 0.5 \text{ and } f(C_i) = f_0 \text{ otherwise} \tag{2.88}$$

- arithmetic averaging:

$$f(C_i) = \sum_{i,i=1,2} C_i f_i + (1 - \sum_{i,i=1,2} C_i) f_0 \tag{2.89}$$

- harmonic averaging:

$$f(C_i) = \frac{\Pi_{i,i=0,2} f_i}{\Pi_{i,i=1,2} C_i f_0 + \Pi_{j,j \neq i}(1 - C_j) f_i} \tag{2.90}$$

The logic about considering harmonic averaging as an alternative to discontinuous or arithmetic averaging have been discussed in a former work by Ritz and Caltagirone [109]. Using different schemes for density and viscosity yield many potential combinations. We have reduced the number of possibilities to a set of four methods that are commonly used in multi-phase flow solvers.

These are, respectively, $M1$: discontinuous for both density and viscosity, $M2$: arithmetic averaging for both density and viscosity, $M5$: arithmetic averaging for density and harmonic averaging for viscosity, and finally $M7$: arithmetic averaging for density, arithmetic averaging when using diagonal components of the viscous stress tensor, and harmonic averaging when using extra-diagonal components of the viscous stress tensor. Obviously, using harmonic averaging for $\rho = f(C_i)$ is physically meaningless, and the additive property of density is perfectly suited to linear interpolation.

Figure 2.17 summarizes the terminal bubble shape predicted with those four methods in comparison to experimental [110] and front-tracking [111] simulation results from literature.

Figure 2.18 presents terminal bubble velocities with corresponding error levels in comparison to experimental velocities.

The best overall results have been achieved with the linear interpolation for both density and viscosity ($M2$). This technique combined with interface smoothing for capillarity exhibits excellent agreement with both experimental and front-tracking simulation results. Arithmetic averaging on viscosity yields non physical surface tension features, made clear from cases $A1 + M5$, $A2 + M5$, and $A3 + M5$. One problematic artifact is highlighted along the symmetry axis when using methods $M5$ or $M7$, as the interface presents a non physical angle inducing over-estimated terminal velocities. The $M1$ technique is globally relevant considering general shapes and terminal velocities, but discontinuous averaging techniques clearly increase the numerical fragmentation of the interface. Special situations of spherical cap shaped bubbles are notably complicated to simulate, as the bubble vertical thickness almost vanishes during its acceleration so that a sub-resolved gas layer on the bubble axis makes it split into toroidal bubbles instead. $A5$ cases are generally stable with our method but non-negligible fragmentation into secondary macro-bubbles are eventually observed. A6 cases have been obtained from one ellipsoidal starting shape but produced very unstable results in general. Consequently, M2 models will be used in this work, as a very good physical consistency is achieved. Moreover, the combination of VOF-PLIC, $M2$ averaging and interface smoothing for capillarity, exhibits excellent interface cohesion as very little numerical fragmentation is observed compared to other combinations.

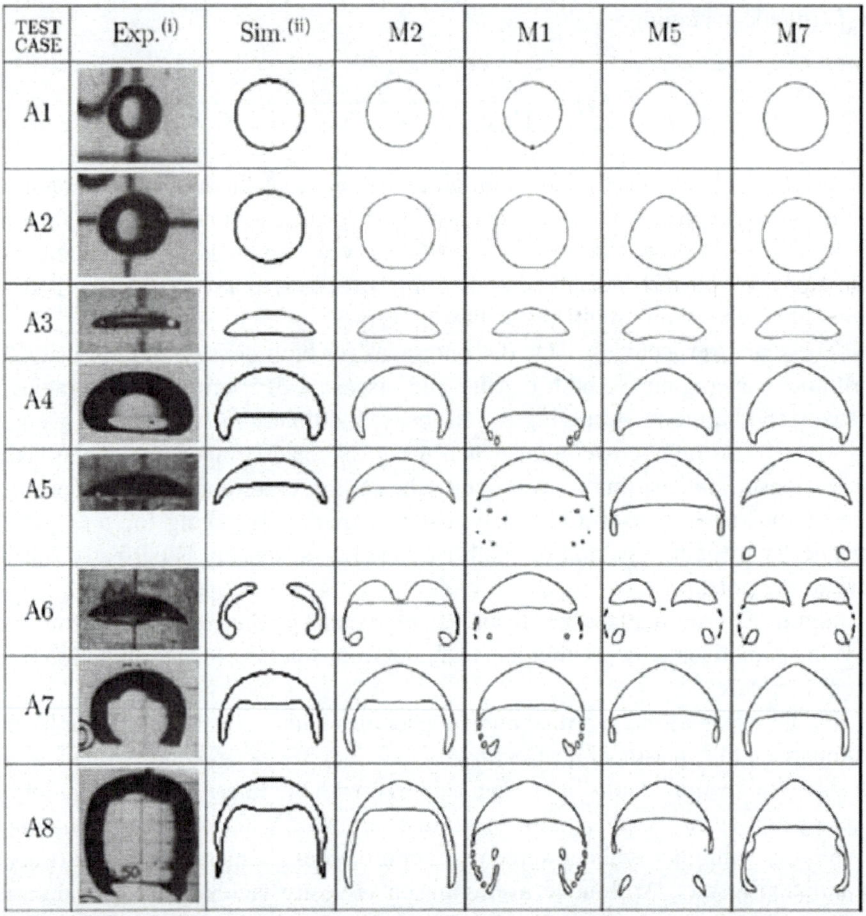


**Fig. 2.17** Terminal bubble shape for $M2$, $M1$, $M5$, and $M7$ averaging methods. Comparison to (i) experiments of Bhaga and Weber [110] and (ii) front-tracking simulations of Hua and Lou [111]

## *2.5.3 Capillary Forces*

The Laplace-Young law is now considered. A drop whose radius is $R$ is centered in a $L_b \times L_b$ square box. Gravity is zero. Initial velocity is zero. Inside the drop, the density and viscosity are written $\rho_{\text{int}}$ and $\mu_{\text{int}}$. Outside, they are written $\rho_{\text{ext}}$ and $\mu_{\text{ext}}$. Numerical parameters are as follow

| Test-Case | Run# | Interpolation | $U_{if}$ ($\mathrm{m\,s^{-1}}$) | $U_{Exp}$ ($\mathrm{m\,s^{-1}}$) | Error (%) | Best overall |
|---|---|---|---|---|---|---|
| A1 | 71 | M1 | 0.0364 | 0.0353 | 3.1 | |
| - | 51 | M2 | 0.0356 | - | 0.8 | × |
| - | 81 | M5 | 0.0573 | - | 62.4 | |
| - | 91 | M7 | 0.0418 | - | 18.4 | |
| A2 | 72 | M1 | 0.0613 | 0.0615 | 0.3 | |
| - | 52 | M2 | 0.0599 | - | 2.6 | × |
| - | 82 | M5 | 0.0890 | - | 44.7 | |
| - | 92 | M7 | 0.0705 | - | 14.6 | |
| A3 | 73 | M1 | 0.3139 | 0.3066 | 2.3 | |
| - | 53 | M2 | 0.3031 | - | 1.1 | × |
| - | 83 | M5 | 0.3132 | - | 2.1 | |
| - | 93 | M7 | 0.3103 | - | 1.2 | |
| A4 | 74 | M1 | 0.2465 | 0.2258 | 9.1 | |
| - | 54 | M2 | 0.2379 | - | 5.3 | × |
| - | 84 | M5 | 0.2858 | - | 26.5 | |
| - | 94 | M7 | 0.2804 | - | 24.2 | |
| A5 | 75 | M1 | 0.3144 | 0.3093 | 1.7 | |
| - | 55 | M2 | 0.2908 | - | 10.2 | × |
| - | 85 | M5 | 0.2280 | - | 26.3 | |
| - | 95 | M7 | 0.2999 | - | 3.0 | |
| A7 | 77 | M1 | 0.2877 | 0.2629 | 9.5 | |
| - | 57 | M2 | 0.2774 | - | 5.5 | × |
| - | 87 | M5 | 0.3236 | - | 23.1 | |
| - | 97 | M7 | 0.3185 | - | 21.2 | |
| A8 | 78 | M1 | 0.3016 | 0.2699 | 11.7 | |
| - | 58 | M2 | 0.2903 | - | 7.5 | × |
| - | 88 | M5 | 0.3373 | - | 24.9 | |
| - | 98 | M7 | 0.3411 | - | 26.4 | |

**Fig. 2.18** Terminal bubble rising velocity for $M2$, $M1$, $M5$, and $M7$ averaging methods. Comparison to the experimental values $U_{Exp}$ from the work of Bhaga and Weber [110]

$$L_b = 0.04\,m \tag{2.91}$$

$$R = 0.01\,m \tag{2.92}$$

$$\rho_{\text{int}} = 1000\,\text{kg m}^{-3} \tag{2.93}$$

$$\rho_{\text{int}} = 10^{-2}\text{Pa s} \tag{2.94}$$

$$\rho_{\text{int}} = 1\,\text{kg m}^{-3} \tag{2.95}$$

$$\rho_{\text{int}} = 10^{-4}\text{Pa s} \tag{2.96}$$

$$\sigma = 0.1\,\text{N m}^{-1} \tag{2.97}$$

The theoretical solution of the problem follows the Laplace solution (written in 2D):

$$\Delta p = p_{\text{int}} - p_{\text{ext}} = \frac{\sigma}{R} \tag{2.98}$$

where $\delta P$ is the pressure jump across the interface. Theoretically, the solution (2.98) is stationary and the velocity field should remain zero. However, local velocities appear in the vicinity of the interface because of numerical errors. They are denoted spurious currents [17]. Tough conditions are found: high density ratio (1000kg $\mathrm{m}^{-3}$) and high curvatures (drop with small radius). The characteristic dimensionless Ohnesorge (Oh) and Laplace (La) numbers are given by:

$$Oh = \frac{\mu_{\text{int}}}{\sqrt{2\sigma\rho_{\text{int}}R}} \tag{2.99}$$

$$La = \frac{1}{Oh^2} \tag{2.100}$$

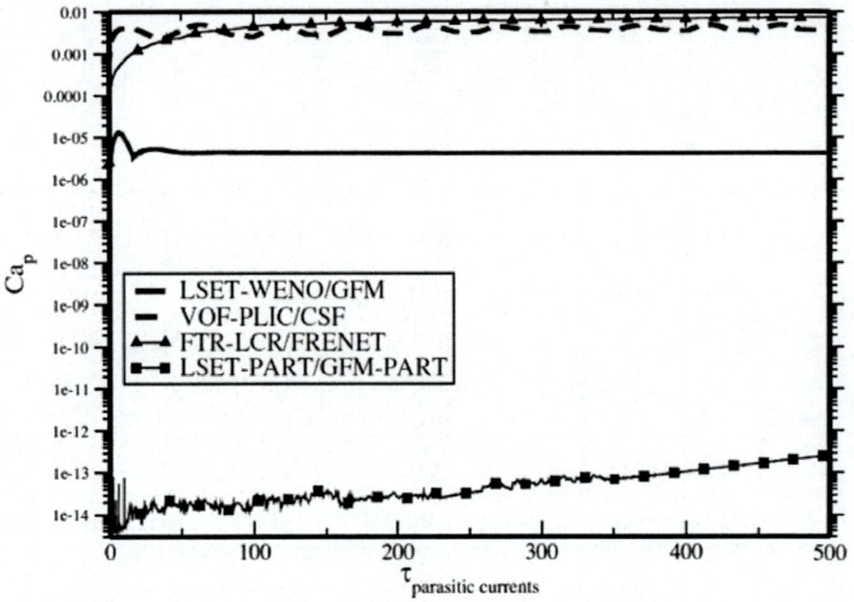


**Fig. 2.19** Spurious currents test case. Time evolution of $Ca_p$. A $32^2$ grid is used [106]

For the drop, $La = 20{,}000$ and $Oh = 7 \times 10^{-3}$, corresponding to flows dominated by capillary effects. The magnitude of the spurious currents is given by the capillary number $Ca_p(t)$ [17]:

$$Ca_p(t) = \frac{\mu_{\text{int}} \|\boldsymbol{u}\|_\infty}{\sigma} \tag{2.101}$$

where $\|\boldsymbol{u}\|_\infty$ is the maximum norm of the velocity field taken at time $t$. In Fig. 2.19, the time evolution of $Ca_p(t)$ is given vs $\tau_{\text{spurious currents}} = \frac{t\sigma}{2\mu_{\text{int}} R}$ ($t$ is the time and $\tau_{\text{spurious currents}}$ is a dimensionless time for the spurious velocities) for a $32^2$ grid size, which corresponds to 16 grid cells into the drop. It can be observed in Fig. 2.19 that the Level Set method, coupled to a ghost-fluid method (GFM) [45], gives errors two orders of magnitude lesser than the VOF-PLIC (coupled to the CSF treatment of capillary forces [44]) and the Front Tracking method (with Frenet approximation of surface tension forces [82]) concerning spurious currents test case.

For the LSET-PART method, also coupled to the ghost fluid approach, jumps are performed in a quasi-analytical way exactly on the interface without any extrapolation. Therefore, spurious currents are reduced up to the machine error. The differences observed in Fig. 2.19 between VOF-PLIC/CSF, FTR-LCR/Frenet and LSET-WENO/GFM are due to:

- The intrinsic sharp or volume character of the methods. The volume methods induce a regularization of the volume fraction before the estimate of the curvature (VOF-PLIC/CSF). The sharp front-tracking approach is based on a regularization of the exact surface tension force estimated by the Frenet theorem (FTR-

LCR/FRENET). In the sharp ghost-fluid approach, the exact surface tension force is distributed on the neighboring discrete points (LSET-WENO/GFM). All the methods generate spurious currents due to the regularization or redistribution steps required due to the non-conformity between the interface and the flow calculation grid.

- The numerical estimate of the curvature itself, which can be achieved through an Eulerian or Lagrangian procedure. A specific order of accuracy exists concerning this geometric parameter of the interface.

# Chapter 3
# Interface Tracking

In the fictitious domain approach, the building of the properties of the equivalent one-fluid model (2.4–2.5) requires the definition of the local values of densities and viscosities. These values are known as soon as the local volume fraction, or equivalently the level set function, has been advected through Eq. (2.3). Here, we discuss different ways of tracking these characteristic functions, and so the interface.

The definition and time evolution of $C_i$ characteristic functions can be satisfied by using and solving Eq. (2.3), what is called the Volume of Fluid method [18, 77]. The main advantage of these approaches, as soon as geometrical or high order fluxes are numerically used to approximate the advection equation, is that the volume, and so the mass, is nicely conserved by the method. The main drawback lies in the sharp $C_i$ profiles that the method considers. Indeed, by looking at the surface tension force that has to be estimated in the momentum equations (2.7), second order derivatives of $C_i$ have to be estimated to approximate the local curvature of the interface. By considering sharp VOF functions, these derivative are poorly estimated and additional smoother functions have to be calculated, based on $C_i$, to calculate the surface tension force.

To circumvent the regularity of the $C_i$ function of the VOF, an other approach consists in using signed distance functions $\phi_i$ to the interfaces instead of the VOF [112, 113]. The distance function is regular across the interface, which is defined as $\phi_i = 0$. Equation (2.3) is applied to $\phi_i$ instead of $C_i$. The major drawback of the level set is the mass conservation that is not nicely satisfied, due to approximation errors of hyperbolic schemes used to discretize the advection equation on $\phi_i$. To improve the method, redistance equations are solved. The main interest of the Level Set approach is that it provides an accurate representation of the normal and curvature of the interface, through the gradients of the Heaviside function (equivalent to VOF) defined according to $\phi_i$.



S. Vincent et al., *Small Scale Modeling and Simulation of Incompressible Turbulent Multi-Phase Flow*, CISM International Centre for Mechanical Sciences 607,
https://doi.org/10.1007/978-3-031-09265-7_3

## 3.1 VOF

### 3.1.1 Introduction to VOF Methods

The Volume-of-Fluid (VOF) method is a popular numerical technique to follow the evolution of interfaces in two-phase and free-surface flows within the framework of direct numerical simulation (DNS). The method is based on the phase indicator or characteristic function $\chi(\boldsymbol{x}, t)$, defined over a computational domain $V$, that represents a multidimensional Heaviside step function with value 1 in the volume $V_1$ occupied by the reference phase, and value 0 in the volume $V - V_1$ with the secondary phase or vacuum. A two-dimensional schematic representation of an initial configuration for a typical two-phase flow simulation is shown in Fig. 3.1.

The function $\chi$ can also be expressed as the integral over the volume $V_1$ of the Dirac delta function

$$\chi(\boldsymbol{x}, t) = \int_{V_1} \delta(\boldsymbol{x} - \boldsymbol{x}')\, d\,V' \tag{3.1}$$

The volume $V_1$ is bounded by the interface $S_1$, here supposed to be continuous and differentiable, where both $V_1$ and $S_1$ can change with time $t$. With this formulation we can compute the outward unit normal $\boldsymbol{n}$ by taking the gradient of (3.1) and transforming a volume integral into an integral over the interface

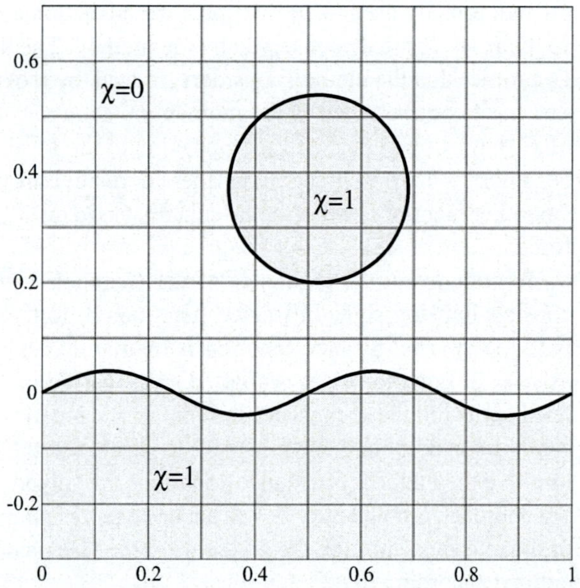


**Fig. 3.1** Two-dimensional initialization of a circular droplet impacting on a wavy gas/liquid interface

$$\nabla \chi(\boldsymbol{x}, t) = - \int_{V_1} \nabla'\big(\delta(\boldsymbol{x} - \boldsymbol{x}')\big)\, d\,V' = - \int_{S_1} \boldsymbol{n}'\, \delta(\boldsymbol{x} - \boldsymbol{x}')\, d\,S' \tag{3.2}$$

The function $\nabla \chi$ is singular over the domain $V$ in the sense that is different from zero only over sets of zero measure. In addition to its position and local outward unit normal, another important geometrical property of the interface is the mean curvature $\kappa$, which is related to the spatial variation of the normal vector.

Since the fluid type does not change following the fluid paths, as we are not considering here phase changes, the indicator function $\chi$ behaves like a passive scalar and satisfies the following advection equation

$$\frac{d\,\chi}{d\,t} = \frac{\partial\,\chi}{\partial\,t} + \boldsymbol{u} \cdot \nabla \chi = 0 \tag{3.3}$$

moreover if $\boldsymbol{u}$ is a divergence-free flow field, $\nabla \cdot \boldsymbol{u} = 0$, then

$$\frac{\partial\,\chi}{\partial\,t} + \nabla \cdot (\chi\, \boldsymbol{u}) = 0 \tag{3.4}$$

Here we use a weak formulation of this partial differential equation, since the partial derivatives of the discontinuous function $\chi$ are singular. In the weak formulation, the differential equation is interpreted by space integrals, which are well-defined.

We now consider a Cartesian domain subdivided by cubic cells of side $\Delta x = \Delta y = \Delta z = h_0$ and define the color or volume fraction function $C_i(t)$, in the $i$-th grid cell of volume $V_i = h_0^3$, by

$$C_i(t) = \frac{1}{V_i} \int_{V_i} \chi(\boldsymbol{x}, t)\, d\,V \tag{3.5}$$

Hence, the volume fraction $C$ is a discrete version of $\chi$, and satisfies $0 \leq C \leq 1$, with $C = 0$ in the empty cells, i.e. with no reference phase inside, and $C = 1$ in the full cells. We now integrate (3.4) over the cell volume $V_i$, and use Gauss's theorem, to get

$$V_i\, \frac{\partial\, C_i(t)}{\partial\,t} + \int_{V_i} \chi\, \boldsymbol{u} \cdot \boldsymbol{n}_i\, d\,V = 0 \tag{3.6}$$

where $\boldsymbol{n}_i$ is the cell outward unit normal. Strictly speaking, (3.6) cannot be written only as a function of $C_i$. In fact the color function does not contain the precise information on the interface location which are necessary to compute the reference phase fluxes across the cell boundary [114]. Classical geometric VOF algorithms base their computations only on the color function $C$. They actually solve an inverse problem, since from the known color function $C$ an estimate $\widetilde{\chi}$ of $\chi$ is found that satisfies (3.5) and approximates the interface in the $i$-th cell. After the reconstruction, the reference phase fluxes can be calculated with a geometric approach. The advection equation for $C$ is often written more simply as

$$\frac{\partial C(t)}{\partial t} + \nabla \cdot (C\,\boldsymbol{u}) = 0 \tag{3.7}$$

and standard algebraic methods for hyperbolic equations, such as upwind, ENO and flux-corrected transport schemes can also be used.

### 3.1.2 Initialization of the Color Function C

The equation of an interface separating two immiscible phases can be written in the implicit form $f(\boldsymbol{x}) = 0$, and the reference phase can be arbitrarily located in the region where $f(\boldsymbol{x}) < 0$ and the secondary phase where $f(\boldsymbol{x}) > 0$, or viceversa.

A straightforward initialization of the color function in each cell of the computational domain can be easily attained by considering an arbitrary number of internal nodes, either on a regular submesh or randomly generated. The local value $C$ is then given by the ratio of the number of nodes inside the reference phase to the total number of nodes. For the two-dimensional case of Fig. 3.2 we assume that the reference phase is where $f(\boldsymbol{x}) < 0$ and the internal nodes are the dark dots. More advanced techniques for multidimensional integration include Monte Carlo methods with different sampling strategies to improve the efficiency of the method [115].

A second possibility is to initialize the color function with a recursive local mesh refinement in the cells cut by the interface. At the finest level the intersections of the interface with the refined boundary should be computed and connected to form in two-dimensions a piecewise-linear approximation of the interface in each cut cell. The area comprised between the approximated interface and the cell boundary can

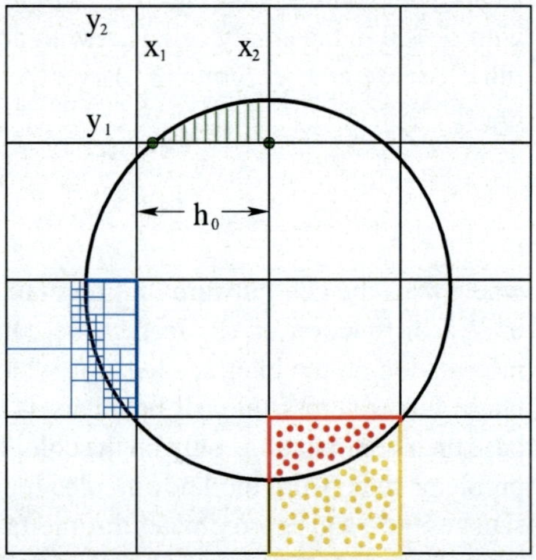


**Fig. 3.2** Different initialization of the color function: Monte Carlo approach (bottom cell with dark/light dots), recursive local mesh refinement (left cell) and analytical integration (top cell) with local height $h$

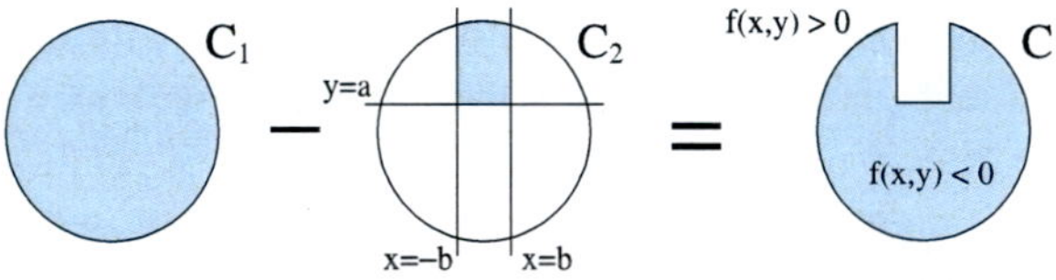


**Fig. 3.3** Initialization of the Zalesak's disk: color functions $C_1$ for the circle and $C_2$ for the notch, and color function $C$ for the the notched disk from their difference, $C = C_1 - C_2$

then be easily computed. Four levels of mesh refinement are used in the left cell of Fig. 3.2, and the method has been shown to be globally second-order accurate [116].

A very accurate initialization of the color function may require an integration of the analytical expression of the interface. For the top cell of Fig. 3.2, from the explicit representation of the interface line $y = g(x)$, we have

$$C_i(t) = \frac{1}{V_i} \int_{V_i} \chi(\boldsymbol{x}, t)\, dV = \frac{1}{h_0^2} \int_{x_1}^{x_2} \min\big(h_0, \max(0, g(x) - y_1)\big)\, dx \tag{3.8}$$

The integrand $h(x) = \min\big(h_0, \max(0, g(x) - y_1)\big)$ in (3.8) represents the local height that must satisfy the relation $0 \le h \le h_0$.

### 3.1.3 *A Library to Initialize the Volume Fraction Field*

The volume fraction field can be initialized with the open-source library VOFI from a given implicit equation of the interface, $f(\boldsymbol{x}) = 0$, which is particularly convenient for closed surfaces [117, 118]. The algorithm locates the reference phase in the points $\boldsymbol{x}$ where the function is negative. With this approach is relatively easy to determine the color function for complex figures, by subdividing each figure in simpler shapes and by adding algebraically the corresponding color functions, as for the case of the Zalesak's notched disk of Fig. 3.3.

The first step of the initialization of the volume fraction field is the computation of a cutoff function value $f_h$ which is used to determine if a grid cell is either full or empty, or if it requires a more detailed analysis. From an initial position $\boldsymbol{x}_0$ the library computes the function value $f(\boldsymbol{x}_0)$ and the local gradient $\nabla f(\boldsymbol{x}_0)$ with centered finite differences. The search algorithm is based on an iterative procedure that moves along the gradient to a new point where the values of the function and of its gradient are updated. When the interface is finally crossed the intersection point $\boldsymbol{x}_I$ is computed and the cutoff value $f_h$ is given by the expression

$$f_h = \max\left(|f(\boldsymbol{x}_I + \rho\, h_0\, \boldsymbol{n})|, |f(\boldsymbol{x}_I - \rho\, h_0\, \boldsymbol{n})|\right) \tag{3.9}$$

where $\boldsymbol{n}$ is the interface unit normal and $\rho = \sqrt{2}/4$.

Each cell of the computational grid is then examined with the computation of the function value $f_j$ at $N_t$ points of a local submesh with step size $h_0/2$ along each coordinate direction, hence $N_t = 9$ in 2D and $N_t = 27$ in 3D. If all the values $f_j$ have the same sign and satisfy $|f_j| > f_h$, then the cell is full if $f_j < 0$, and $C = 1$, otherwise it is empty and $C = 0$. In all the other cases the library computes the gradient with centered finite differences in the points of the submesh. The cell-averaged gradient component with the greatest absolute value determines the main coordinate direction $x_1$, while the second and third directions, $x_2$ and $x_3$ respectively, are associated with the next largest and smallest components. The gradients computed in the subgrid points are also used to determine a tentative number of nodes $n_*$ to be used in the numerical integration.

A first major assumption in the library design is that in each cut cell the implicit equation of the interface can be locally written in the explicit form $x_1 = g(x_2, x_3)$, where $x_1$ is the main coordinate direction and each $x_i$ is one of the three Cartesian coordinates. In the numerical integration to compute the volume fraction $C$, values of the local height function $h(x_2, x_3)$ are calculated along the main coordinate direction $x_1$, while the internal and external limits of integration are calculated along the second and third directions. A second major assumption is that the interface cannot intersect a cell side more than twice, since this would imply a characteristic lengthscale smaller than the grid spacing $h_0$. This assumption simplifies considerably the algorithm.

Furthermore, when an interface intersects the sides of a two-dimensional cell, as shown in Fig. 3.4, the integrand $h(x)$ has a discontinuous first derivative in the intersection point $(x_{i1}, y_1)$. To limit the integration error near the discontinuity, the library explicitly computes the abscissa $x_{i1}$ of the zero and performs the integration separately in the two subintervals $[x_1, x_{i1}]$ and $[x_{i1}, x_2]$. A hybrid root-finding algorithm, based on the secant and bisection methods, has been implemented to compute both the intersections of the interface with the cell sides and the local heights $h(x)$ for the numerical integration, as shown in Fig. 3.4. The local height is computed in the region where $f(\mathbf{x}) < 0$ and satisfies $0 \le h \le h_0$.

In three dimensions the interface can also intersect the cell face on a closed line, as shown on the left of Fig. 3.5, while in two-dimensions there can be a maximum of two intersections with a cell side, as shown on the right of the same figure. In the first case, in order to determine the two external limits of integration $x_{31}$ and $x_{32}$, a minimum search routine on a cell face is first used to detect a change of sign of the function $f(\mathbf{x})$, then the root-finding routine is called several times to get a convergent sequence of approximations to the two external limits $x_{31}$ and $x_{32}$. In the second case, a minimum search routine on a cell segment is used to detect the function change of sign. The minimum search on a segment is based on Brent's method [119], while a preconditioned conjugate gradient method, with the Polak-Ribiere's expression for the parameter used to update the conjugate direction, is considered for the minimum search on a cell face [120].

The numerical calculation of the volume fraction in a cubic cell cut by the interface is performed by subdividing the cube into rectangular hexahedra and by considering a double Gauss-Legendre integration in each hexahedron cut by the interface. As an example, the cell of Fig. 3.5 is subdivided into three hexahedra, with the interface

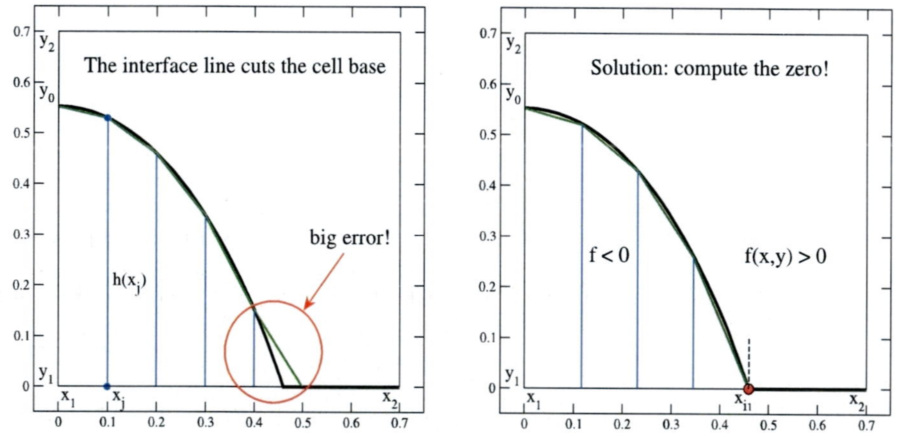


**Fig. 3.4** Most of the integration error of a numerical method is located near the intersection $x_{i1}$ with the cell base (left), this error is greatly reduced with two separate integrations in the subintervals $[x_1, x_{i1}]$ and $[x_{i1}, x_2]$ (right)

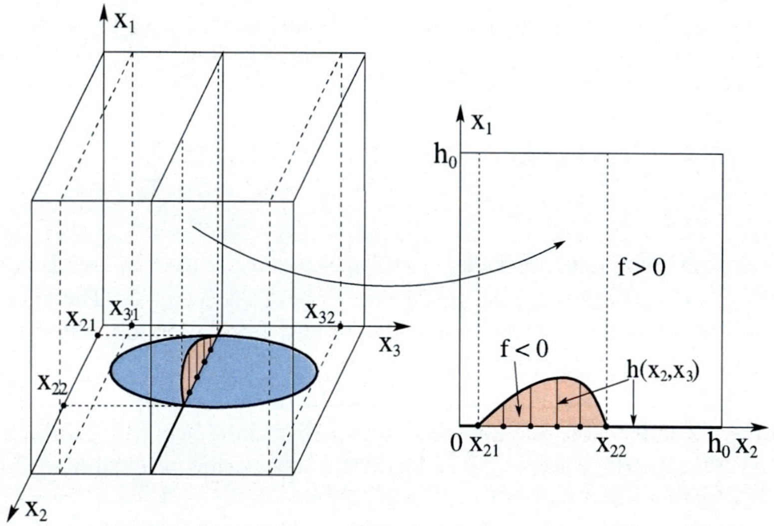


**Fig. 3.5** For the initialization of the volume fraction in a cut cell in three-dimensions both the external limits of integration $x_{31}$ and $x_{32}$ (left) and the internal limits $x_{21}$ and $x_{22}$ (right) must be computed

being present only in the middle one. Therefore, the algorithm computes the volume fraction with the following numerical integration along the third direction $x_3$

$$C = \frac{1}{h_0} \int_{x_{31}}^{x_{32}} A(x_3)\, dx_3 \simeq \frac{x_{32} - x_{31}}{2\, h_0} \sum_{k=1}^{n_e} \omega_k\, A(x_{3k}) \tag{3.10}$$

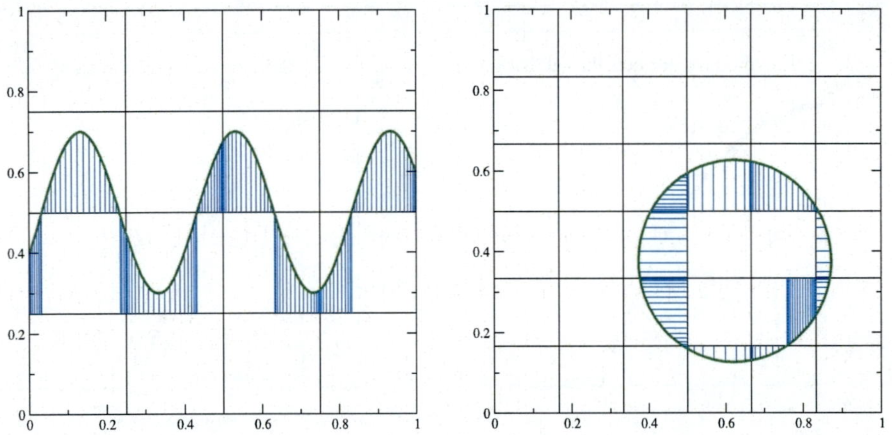


**Fig. 3.6** Initialization in the unit square of a sinusoisal wave with grid resolution $m = 4$ (left) and of a circle with $m = 6$ (right)

where the coefficients $\omega_k$ are the integration weights and $A(x_{3k})$ the normalized areas that come from the integration of the local height function $h(x_2, x_{3k})$ along the second direction $x_2$

$$A(x_{3k}) = \frac{1}{h_0^2} \int_{x_{21}(x_{3k})}^{x_{22}(x_{3k})} h(x_2, x_{3k})\, dx_2 \simeq \frac{x_{22}(x_{3k}) - x_{21}(x_{3k})}{2\, h_0^2} \sum_{j=1}^{n_i} \omega_j\, h(x_{2j}, x_{3k}) \tag{3.11}$$

To determine the number of nodes $n_i$ and $n_e$ we consider the two non-dimensional lengths $\delta_i = (x_{22}(x_{3k}) - x_{21}(x_{3k}))/h_0$ and $\delta_e = (x_{32} - x_{31})/h_0$. The number of nodes $n_i$ of the internal integration is determined from a lookup table as a function of the length $\delta_i$ and of the tentative number of nodes $n_*$, while the number of nodes $n_e$ of the external integration is only a function of the length $\delta_e$ [117].

For an accurate reconstruction and advection of an interface, with a linear approximation of the interface in each cut cell, the ratio between the local radius of curvature $r_0$ of the interface and the grid spacing $h_0$ should satisfy $r_0/h_0 \geq 4$. In Fig. 3.6 we show the initialization of a sinusoidal wave and of a circle in the unit square, with grid resolution $m = 4$ and $m = 6$ respectively. In each cut cell are also shown the local heights that are used in the numerical integration. The initialization is very accurate as long as the two basic assumptions are satisfied even if the ratio $r_0/h_0$ is locally smaller than four.

In particular, the initialization of the volume fraction field inside the circle shows how the numerical algorithm selects the number of integration nodes according to the local variation of the interface line. The relative error of the initialization is defined by $E_1 = |A_a - A_n|/A_a$, where $A_n$ is the numerical area computed with the VOFI library and $A_a$ its analytical value. The results of Table 3.1 show that for the circle test the relative error with the proposed algorithm with a variable number of nodes corresponds to that with a fixed number of nodes comprised between 12 and 16.

**Table 3.1** Grid resolution $m$ of the unit square for the initialization of a circle, relative error $E_1$ with a fixed number of integration nodes: 8,12,16,20, relative error $E_1$ with the average number of nodes $n_{ave}$ selected by the VOFI library

| $m$ | $E_1$ (8) | $E_1$ (12) | $E_1$ (16) | $E_1$ (20) | $E_1$ ($n_{ave}$) |
|---|---|---|---|---|---|
| 4 | 3.89e−11 | 1.98e−15 | 2.83e−16 | 1.41e−16 | 1.41e−15 (16) |
| 6 | 6.47e−10 | 1.86e−13 | 2.83e−16 | 2.83e−16 | 8.00e−14 (9.7) |
| 10 | 1.03e−11 | 8.48e−16 | 1.41e−16 | 2.82e−16 | 7.07e−16 (8.4) |

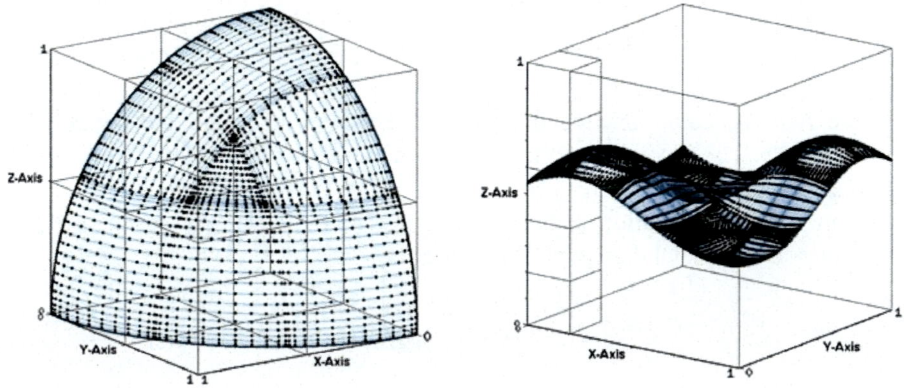


**Fig. 3.7** End points of the local height for the initialization in the unit cube of an eighth of a sphere with grid resolution $m = 2$ (left) and of a sinusoisal wave with $m = 5$ (right)

However, the average number of integration nodes in a cut cell diminishes with the grid resolution, and with $m = 10$ this number is $n_{ave} = 8.4$. For the two resolutions $m = 4$ and $m = 6$ the results are systematically better at the lower resolution. This is due to the symmetrical alignment of the circle with respect to the grid lines when $m = 4$, since the circle center almost coincides with a cell center.

In Fig. 3.7 we show a portion of two interfaces that are commonly used to initialize a two-phase flow simulation with the VOF method in 3D. The end points of the local heights, which are on the interface and are used in the 2D integration along the second direction $x_2$, are connected by a line. The grid spacing for the eighth of a sphere in the unit cube is $h_0 = r_0/2$ with a radius $r_0 = 1$, while the grid resolution for the sinusoidal wave is $m = 5$. In both cases the relative error $E_1 = |V_a - V_n|/V_a$, where $V$ is now the volume, is very small, $E_1 \simeq 1.\,10^{-15}$.

### 3.1.4 *Algebraic Methods for the Advection of the Color Function*

Let us consider a one dimensional discrete version of (3.7) with staggered spatial variables, i.e. the scalar variable $C$ is positioned at the cell center and the velocity $u$ at the cell face, as shown in Fig. 3.8

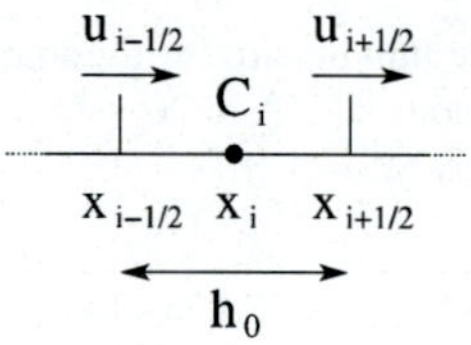


**Fig. 3.8** One-dimensional staggered grid with the color function $C$ at the cell center and velocity $u$ at the cell faces

$$\frac{\partial C_i}{\partial t} = \frac{1}{h_0}\left(u_{i-1/2}\,C_{i-1/2} - u_{i+1/2}\,C_{i+1/2}\right) = \frac{1}{h_0}\left(F_{i-1/2} - F_{i+1/2}\right) \tag{3.12}$$

where $F$ is the color function flux, $u$ is a face-averaged value and the cell face area is unity. The incompressibility condition $\nabla \cdot \boldsymbol{u} = 0$ in one dimension has the trivial solution $u = u_0$, where $u_0$ is a constant value. Finally we consider a first-order forward time-discretization scheme from time $t^n$ to $t^{n+1}$, with $\Delta t = t^{n+1} - t^n$, and variables in the spatial derivative computed at $t^n$

$$C_i^{n+1} = C_i^n + \frac{u_0\,\Delta t}{h_0}\left(C_{i-1/2}^n - C_{i+1/2}^n\right) \tag{3.13}$$

This equation has been written in a discretely conservative form, since the fluxes at the boundary between two cells cancel out exactly. The ratio on the right hand side is the Courant-Friedrich-Levy (CFL) number that for stability should be smaller than one. In the following simulations we assume $u_0\,\Delta t/h_0 = 1/4$.

In one dimension the interface is completely defined by its position $x_I$ and at the initial time $t_0$ we assume $x_I = 1/2$ and the following Heaviside step function for the initial $\chi$ profile

$$\chi(x, t_0) = \begin{cases} 1 & \text{if } x < 1/2 \\ 0 & \text{if } x > 1/2 \end{cases} \tag{3.14}$$

The computational domain is the unit segment, $0 \le x \le 1$, subdivided with $m = 100$ equal subintervals. Furthermore, the following condition is applied on the left boundary $C(0, t) = 1$, since $u_0 > 0$.

Moving around the piecewise-constant characteristic function with a constant velocity $u_0$ is actually a rather difficult task, but at the same time this difficulty will point out very quickly the problems related to the numerical scheme under consideration. The evolution equation (3.13) has been derived for the cell-centered value $C_i$ but we need to compute the face-averaged values $C_{i\pm1/2}$ as well.

A first choice is the centered scheme

$$C_{i+1/2} = \frac{C_{i+1} + C_i}{2} \tag{3.15}$$

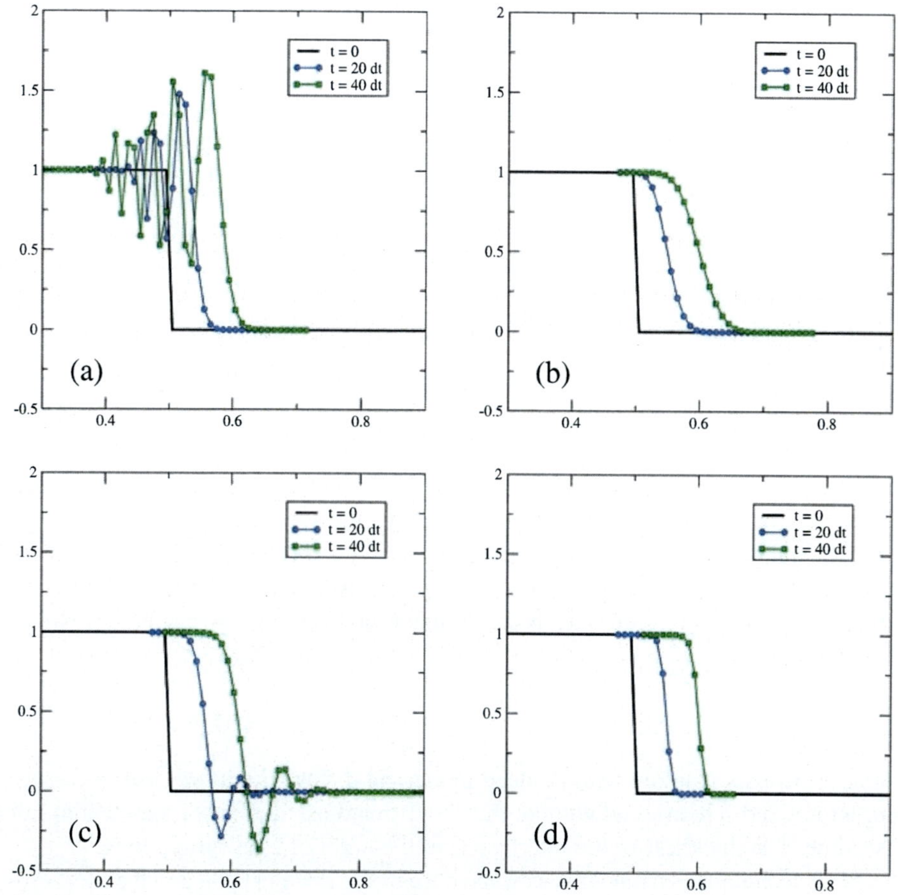


**Fig. 3.9** Advection of a piecewise constant function: **a** centered scheme, **b** first-order upwind scheme, **c** second-order upwind scheme, **d** slope-limter scheme

which is exact for a linear profile and second-order accurate in space, $O(h_0^2)$. This method is known to be unstable as it develops oscillations of increasing amplitude, as seen in Fig. 3.9a, and also shows that the CFL condition is not sufficient for stability.

A second choice is the first-order upwind scheme, an asymmetric flux that takes into account the direction of propagation. In this case $u_0 > 0$ and the approximation becomes

$$C_{i+1/2} = C_i \tag{3.16}$$

which is exact for a constant profile and first-order accurate, $O(h_0)$. The solution remains bounded in time and within the prescribed limits, $0 \leq C \leq 1$, but the interfaces thickens in time, as shown in Fig. 3.9b. This is a clear indication that the method is numerically diffusive.

A second-order accurate upwind scheme requires a linear extrapolation of the profile from the upwind direction

$$C_{i+1/2} = C_i + \frac{C_i - C_{i-1}}{2} \tag{3.17}$$

The method appears to be less diffusive near the interface, but it develops large oscillations downstream the interface with increasing amplitude in time, see Fig. 3.9c.

The linear extrapolation (3.17) can be generalized in the following way

$$C_{i+1/2} = C_i + \phi(r)\,\frac{C_i - C_{i-1}}{2} \tag{3.18}$$

where $\phi(r)$ is the slope limiter function and $r$ is the ratio of successive gradients on the mesh

$$r_i = \frac{C_i - C_{i-1}}{C_{i+1} - C_i} \tag{3.19}$$

The limiter function must satisfy the constraint $\phi(r) \geq 0$. In the regions of sharp gradients or opposite slopes, where $\phi(r) \approx 0$, the face-averaged value $C_{i+1/2}$ is approximated by the stable first-order scheme, on the other hand in the smooth regions, where $\phi(r) \approx 1$, $C_{i+1/2}$ will be given by the second-order scheme. The results shown in Fig. 3.9d have been obtained with the van Leer symmetric limiter function [121]

$$\phi(r) = \frac{r + |r|}{1 + |r|} \tag{3.20}$$

that provides a solution with reduced numerical diffusion with respect to the first-order upwind scheme, and without the downstream oscillations of the second-order scheme. The thickness of the interface is of the order of a few grid points.

This technique is called flux-corrected transport (FCT) [122], and the flux across a cell boundary is given by a weighted average of a flux computed by a low order scheme and a flux computed by a high order scheme. The weighting procedure tries to use the high order flux as much as possible without introducing oscillations and to limit the low order diffusive scheme in the regions near steep gradients. These algorithms can resolve contact discontinuities over a few grid points. Other high-order schemes such as Essentially non-oscillatory (ENO) and Weighted ENO (WENO) have been extensively and successfully used for the advection of the level set function, which is a smooth function.

For a one-dimensional incompressible flow it is possible to devise a scheme that updates the color function without any thickening of the interface or the appearance of oscillations. This is the geometric VOF method and it is based on the fact that the only relevant parameter is the position $x_I(t)$ of the interface. With reference to (3.12) and with $u_0 > 0$, the color function flux through the right face of the $i$-th cell is $F_{i+1/2} = 0$ if $C_i = 0$, and $F_{i+1/2} = u_0$ if $C_i = 1$. On the other hand if the $i$-th cell is cut by the interface, its position is simply $x_I = x_{i-1/2} + C_i\, h_0$, for the given initial distribution $\chi(x, t_0)$. In this case the reference phase flux is $F_{i+1/2} = 0$ when $u_0\, \Delta t \leq (1 - C_i)\, h_0$, as shown in Fig. 3.10a, and $F_{i+1/2} = u_0 - h_0\,(1 - C_i)/\Delta t$ in the case of Fig. 3.10b. Considering the fact that the ratio $u_0\, \Delta t / h_0$ is the CFL number, the complete fluxing scheme for all values of $C_i$ becomes

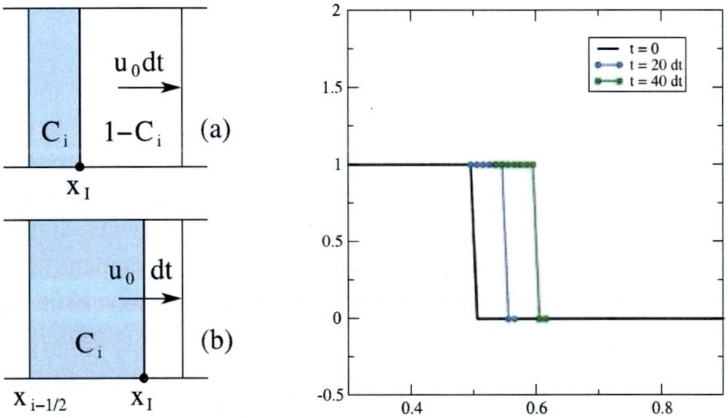


**Fig. 3.10** Geometric VOF advection scheme in a cut cell for two different interface positions: **a** there is no flux of the reference phase through the right side (top-left), **b** the flux is $F_{i+1/2} = u_0 - h_0\,(1 - C_i)/\Delta t$ (bottom-left); advection of a piecewise constant function with the VOF scheme (right)

$$F_{i+1/2} = \begin{cases} 0 & \text{if } CFL \le (1 - C_i) \\ u_0 - \dfrac{h_0}{\Delta t}(1 - C_i) & \text{otherwise} \end{cases} \tag{3.21}$$

The advection of the initial $\chi$ profile (3.14) with this scheme is shown on the right of Fig. 3.10, that demonstrates that the one-dimensional geometric VOF method is conservative, not diffusive and with no oscillation.

Algebraic VOF algorithms are still rather common both in research and commercial codes since they do not require a geometrical reconstruction and it is somewhat easier to implement them in unstructured grids. However, since the interface is not reconstructed, its local unit normal vector and curvature must be computed in some other way. For a multidimensional VOF calculation a split technique is often used, that is characterized by a sequence of one-dimensional sweeps along one coordinate direction at a time. In this case, even if the velocity field $\boldsymbol{u} = (u, v, w)$ is divergence-free, the sweep along one coordinate direction is not, and the advection equation (3.7), limited to the $x$-axis, should be written as

$$\frac{\partial C}{\partial t} + \frac{\partial}{\partial x}(C\,u) - C\,\frac{\partial u}{\partial x} = 0 \tag{3.22}$$

with the following explicit finite-volume discretization and the velocity at time $t^n$

$$C_i^{n+1} = C_i^n + \frac{\Delta t}{h_0}\left(F_{i-1/2}^n - F_{i+1/2}^n\right) + C_i^n\,\frac{\Delta t}{h_0}\left(u_{i+1/2} - u_{i-1/2}\right) \tag{3.23}$$

We now discuss briefly a few algebraic VOF schemes. In the SURFER code [17] both the first-order upwind scheme (3.16) and the first-order downwind scheme are considered, therefore when $u > 0$

$$C_{i+1/2} = C^{up} = C_i \ ; \quad C_{i+1/2} = C^{down} = C_{i+1} \tag{3.24}$$

Furthermore, in the downwind scheme the flux is limited to a clean sweep of the upwind cell and is made symmetric to the transport of one phase or the other as well. While the upwind scheme is stable but diffusive, the downwind one is unstable but it sharpens the interface. The front sharpening feature of the downwind scheme is selected when the flow is mainly perpendicular to the flow, however if the interface is parallel to the flow this scheme creates spikes on the interface and the overall scheme would also be unstable. Therefore the upwind scheme is selected in the second case. The method requires the computation of a local approximation of the interface unit normal

$$\boldsymbol{n} = (n_x, n_y, n_z) = \nabla^h C / |\nabla^h C| \tag{3.25}$$

where $\nabla^h$ is a finite difference approximation of the gradient operator, and the definition of a critical angle $\theta_c$. For a 1D flow along the $x$-direction the angle $\theta_x = \arccos(n_x)$ is computed and the upwind scheme is selected when $\theta_c < \theta_x$, the downwind scheme in the other cases.

It is also possible to avoid the need to devise different schemes for convection by adding an additional convective term to the transport equation for the volume fraction $C$ [123]

$$\frac{\partial C(t)}{\partial t} + \nabla \cdot (C\,\boldsymbol{u}) + \nabla \cdot \big(C\,(1 - C)\,\boldsymbol{u}_c\big) = 0 \tag{3.26}$$

This term is clearly active only near the interface, where the product $C\,(1 - C)$ is different from zero, and contributes to a higher interface resolution. The velocity $\boldsymbol{u}_c$ is actually the relative velocity between the two phases that "compresses" the interface and keeps it sharp and can be written as a function of the volume fraction gradient $\nabla^h C$, the local velocity $\boldsymbol{u}$ and a parameter that controls the amount of interface compression.

The 1D THINC scheme [124] computes the solution $C_i^{n+1}$ of (3.23) by approximating the discontinuity at the interface of the phase indicator function in the $i$-th cut cell with a hyperbolic tangent profile

$$\Phi_i(x) = \frac{1}{2}\Big(1 + \gamma \tanh\Big(\beta\Big(\frac{x - x_{i-1/2}}{h_0} - X_i\Big)\Big)\Big) \tag{3.27}$$

where $\gamma = 1$ if $C_{i-1}^n < C_{i+1}^n$, otherwise $\gamma = -1$, and $\beta$ is a parameter that controls the slope and thickness of the jump. The position of the jump center $X_i$ is uniquely determined by the integral condition (3.5)

$$C_i^n = \frac{1}{h_0} \int_{x_{i-1/2}}^{x_{i+1/2}} \Phi_i(x)\, dx \tag{3.28}$$

The reference phase flux $F_{i+1/2}$ across the cell boundary at $x = x_{i+1/2}$ is computed with a first-order upwind scheme, for example if $u_{i+1/2} = u_+ > 0$ then

$$F_{i+1/2} = \int_{t^n}^{t^{n+1}} \Phi_i\big(x_{i+1/2} - u_+(t - t^n)\big)\, u_+\, dt \tag{3.29}$$

If $u_+ < 0$ the profile $\Phi_{i+1}(x)$ of the upwind cell should be considered. It is also possible to derive closed formulae for the jump center and the reference phase flux. For $u_+ \geq 0$ the upwind cell is the $i$-th cell and the parameter $\lambda = 1$, otherwise the upwind cell is the $i+1$-th cell and the parameter $\lambda = 0$. The jump center $X^{up}$ of the upwind cell is

$$X^{up} = \frac{1}{2\beta} \ln\left(\frac{\exp\big(\beta\,(1+\gamma-2\,C^{up})/\gamma\big) - 1}{1 - \exp\big(\beta\,(1-\gamma-2\,C^{up})/\gamma\big)}\right) \tag{3.30}$$

where the volume fraction $C^{up}$ is the upwind value, and the reference phase flux $F_{i+1/2}$ is

$$F_{i+1/2} = \frac{1}{2}\left( - u_+\,\Delta t + \frac{\gamma\, h_0}{\beta} \ln\left(\frac{\cosh\big(\beta(\lambda - X^{up} - u_+\,\Delta t/h_0)\big)}{\cosh\big(\beta(\lambda - X^{up})\big)}\right)\right) \tag{3.31}$$

The thickness of the jump transition is controlled by the parameter $\beta$, and for $\beta = 2.3$ the jump is resolved by approximately 2–3 mesh cells.

In a multi-dimensional advection a constant value of $\beta$ will tend to ruffle the interface when it is aligned to the velocity direction. A more general approach would be to adapt the value of $\beta$ to the orientation of the interface. Parker and Youngs' algorithm can be used to compute the interface unit normal $\boldsymbol{n} = (n_x, n_y, n_z)$, then an estimate of $\beta_k$ for a 1D advection along the $k$-direction is

$$\beta_k = 2.3\,|n_k| + 0.01 \tag{3.32}$$

### 3.1.5 Simple Geometric Methods for the Advection of the Color Function

The geometric VOF method (3.21) is exact for a one-dimensional and incompressible flow, but its extension to multi-dimensional problems is not straightforward. The simplest interface reconstruction in each cut cell is to align the interface to a coordinate axis in two dimensions and to a coordinate plane in three dimensions, hence the reconstruction is piecewise constant.

The SLIC method (Simple Line Interface Calculation) [125] is a direction-split method where only cell neighbours along the sweep direction are used to determine

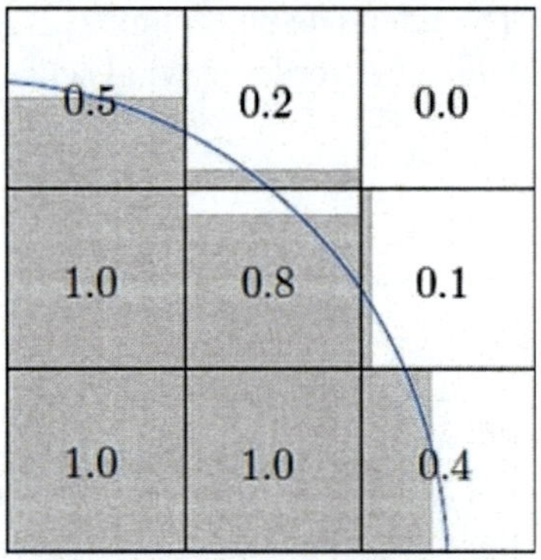


**Fig. 3.11** SLIC method: **a** interface line; **b** SLIC reconstruction for the $x$-sweep; **c** SLIC reconstruction for the $y$-sweep

the interface reconstruction. There are three basic cases and because the reconstruction considers neighbours only in the fluxing direction, a cut cell can have a different representation for each sweep direction, as shown in Fig. 3.11. The interface is reconstructed cell by cell before advection and fluxes are computed geometrically. The order of the 1D advection sweeps is changed every timestep to avoid a directional bias and, because of the split technique, Eq. (3.22) should be considered.

In the original VOF method [73] the interface reconstruction in 2D is again parallel to a coordinate axis in each cell, but it does not change with the sweep direction. A block of $3 \times 3$ cells, centered on the cell under investigation, is used to estimate the local normal $\boldsymbol{m} = \nabla^h C$ and the interface is specified as horizontal or vertical depending on the relative magnitude of the normal components. The extension to 3D is straightforward, and it requires a surrounding block with 27 cells. For a 1D advection parallel to the interface reconstruction, a first-order upwind flux is considered. When the propagation is perpendicular to the interface, as shown in Fig. 3.12 for a propagation in the positive $x$-direction, a combination of upwind and downwind fluxes is considered. The resulting flux $F_{i+1/2}$ through the side at $x_{i+1/2}$ is limited so that the fluxes of the reference and secondary phases leaving the $i$-th cell do not exceed the amount of each phase present in the cell

$$F_{i+1/2} = \min\left(C_i \frac{h_0}{\Delta t}, u_{i+1/2}\, C_{i+1} + \max\left(0, u_{i+1/2}\,(1 - C_{i+1}) - (1 - C_i)\frac{h_0}{\Delta t}\right)\right)$$

An alternative way to discretize the one-dimensional advection equation (3.22) is to compute separately the change of the reference phase volume and of the cell volume

$$\Delta C = C_i^n h_0^2 + h_0\, \Delta t\, \left(F_{i-1/2}^n - F_{i+1/2}^n\right) \;\; ; \;\; \Delta V = h_0^2 + h_0\, \Delta t\, \left(u_{i-1/2} - u_{i+1/2}\right)$$

The updated value of the color function is then

$$C_i^{n+1} = \frac{\Delta C}{\Delta V} \tag{3.33}$$

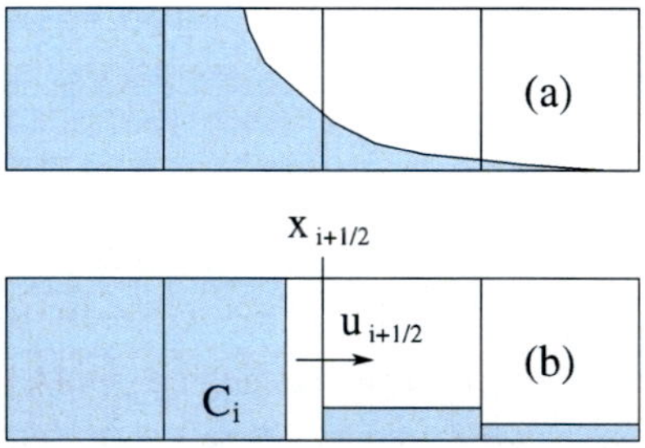


**Fig. 3.12** Original VOF method: **a** interface line; **b** VOF reconstruction with the interface line in the $i$th-cell perpendicular to the velocity component $u_{i+1/2}$

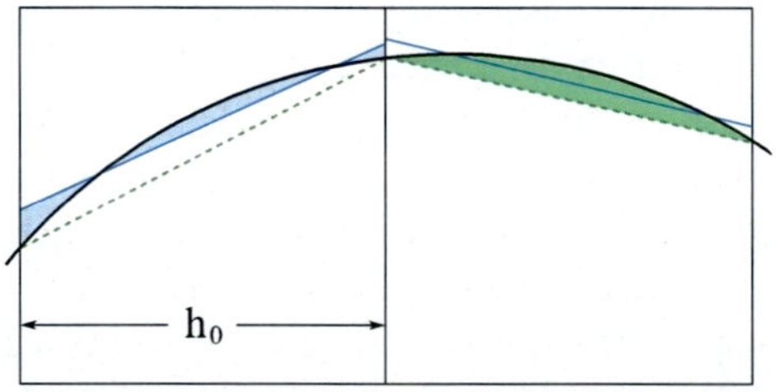


**Fig. 3.13** In the trapezoidal rule the interface line (thick solid line) is approximated by a continuous piecewise linear line (dashed line) that connects consecutive points on the line, while a VOF-PLIC reconstruction (thin solid lines) is not continuous across the cell boundary

It is straightforward to demonstrate that this discretization is equivalent to an implicit discretization of the color function in the compressible term of (3.22). However, we anticipate that the reference phase fluxes should be defined differently in the two schemes. Simple advection tests, such as translations and rotations, show that these geometric methods tend to generate isolated blobs of reference phase, usually called "flotsam" and "jetsam", and that this feature is more appreciable for a multidimensional scheme rather than for a sequence of one-dimensional sweeps [126].

### *3.1.6 VOF-PLIC Methods: Interface Reconstruction*

In the PLIC method (Piecewise Linear Interface Calculation) the interface line in 2D is approximated by a line segment in each cell, but the segment can be oriented arbitrarily with respect to the coordinate axes. The orientation of the line is determined by the normal $\boldsymbol{m} = \nabla^h C$ to the interface, which is computed by considering the values of $C$ in a neighbourhood of the cell under consideration. Once the interface has been reconstructed, the reference phase fluxes from one cell to another are computed by geometric considerations.

In Fig. 3.13 we compare the trapezoidal rule, that computes an approximated value of the area under the interface line $y = f(x)$, and a VOF-PLIC reconstruction, that

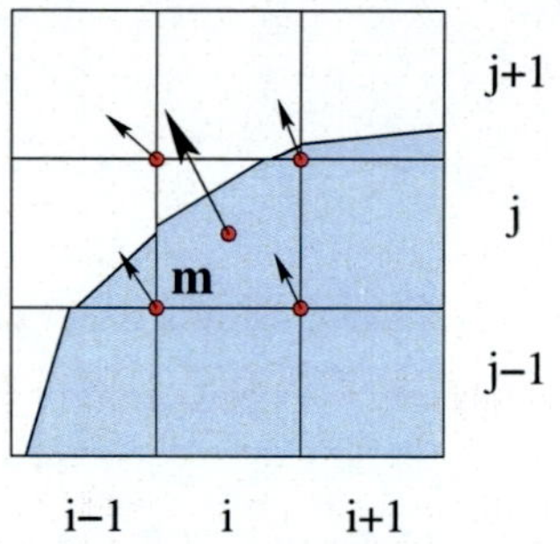


**Fig. 3.14** In Youngs' method the normal $\boldsymbol{m}$ is computed at the four corners of the central cell and then at its center by averaging

in each cell satisfies the area conservation constraint. In the trapezoidal rule the area under the function is approximated by that of a sequence of right trapezoids. The integration error in the evaluation of the area, if the function has a continuous second derivative, is second-order accurate with respect to the grid spacing $h_0$, $\mathcal{O}(h_0^2)$, and a straight line is reproduced exactly. A VOF-PLIC reconstruction actually solves the inverse problem: the area under the interface line in each cut cell is known, and the function should now be approximated. The inverse problem has clearly an infinite number of solutions. The reconstruction is in general not continuous across the cell boundary, but it is reasonable to require that a second-order accurate reconstruction algorithm should reproduce exactly a straight line, as for the trapezoidal rule. Reconstruction techniques that do not satisfy this requirement are expected to be $\mathcal{O}(h_0^n)$, with $1 \leq n \leq 2$.

The reconstruction is basically a two-steps procedure. In any cut cell the normal $\boldsymbol{m}$ (the unit normal is $\boldsymbol{n} = \boldsymbol{m}/|\boldsymbol{m}|$) is first determined from the knowledge of the color function in the given cell and in the neighbouring ones. The equation of the interface is then written in 2D as

$$\boldsymbol{m} \cdot \boldsymbol{x} = m_x\, x + m_y\, y = \alpha \tag{3.34}$$

and the value of $\alpha$ is adjusted until the area under the segment equals $h_0^2 C$. The reconstruction methods that are now described are based on a $3 \times 3$ block of cells to compute the normal in the central cell of the block.

In Youngs' method the normal is estimated as a gradient with finite differences. The normal $\boldsymbol{m} = -\nabla^h C$, pointing outwards from the reference phase, is first evaluated at the four corners of the central cell $(i, j)$ of Fig. 3.14, for example its components on the top-right corner are given by

$$m_{x;i+1/2,j+1/2} = -\frac{1}{2h_0}(C_{i+1,j+1} + C_{i+1,j} - C_{i,j+1} - C_{i,j})$$

$$m_{y;i+1/2,j+1/2} = -\frac{1}{2h_0}(C_{i+1,j+1} - C_{i+1,j} + C_{i,j+1} - C_{i,j})$$

and similarly for the other three corners. The cell-centered vector is finally obtained by averaging the four cell-corner values

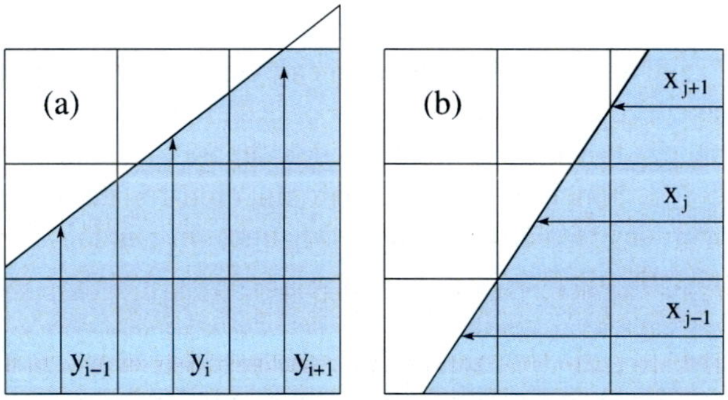


**Fig. 3.15** **a** Volume fractions are added columnwise to define the vertical height y, in **b** rowwise for the horizontal height x; **c** the central scheme gives the best linear approximation for the curved interface

$$\boldsymbol{m}_{i,j} = \frac{1}{4}(\boldsymbol{m}_{i+1/2,j+1/2} + \boldsymbol{m}_{i+1/2,j-1/2} + \boldsymbol{m}_{i-1/2,j+1/2} + \boldsymbol{m}_{i-1/2,j-1/2})$$

This method is rather simple to implement, both in 2D and 3D, but it does not reproduce any straight line exactly. As a matter of fact, the method is asymptotically first-order accurate, as evidenced by numerical tests. However, at low resolution, when the local radius of curvature of the interface is not much bigger than the grid spacing $h_0$, the method is competitive with more accurate schemes.

In order to define a local height function we consider the $3 \times 3$ block of square cells of side $h_0$ of Fig. 3.15, and add the volume fraction values columnwise to define $y = f(x)$ or rowwise for $x = g(y)$. In Fig. 3.15a the height $y_{i-1}$ at the abscissa $x_{i-1}$, placed in the center of the column, is given by the expression $h_0\, y_{i-1} = h_0^2 \sum_{k=-1}^{1} C_{i-1,j+k}$, while in Fig. 3.15b the width $x_{j+1}$ at the ordinate $y_{j+1}$ is $h_0\, x_{j+1} = h_0^2 \sum_{k=-1}^{1} C_{i+k,j+1}$. In the case of Fig. 3.15a we consider the straight line: $\mathrm{sgn}(m_y)\, y = m_x\, x + \alpha'$, where $m_y = -\partial C/\partial y$. The sign of $m_y$ is required because in the integration to compute the local height we lose track of what phase is on the top or on the bottom of the column. The sign can be computed with centered finite differences. The angular coefficient $m_x$ can be computed with backward, centered and forward finite differences, $m_{xb} = -(y_i - y_{i-1})/h_0, m_{xc} = -(y_{i+1} - y_{i-1})/2\,h_0$ and $m_{xf} = -(y_{i+1} - y_i)/h_0$, respectively. The local height $y$ is placed exactly on the interface if the interface is a straight line that cuts the two vertical sides of the column, as for the heights $y_{i-1}$ and $y_i$, but not for $y_{i+1}$. As a result the backward estimate $m_{xb}$ is correct, while the other two underestimate the value of the angular coefficient. Therefore, if we have two estimates of the same angular coefficient, say $m_{x1}$ and $m_{x2}$, we select one of them with the following criterion based on the recostruction of a linear interface

$$|m^*| = \max\left(|m_{x1}|, |m_{x2}|\right) \tag{3.35}$$

For the almost vertical line of Fig. 3.15b it is convenient to consider the straight line: $\text{sgn}(m_x)\, x = m_y\, y + \alpha''$, and any finite difference scheme calculates the correct value $m_y$, as the interface line cuts the two opposite sides of each row. On the other hand if we use the vertical height function $y$ any discrete estimate of the angular coefficient $m_x$ satisfies $|m_x| \le 3$, which is a crude approximation for an almost vertical line. Therefore between any two discrete approximations $m_{x1}$ and $m_{y1}$ for the angular coefficients we now choose

$$|m^*| = \min\left(|m_{x1}|, |m_{y1}|\right) \tag{3.36}$$

In this way we use the form $y = f(x)$ if the interface is almost horizontal and $x = g(y)$ when it is about vertical. These two criteria optimize the reconstruction of a linear interface, but they may not select the best guess for a curved line. In Fig. 3.15c the line has a vertical axis of symmetry through the middle of the central column, where $f'(x) = 0$. The best approximation is $m_{xc} = 0$, but this would require the minimum value in (3.35). However, numerical tests show that these two criteria are the most efficient ones and can be easily extended to 3D, where we now have three angular coefficients, $m_x$, $m_y$ and $m_z$.

The mixed Youngs-centered method (MYC) in 3D is based on these two criteria and computes four normal vectors including Youngs' scheme, for its good behavior at low resolution, and three centered column schemes (CC), for their better behavior at high resolution [88]. In the CC scheme the function $z = f(x, y)$ in the central cell of a $3 \times 3 \times 3$ block of square cells is approximated by the linear equation: $\text{sgn}(m_z)\, z = m_x\, x + m_y\, y + \alpha$, where the angular coefficients $m_x$ and $m_y$ are computed with finite differences from the vertical height function $z$

$$m_x = \frac{z_{i+1,j} - z_{i-1,j}}{2\,h_0} = \frac{1}{2}\left(\sum_{l=-1}^{1} C_{i+1,j,k+l} - \sum_{l=-1}^{1} C_{i-1,j,k+l}\right)$$

$$m_y = \frac{z_{i,j+1} - z_{i,j-1}}{2\,h_0} = \frac{1}{2}\left(\sum_{l=-1}^{1} C_{i,j+1,k+l} - \sum_{l=-1}^{1} C_{i,j-1,k+l}\right)$$

We also consider the two functions $x = h(y, z)$ and $y = g(x, z)$ and their linear approximation to calculate with finite differences the corresponding normal vectors. We select one candidate among the three approximations with the CC scheme by using the 3D equivalent of (3.36) and then between the chosen candidate and Youngs' approximation with (3.35). For the selection it is convenient to normalize the angular coefficients so that $|m_x| + |m_y| + |m_z| = 1$ [88].

The ELVIRA method in 2D considers both the vertical height $y = f(x)$ and the horizontal height $x = g(y)$ and their linear approximations [127]. The two angular coefficients $m_x$ and $m_y$ are computed with backward, centered and forward finite differences. There are now six different choices and a selection is done by minimizing a measure of the error between the volume fractions given by the actual and approximate interfaces. For each of the six normal vectors $\boldsymbol{m}_n$, with $1 \le n \le 6$, the

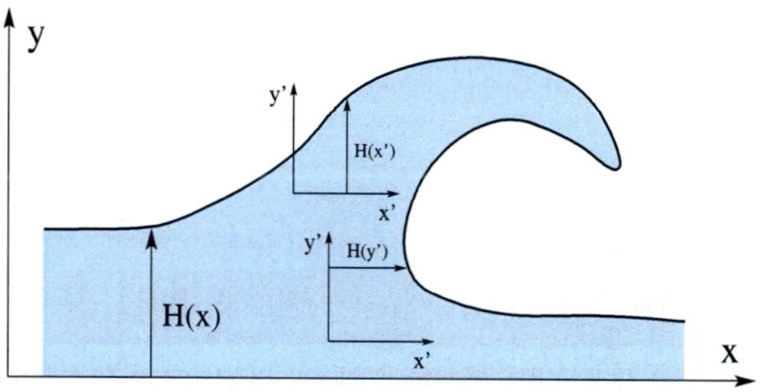


**Fig. 3.16** Fixed coordinate system $x$-$y$ with height $H(x)$, and local coordinate systems $x' - y'$ with horizontal and vertical heights $H(y')$ and $H(x')$

corresponding value of the line constant $\alpha_n$ is determined. Then the reconstructed linear interface is drawn across the whole $3 \times 3$ block of cells, defining a tentative volume fraction value $\widetilde{C}$ in each of the surrounding cells. The area error $E_n$ in $L_2$ between the reconstruction with normal $\boldsymbol{m}_n$ and the true values is

$$E_n = h_0^2 \sum_{k=-1}^{1} \sum_{l=-1}^{1} \left( \widetilde{C}_{i+k,j+l}(\boldsymbol{m}_n) - C_{i+k,j+l} \right)^2 \tag{3.37}$$

The selected value of $\boldsymbol{m}_n$ is the one that minimizes the error $E$. This method reconstructs exactly any linear interface, but its extension to 3D is computationally very expensive, because of the number of angular coefficients that must be considered and the fact that a $5 \times 5 \times 5$ block of cells is required in order to reconstruct exactly any linear interface.

The height function $H(x)$ in 2D represents the distance of the points of the interface line from a reference coordinate axis. This feature limits the possibility to follow the full non-linear evolution of an unstable wave: we have to stop the simulation when the interface "overhangs" and becomes a multi-valued relation. However, we are not obliged to consider a fixed coordinate system, but we can use local coordinate systems in order to define a local vertical height $H(x')$ or a horizontal height $H(y')$, as shown in Fig. 3.16.

We now consider a continuous function $f(x)$ with continuous derivatives, $f^{(n)}(x) = d^n f(x)/dx^n$, and define the continuous height function

$$H(x; h_0) = \frac{1}{h_0} \int_{x-h_0/2}^{x+h_0/2} f(t)\, dt \tag{3.38}$$

where $h_0$ is the grid spacing. Furthermore we define $H_{\pm n} = H(x \pm n\, h_0; h_0)$ and approximate the first and second derivatives of the function $f(x)$ with the finite differences of the height function

$$H^{(1)}(x; h_0) = \frac{H_{+1} - H_{-1}}{2\, h_0}\,, \quad H^{(2)}(x; h_0) = \frac{H_{+1} + H_{-1} - 2\, H_0}{2\, h_0} \tag{3.39}$$

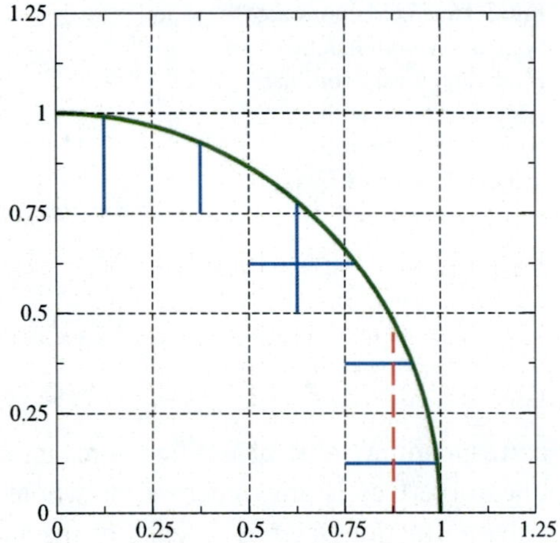


**Fig. 3.17** Horizontal and vertical heights for the circle with center at (0, 0) and radius $r_0 = 1$ in the quadrant $x \geq 0$, $y \geq 0$, with grid spacing $h_0 = 1/4$

Then with an expansion in Taylor's series it is straightforward to show that these approximations are second-order accurate, in particular

$$f(x) = H(x; h_0) + \mathcal{O}(h_0^2)\,, \quad f^{(1)}(x) = H^{(1)}(x; h_0) + \mathcal{O}(h_0^2) \tag{3.40}$$

The discrete height can be calculated by adding the $C$ data along one coordinate direction. For example in 2D starting from the cut cell with indices $(i, j)$ the vertical height can be defined by the expression

$$H(x_i) = \frac{1}{h_0} \int_{x_i - h_0/2}^{x_i + h_0/2} f(t)\, dt = \frac{1}{h_0} \sum_{k=-n}^{n} C_{i,j+k}\, h_0^2 \tag{3.41}$$

where the values $n = 1, 2, 3$ have been widely used, corresponding to a stencil with 3, 5 and 7 cells. The definition (3.38) is meaningful only if the interface line crosses the column centered at $x$ and with thickness $h_0$. In a 2D grid with square cells this requirement can be satisfied if the stencil, which is used to define the height, is bounded by a full cell on one side and an empty one on the other side.

As an example, the portion of the circle of Fig. 3.17 has center at (0, 0) and radius $r_0 = 4\,h_0$, with $h_0 = 0.25$. The first two vertical heights on the left require three consecutive cells for their correct definition, four cells are required for the third one. In this last case there are two consecutive cut cells, and the height point is in the top cell. In the vertical column with the dashed line there is no full cell, hence it is impossible the define the discrete vertical height. However the horizontal heights in this area are well defined. As shown in Fig. 3.17, near the 45° degree line and at low resolution it is possible that three consecutive heights along the same coordinate direction are not available and the angular coefficient of (3.39) cannot be computed. Furthermore, as previously mentioned, a cell may contribute to the definition of a

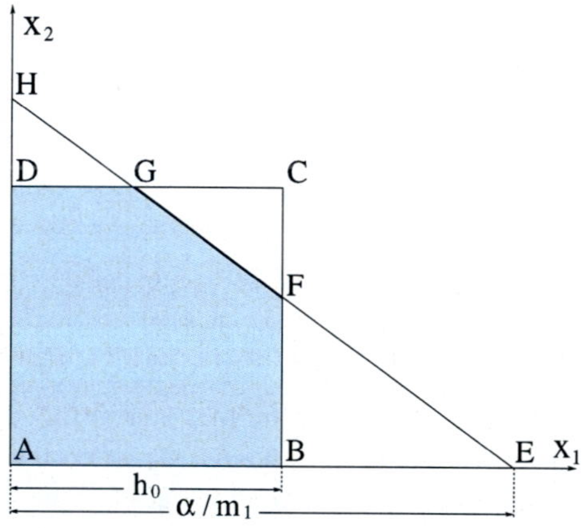


**Fig. 3.18** The reconstructed interface, segment $FG$, and the cell boundary define the polygon $ABFGD$ of area $h_0^2 C$; here both normal components, $m_x$ and $m_y$, are positive

height point which is not inside the cell, or may even be isolated. For all these cases, some other scheme should be devised.

In the hierarchical scheme, the numerical algorithm searches first for three consecutive heights, then mixes vertical and horizontal heights, and then even centroids of a preliminary Youngs' reconstruction in order to compute the angular coefficient and the local curvature as well [128, 129]. In 2D a linear interface reconstruction has also been used as a preliminary step for a quadratic spline reconstruction [130].

Once the normal $\boldsymbol{m}$ has been calculated, the non-homogeneous term $\alpha$ of (3.34) must be determined by enforcing area conservation. In the square cell of side $h_0$ of Fig. 3.18 and for a given $\boldsymbol{m}$, a value of $\alpha$ determines the area $A(\alpha)$ of the polygon $ABFGD$ that should be compared with the value $h_0^2 C$. In other terms we have to find a zero of the nonlinear function $g(\alpha)$

$$g(\alpha) = A(\alpha) - h_0^2 C = 0 \tag{3.42}$$

An alternative method to a root-finding algorithm for the solution of (3.42) relies heavily on the symmetry of a Cartesian cell, so that, with proper mirror reflections about the coordinate directions, the interface line (3.34) has both angular coefficients $m_x$ and $m_y$ positive, as in Fig. 3.18. Furthermore, with $m_1 = \min(m_x, m_y)$ and $m_2 = \max(m_x, m_y)$ the area $A(\alpha)$ is given by the expression [131, 132]

$$A(\alpha) = h_0^2 C = \frac{1}{2 m_1 m_2}\Big(\alpha^2 - F_2\big(\alpha - m_1 h_0\big) - F_2\big(\alpha - m_2 h_0\big)\Big) \tag{3.43}$$

where $F_2(z) = z^2$ when $z > 0$ and zero otherwise. The three contribution to $A(a)$ represent the areas of the similar triangles $AEH$, $BEF$ and $DGH$. Expression (3.43) is a quadratic function of $\alpha$ when the interface reconstruction cuts two consecutive cell sides, and is linear when the two intersections are on opposite sides. Furthermore, it is a continuous, strictly monotonically increasing function of $\alpha$ that can be easily inverted to get $\alpha = \alpha(C)$ [132].

In 3D the interface plane, after mirror reflections, is given by the equation

$$\boldsymbol{m} \cdot \boldsymbol{x} = m_1 x_1 + m_2 x_2 + m_3 x_3 = \alpha \tag{3.44}$$

with $m_i > 0$, $i = 1, 2, 3$, and the cut volume $V(\alpha)$ is

$$V(\alpha) = h_0^3 C = \frac{1}{6 m_1 m_2 m_3} \Big( \alpha^3 - \sum_{i=1}^{3} F_3(\alpha - m_i h_0) + \sum_{i=1}^{3} F_3(\alpha - \alpha_{max} + m_i h_0) \Big) \tag{3.45}$$

with $\alpha_{max} = \sum_{i=1}^{3} m_i h_0$ and $F_3(z) = z^3$ when $z > 0$ and zero otherwise. Geometrically, the first term is the volume of the tetrahedron under the triangle $AEH$ on the left of Fig. 3.19, the first sum removes the volume of the tetrahedra under the two triangles $CEG$ and $BFH$, when the vertices $E$ and $H$ move beyond the cell faces, the second sum adds back the volume of the tetrahedron under the triangle $DFG$, when the line $EH$ is completely outside the cell. The function $V(\alpha)$ is strictly monotonic and varies continuously from the value $V = 0$, when $\alpha = 0$, to $V = h_0^3$, when $\alpha = \alpha_{max}$. The number of intersections of the reconstructed interface with the cell sides may vary from 3 to 6, as shown on the right of Fig. 3.19. The relation (3.45) can be inverted with some algebra to get $\alpha = \alpha(C)$ [132]. This methodology has been extended to triangular and tetrahedral grids, and to arbitrarily convex cells in 2D and 3D.

The interface reconstruction in each cut cell provides a local approximation $\widetilde{\chi}$ of the indicator function $\chi(\boldsymbol{x}, t)$. A natural measure of the difference between the exact and the reconstructed interfaces is the error $E$

$$E = \int \big| \chi(\boldsymbol{x}, t) - \widetilde{\chi}(\boldsymbol{x}, t) \big| \, dV \tag{3.46}$$

which depends clearly on the reconstruction method and the grid spacing $h_0$. The contribution to $E$ of each cut cell in 2D is the measure of the area between the interface line and the reconstructed one, which is given by the shadowed area in the left cell of Fig. 3.13. The order of convergence $\mathcal{O}$ of a reconstruction method can be numerically calculated by considering the errors $E$ with grid spacing $h_0$ and $h_0/2$ and the following definition

$$\mathcal{O} = \frac{\ln \big( E(h_0) / E(h_0/2) \big)}{\ln \big( h_0 / h_0/2 \big)} \tag{3.47}$$

In Table 3.2 we present the results for the reconstruction error and convergence rate for randomly-positioned straight lines and ellipses. The number of interface lines is such that the results are stabilized up to the third significant digit. For the straight lines we consider Youngs' reconstruction and the centerer column scheme (CC), that are asymptotically first-order. For the ellipse we add the ELVIRA reconstruction, that is asymptotically second-order. Finally in Fig. 3.20 we show how the reconstruction error varies with the grid resolution $r_0/h_0$ for randomly-positioned circles. The hierarchical algorithm based on the height function is consistently more performing than Youngs' and ELVIRA reconstructions.

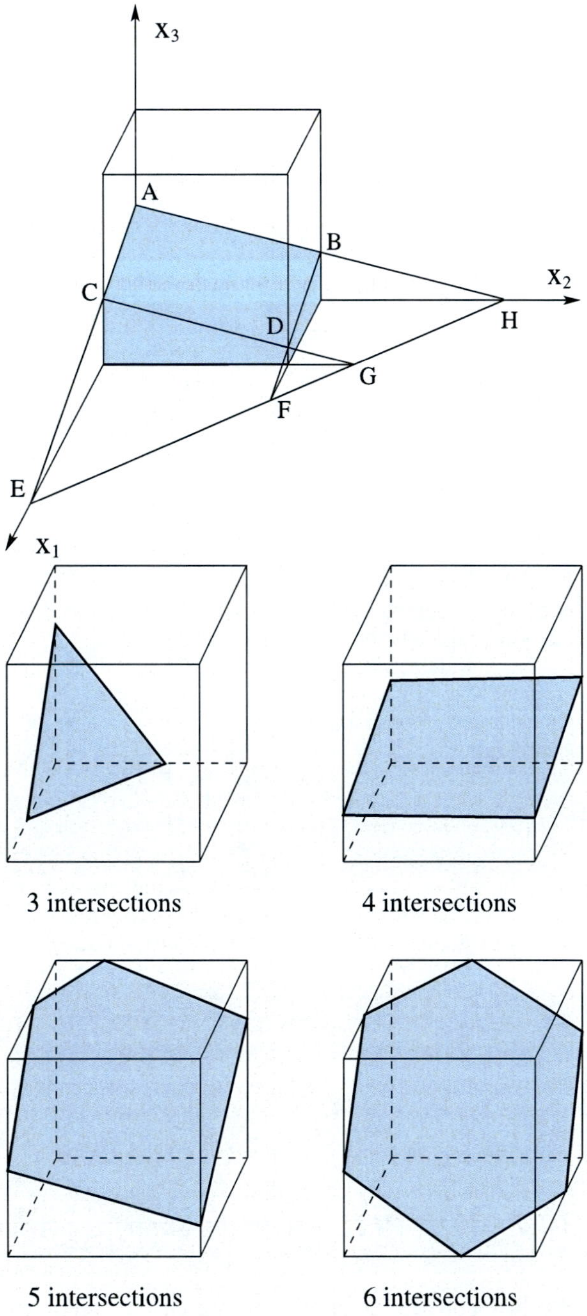


**Fig. 3.19** The reconstructed interface, polygon $ABDC$, and the cell boundary define the cut volume $h_0^3 C$ (left); the number of intersections of a planar reconstruction with the cell sides may vary from 3 to 6 (right)

**Table 3.2** Grid resolution $m$ of the unit square, error $E$ and convergence rate $\mathcal{O}$ for the reconstruction of 1000 straight lines with Youngs' and CC methods (A) and of 150 ellipses with ELVIRA as well (B)

| m | | (A) Youngs | CC | (B) Youngs | CC | ELVIRA |
|---|---|---|---|---|---|---|
| 10 | $E$ | 8.21e−4 | 3.29e−5 | 3.90e−3 | 4.25e−3 | 4.76e−3 |
| | $\mathcal{O}$ | 0.97 | 0.45 | 2.34 | 2.47 | 2.42 |
| 20 | $E$ | 4.18e−4 | 2.40e−5 | 7.68e−4 | 7.65e−4 | 8.87e−4 |
| | $\mathcal{O}$ | 1.01 | 0.65 | 1.84 | 2.27 | 2.33 |
| 40 | $E$ | 2.07e−4 | 1.52e−5 | 2.14e−4 | 1.59e−4 | 1.77e−4 |
| | $\mathcal{O}$ | 1.00 | 0.81 | 1.38 | 2.02 | 2.10 |
| 80 | $E$ | 1.03e−4 | 8.62e−6 | 8.23e−5 | 3.93e−5 | 4.12e−5 |
| | $\mathcal{O}$ | 1.00 | 1.05 | 1.16 | 1.89 | 2.04 |
| 160 | $E$ | 5.16e−5 | 4.15e−6 | 3.69e−5 | 1.06e−5 | 1.00e−5 |
| | $\mathcal{O}$ | 1.01 | 0.96 | 1.06 | 1.73 | 2.00 |
| 320 | $E$ | 2.56e−5 | 2.13e−6 | 1.77e−5 | 3.20e−6 | 2.49e−6 |

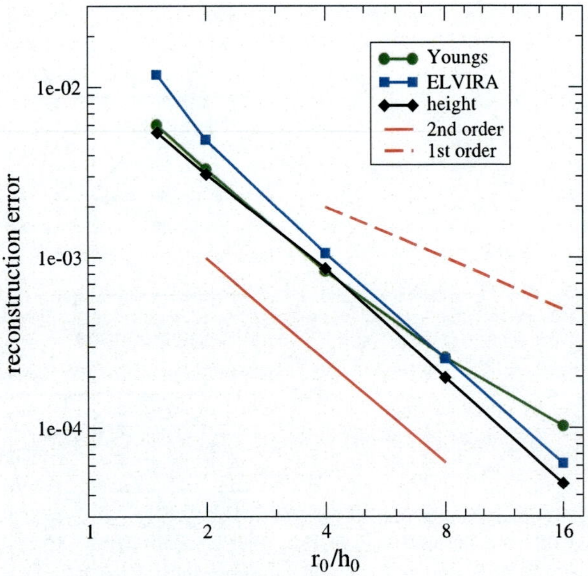


**Fig. 3.20** Reconstruction error and its convergence rate with grid resolution, for a circle with 100 random positions of its center and three different reconstruction algorithms

### 3.1.7 *VOF-PLIC Methods: Interface Advection*

We first consider a multidimensional advection with a split technique, given by a sequence of one-dimensional advection steps. In this case even if the flow is divergence-free, the 1D advection is not and we need to consider again (3.22), with the advection along the $x$-direction

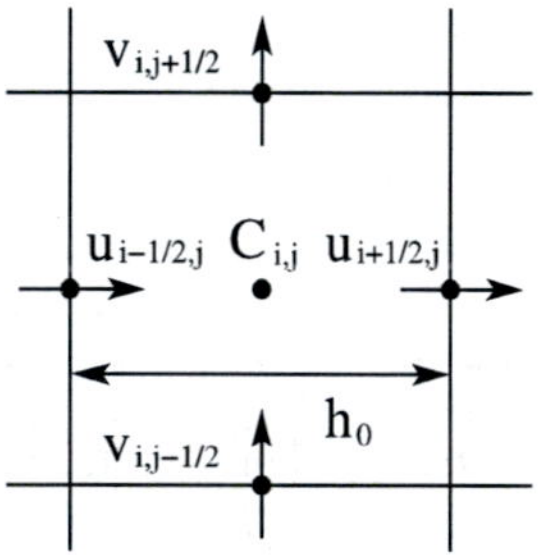


**Fig. 3.21** Two-dimensional staggered grid with the color function $C$ at the cell center and velocity components $u$ and $v$ at the cell faces

$$\frac{\partial C}{\partial t} + \frac{\partial}{\partial x}(C\,u) - C\,\frac{\partial u}{\partial x} = 0$$

This equation is discretized on the staggered grid of Fig. 3.21, for simplicity in two dimensions, with a first-order forward scheme for the time integration and an approximation of the velocity derivative at time $t^n$ with a centered finite-difference scheme, as in (3.23)

$$C_{i,j}^{n+1} = C_{i,j}^{n} + \frac{\Delta t}{h_0}\left(\widetilde{F}_{i-1/2,j} - \widetilde{F}_{i+1/2,j}\right) + \widetilde{C}_{i,j}\,\frac{\Delta t}{h_0}\left(u_{i+1/2,j} - u_{i-1/2,j}\right) \quad (3.48)$$

There are a couple of choices for the term $\widetilde{C}_{i,j}$ that affect the computation of the fluxes $\widetilde{F}$ as well. A deep insight on the flux calculation can be gained by expressing the advection in terms of different mappings of the plane onto itself. Starting with the equation of motion of a particle in a given velocity field $u(x)$, $dx/dt = u(x)$, we can integrate it with an explicit first-order scheme

$$x^{n+1} = x^n + u\left(x^n\right)\Delta t \quad (3.49)$$

and consider a linear interpolation of the velocity between the two values $u_{i-1/2,j}$ on the left edge and $u_{i+1/2,j}$ on the right edge of the cell of Fig. 3.21.

$$u(x) = u_{i-1/2,j}\left(1 - \frac{x}{h_0}\right) + u_{i+1/2,j}\,\frac{x}{h_0} \quad (3.50)$$

By combining (3.49) and (3.50) the updated position of the particle becomes

$$\begin{cases} x' = b\,x + u_{i-1/2,j}\,\Delta t \\ y' = y \end{cases} \quad (3.51)$$

where $b = 1 + (u_{i+1/2,j} - u_{i-1/2,j})\Delta t/h_0$. This system describes a local linear mapping that transforms straight lines into straight lines and the central cell $\Sigma$ of Fig. 3.22 is mapped onto the rectangle $\Gamma$, $\Gamma = T_E(\Sigma)$. The new volume fraction in the central cell is then the sum of three contributions

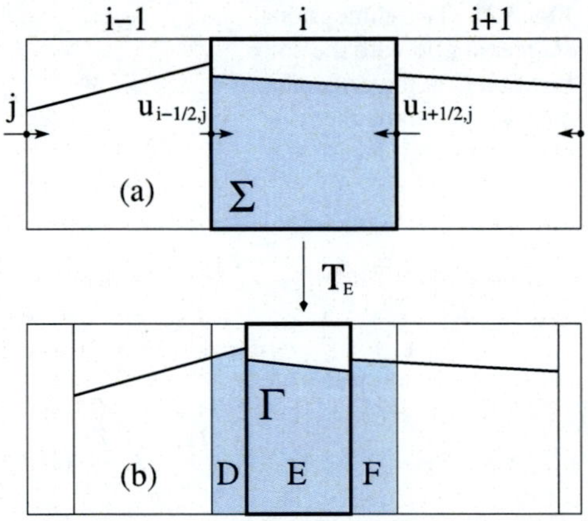


**Fig. 3.22** The horizontal 1D mapping $T_E$ transforms the central cell $\Sigma$ of **a** into the rectangle $\Gamma$ of (**b**), $\Gamma = T_E(\Sigma)$. The reference phase inside $\Sigma$ is mapped to the area $E$, while the mapping of the two lateral cells determines the two contributions $D$ and $F$

$$C_{i,j}^{n+1} = \left(D + E + F\right)/h_0^2 \tag{3.52}$$

There is a different map for each cell of the grid, and the rectangles $\Gamma$ do not overlap or leave empty space by construction, and the updated volume fractions satisfy the condition $0 \le C \le 1$. With a split technique, the interface should be reconstructed after the advection along the $x$-direction and then advected along the $y$-direction. However, the sequence of two $T_E$ mappings completes a 2D advection but does not preserve areas exactly. Furthermore, to reduce possible asymmetries, the direction of the first advection should be alternated in time. The two areas $D$ and $F$ of Fig. 3.22 correspond to the two reference phase fluxes $\widetilde{F}$ of expression (3.48), or (3.23), and should be computed after the interface line has been advected. For this feature it can be called a Lagrangian explicit scheme. In three dimensions the area contributions become right hexahedra.

We now integrate the equation of motion with an implicit first-order scheme

$$x^{n+1} = x^n + u\left(x^{n+1}\right) \Delta t \tag{3.53}$$

and combine (3.53) and (3.50) to get

$$\begin{cases} x' = a\,x + a\,u_{i-1/2,j}\,\Delta t \\ y' = y \end{cases} \tag{3.54}$$

where $a = 1/(1 - (u_{i+1/2,j} - u_{i-1/2,j})\Delta t/h_0)$. This is a different linear mapping that maps the rectangle $\Gamma$ of Fig. 3.23 onto the central cell $\Sigma$, $\Sigma = T_I(\Gamma)$. The updated color function is given by the same expression (3.52), however the two mappings $T_E$ and $T_I$ are different and, as a result, the fluxes through the cell boundary are also different. The sequence of two consecutive $T_I$ mappings along the $x$- and $y$-directions does not conserve the area. The difference between the two computations of the reference phase fluxes across the cell boundary is shown in Fig. 3.24.

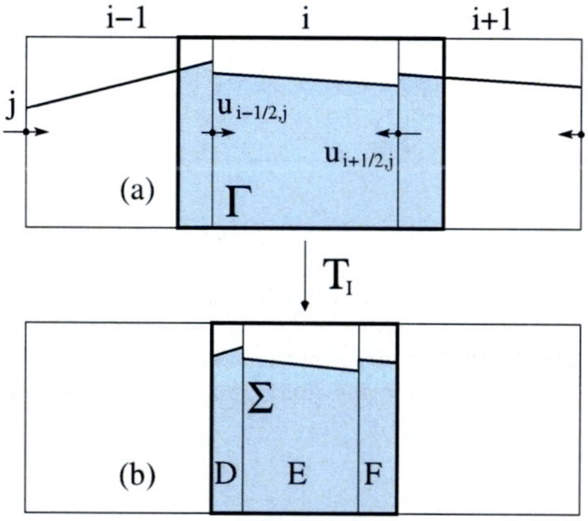


**Fig. 3.23** The horizontal 1D mapping $T_I$ transforms the central rectangle $\Gamma$ of **a** into the square cell $\Sigma$ of **b**, $\Sigma = T_I(\Gamma)$. The reference phase inside $\Gamma$ is mapped into the three contributions $D$, $E$ and $F$

An alternative way to update the volume fraction is suggested by Fig. 3.23, where the three volume fraction contributions can be computed inside $\Gamma$ and their sum multiplied by the coefficient $a$. For this reason, the scheme is sometimes called Eulerian implicit scheme and is somewhat related to the approach that has resulted in (3.33).

A remarkable conservation property is achieved by a 1D advection with the $T_I$ mapping along one coordinate direction followed by the $T_E$ mapping along the other direction. If the first step is along the $x$-direction the initial rectangle $\Gamma_x$ is mapped onto the grid cell $\Sigma$ and it is compressed (or expanded) by the factor $a = 1/(1 - (u_{i+1/2,j} - u_{i-1/2,j})\Delta t/h_0)$. In the second step the cell $\Sigma$ is mapped onto the rectangle $\Gamma_y$ and is compressed (or expanded) by the factor $b = 1 + (v_{i,j+1/2} - v_{i,j-1/2})\Delta t/h_0$. The area of $\Gamma_y = T_{E,y}(\Sigma) = T_{E,y}\, T_{I,x}(\Gamma_x)$ is the area of $\Gamma_x$ multiplied by the factor $a\, b$. For an incompressible flow on a staggered grid with square cells the divergence-free condition is

$$\frac{1}{h_0}\left(u_{i+1/2,j} - u_{i-1/2,j} + v_{i,j+1/2} - v_{i,j-1/2}\right) = 0 \tag{3.55}$$

then $a\, b = 1$. The combined mapping $T_{E,y}\, T_{I,x}$ at one time step should be alternated with $T_{E,x}\, T_{I,y}$ at the next step. The convergence rate of this scheme is intermediate between 1 and 2, and unfortunately it cannot be extended directly to three dimensions.

In order to develop a split three-dimensional mass-conserving scheme we consider three consecutive 1D steps along the coordinate directions, starting from (3.48)

$$C^*_{i,j,k} = C^n_{i,j,k} + \frac{\Delta t}{h_0}\left(\widetilde{F}_{i-1/2,j,k} - \widetilde{F}_{i+1/2,j,k}\right) + \widetilde{C}\,\frac{\Delta t}{h_0}\left(u_{i+1/2,j,k} - u_{i-1/2,j,k}\right)$$
$$C^{**}_{i,j,k} = C^*_{i,j,k} + \frac{\Delta t}{h_0}\left(\widetilde{F}_{i,j-1/2,k} - \widetilde{F}_{i,j+1/2,k}\right) + \widetilde{C}\,\frac{\Delta t}{h_0}\left(v_{i,j+1/2,k} - v_{i,j-1/2,k}\right)$$
$$C^{n+1}_{i,j,k} = C^{**}_{i,j,k} + \frac{\Delta t}{h_0}\left(\widetilde{F}_{i,j,k-1/2} - \widetilde{F}_{i,j,k+1/2}\right) + \widetilde{C}\,\frac{\Delta t}{h_0}\left(w_{i,j,k+1/2} - w_{i,j,k-1/2}\right)$$

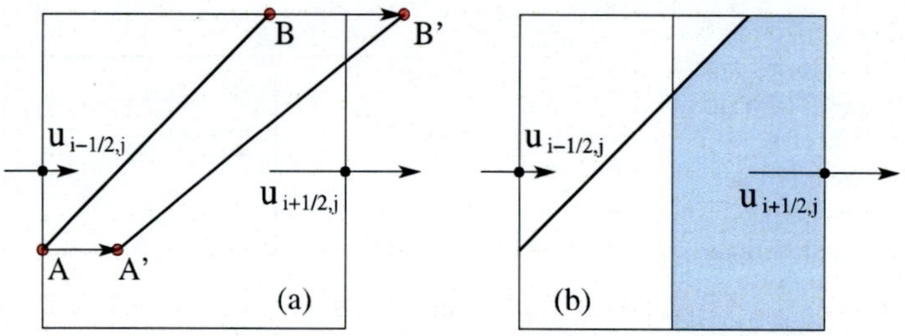


**Fig. 3.24** **a** In the Lagrangian method the end points $A$ and $B$ of the interface segment are advected by the flow; **b** in the Eulerian method the area fluxing through the cell right side is the rectangle with width $u_{i+1/2,j}\,\Delta t$

In the last term on the right hand side the unknown value $\widetilde{C}$ must be kept constant during the three advection steps in order to use the divergence-free condition

$$\frac{1}{h_0}\left(u_{i+1/2,j,k} - u_{i-1/2,j,k} + v_{i,j+1/2,k} - v_{i,j-1/2,k} + w_{i,j,k+1/2} - w_{i,j,k-1/2}\right) = 0 \tag{3.56}$$

Then the sum of the three equations gives

$$C^{n+1}_{i,j,k} = C^{n}_{i,j,k} + \frac{\Delta t}{h_0}\big(\widetilde{F}_{i-1/2,j,k} - \widetilde{F}_{i+1/2,j,k} + \widetilde{F}_{i,j-1/2,k} - \widetilde{F}_{i,j+1/2,k} + \widetilde{F}_{i,j,k-1/2} - \widetilde{F}_{i,j,k+1/2}\big) \tag{3.57}$$

where the fluxes are computed with the Eulerian scheme. It can be demonstrated that the correct choice for $\widetilde{C}$ is the value of the phase indicator $\chi$ in the cell center at the beginning of the sequence of three advection steps [133]

$$\widetilde{C} = \begin{cases} 1 & \text{if } C^{n}_{i,j,k} > 1/2 \\ 0 & \text{otherwise} \end{cases} \tag{3.58}$$

With this choice the fluxing scheme is conservative, since whatever goes out from a cell enters the next one. The divergence term, which is necessary in each advection step to avoid undershoots or overshoots, sums exactly to zero. The consistency condition, $0 \le C \le 1$ is satisfied at each intermediate step, hence no clipping or filling is needed. However, roundoff errors are unavoidably present and in some instances should be considered carefully.

We now illustrate a few examples of two-dimensional advection schemes with an unsplit technique. In the concurrent split scheme of Fig. 3.25a the reference phase fluxes $\widetilde{F}$ are computed with the Eulerian method of Fig. 3.24b and the volume fraction field is updated in the following way

$$C^{n+1}_{i,j} = C^{n}_{i,j} + \frac{\Delta t}{h_0}\left(\widetilde{F}_{i-1/2,j} - \widetilde{F}_{i+1/2,j}\right) + \frac{\Delta t}{h_0}\left(\widetilde{F}_{i,j-1/2} - \widetilde{F}_{i,j+1/2}\right) \tag{3.59}$$

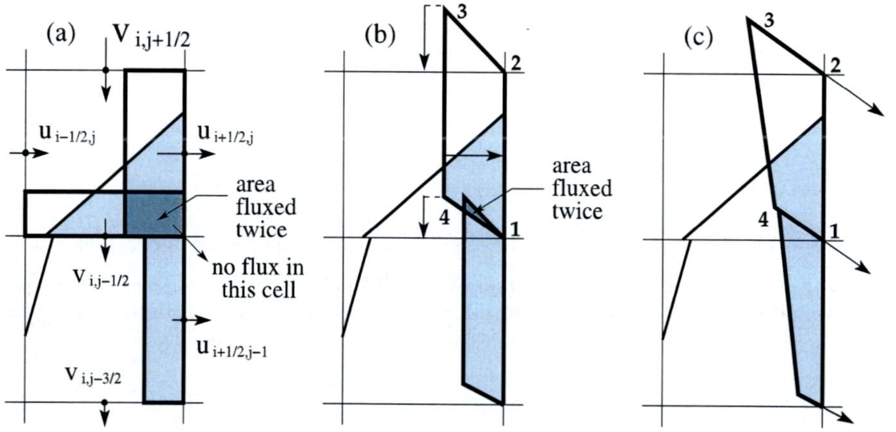


**Fig. 3.25** Unsplit Eulerian advection: (a) a concurrent split method, (b) a trapezoidal flux polygon with staggered velocity components [75], (c) a trapezoidal flux polygon with interpolated velocity field [134]

This scheme conserves mass to machine accuracy, however there are two major flaws. First the consistency condition, $0 \leq C \leq 1$, may not be satisfied in a few cells. As seen in Fig. 3.25a the reference phase located near the bottom-right corner is fluxed twice and the volume fraction update may lead to a negative value in this cell. A second issue is that the mass flows to the neighboring cells only horizontally or vertically, but no flux is trasferred to the diagonal cell. As a result the interface line rapidly develops spurious oscillations even in very smooth divergence-free flows. To overcome this problem the fluxing area through a cell side should be deformed from a rectangle to a trapezoidal polygon at least. In Fig. 3.25b the width of the fluxing area is still $u_{i+1/2,j}\,\Delta t$, and two triangular areas have been considered to take into account the multidimensionality of the flow. The height of the triangle added near the top side is $v_{i,j+1/2}\,\Delta t$, while the height of the triangle removed near the bottom side is $v_{i,j-1/2}\,\Delta t$. The two areas are different, hence the total fluid flux through the cell boundary may not satisfy the divergence-free condition. Furthermore there is still some area that may be fluxed twice. To reduce the negative impact of these inconsistencies it was proposed to consider the 2D version of (3.22) [75], with the following discrete form

$$C_{i,j}^{n+1} = C_{i,j}^{n} + \frac{\Delta t}{h_0}\left(\widetilde{F}_{i-1/2,j} - \widetilde{F}_{i+1/2,j}\right) + \frac{\Delta t}{h_0}\left(\widetilde{F}_{i,j-1/2} - \widetilde{F}_{i,j+1/2}\right) + \frac{C_{i,j}^{n+1} + C_{i,j}^{n}}{2}\,\Delta t\,(\nabla \cdot \boldsymbol{u})_{i,j}\,, \tag{3.60}$$

where the term $(\nabla \cdot \boldsymbol{u})_{i,j}$ represents the net balance between the incoming and outgoing fluid fluxes.

The velocity components in a staggered grid can also be interpolated in the cell vertices [134], as shown in Fig. 3.25c. The velocity vectors in the cell vertices define the slope of side 3-4

$$\left(\frac{dx}{dy}\right)_{3-4} = \frac{x_2 - x_1 - (u_2 - u_1)\Delta t}{y_2 - y_1 - (v_2 - v_1)\Delta t}$$

Side 3–4 should be moved parallel to itself until the area of the trapezoid is equal to $u_{i+1/2,j}\,\Delta t\, h_0$. In this scheme no fluid area is fluxed twice.

We have pointed out a several inaccuracies present in the advection schemes that may lead to undershoots, $C < 0$, or overshoots, $C > 1$. If they are small in magnitudo and exact mass-conservation is not a major issue they can simply be ignored

$$C_{i,j} = \min\big(1, \max(C_{i,j}, 0)\big)$$

If this solution is not sufficient, the $C$ data should redistributed locally [135]. Suppose $C_{i,j} > 1$, then we define $\varepsilon = C_{i,j} - 1$ and look in the surrounding 8 cells for $C_{i',j'} < 1$. We let $C_T = C_{i',j'} + \varepsilon$ and redefine the $C$ value in the 2 cells $(i, j)$ and $(i', j')$ as $C_{i',j'} = \min(C_T, 1)$ and $C_{i,j} = \max(C_T, 1)$. If the new value $C_{i,j} \leq 1$ the procedure stops, otherwise the algorithm will look for the next maximum value in the neighboring cells.

Typical advection tests involve translations, solid body rotations and more complex divergence-free velocity fields that deform the interface. The first two tests are particularly suited to evaluate the basic properties of a combined reconstruction and advection algorithm, since a fluid body in these flows will preserve its shape. In the translation test we consider a unit square with $80 \times 80$ square cells of side $h_0$ and three circles with different radius $r_0/h_0 = 2.5, 4, 5.5$, as shown in Fig. 3.26a. We consider the ELVIRA reconstruction algorithm and the sequence of two 1D advections, more precisely the $T_I$ mapping followed by the $T_E$ mapping.

A translation along a coordinate axis is well suited to study the performance of the reconstruction error, as the fluid fluxes are exact. We expect the error to be zero at CFL=1, as the volume fraction is simply shifted from one cell to the next one, and it should increase by lowering the CFL number, as a small amount of error is added at each reconstruction. The translation test brings the figure back to its initial position at time $t = T$, and the geometrical error E can be defined by an $L_1$ norm as

$$E = \sum_{i,j} h_0^2 |C_{i,j}^T - C_{i,j}^0|$$

With expect this error to reach an asymptotic value as CFL $\rightarrow$ 0, since the two reconstructions will be very close to each other. These features are shown very well in Fig. 3.27 for the circle with $r_0/h_0 = 5.5$. Notice that in the asympotic region at fixed CFL the convergence rate is about first order at the lowest resolutions then slowly increases to second order.

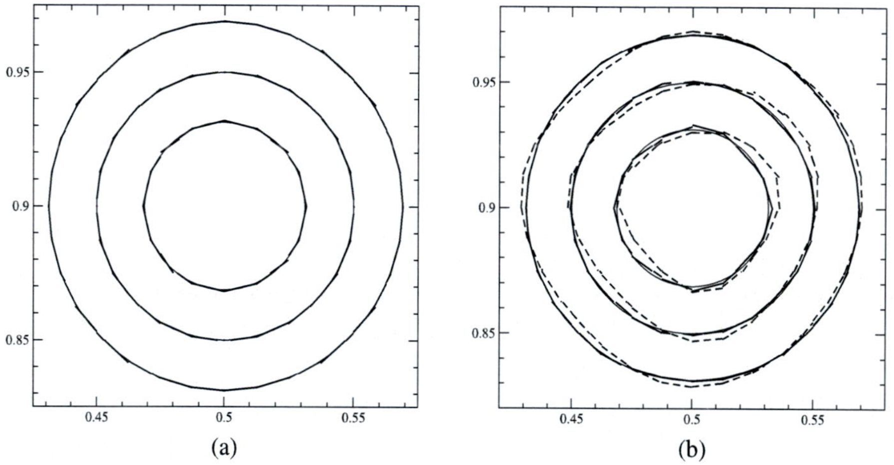


**Fig. 3.26** Translation test: **a** three circles of different radii and their initial reconstruction, **b** their reconstruction at the end of the translation test with CFL = 1 (solid segments) and CFL = 0.1 (dashed segments)

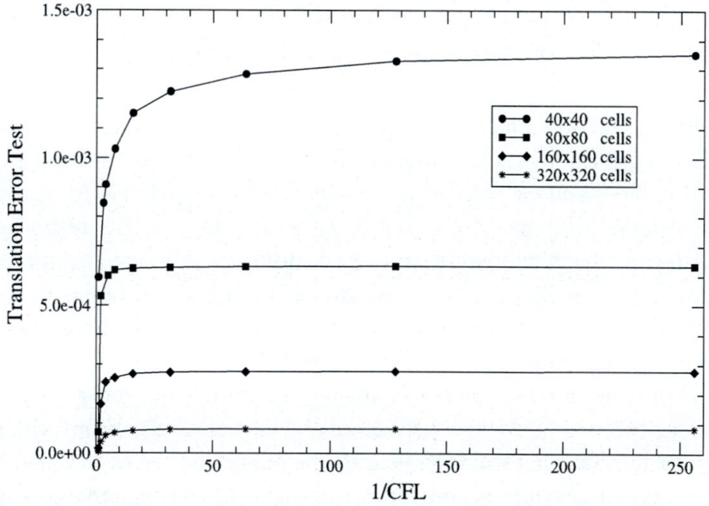


**Fig. 3.27** Translation errors as a function of the CFL number and grid resolution for the circle with $r_0/h_0 = 5.5$

In Fig. 3.26b we show the reconstruction at the end of a translation along a diagonal direction. The figure becomes flat in the rear partand with a higher curvature in the front part. We also notice that a minimum ratio of $r_0/h_0 \approx 4$ is required for a reasonable well-resolved interface with PLIC and simple advection tests.

The vortex-in-a-box test is often used to study the combined performance of the reconstruction and advection algorithms. The velocity field is determined by the stream function

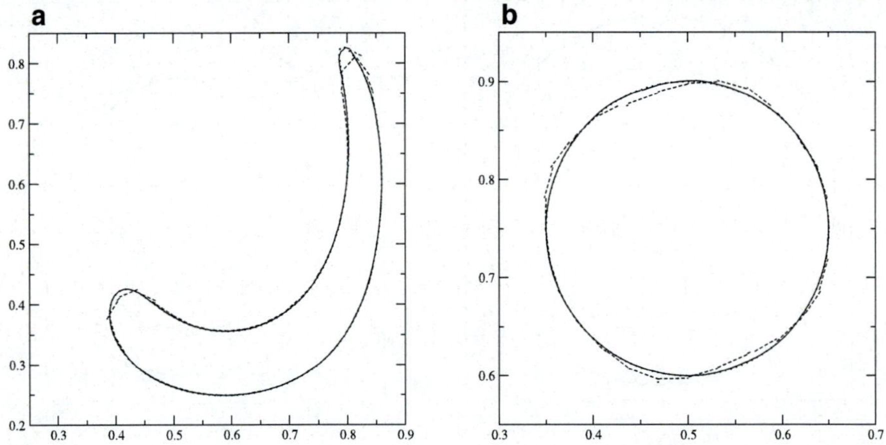


**Fig. 3.28** Vortex-in-a-box test with period $T = 2$: **a** the exact and reconstructed interface at $t = T/2$ (left) and $t = T$ (right), with grid resolutions $n = 32$ (dashed segments) and $n = 128$ (solid segments)

$$\Psi(x, y, t) = \frac{1}{\pi} \cos\left(\pi t/T\right) \sin^2\left(\pi x\right) \sin^2\left(\pi y\right)$$

The computational domain is a unit box with square cells of side $h_0 = 1/n$ and $n$ is the number of cells along each coordinate direction. The stream function is computed on the cell vertices and the velocity components on a staggered MAC grid are found by centered finite differences of $u = \partial\Psi/\partial y$, $v = -\partial\psi/\partial x$. The resulting velocity field satisfies the discrete divergence-free condition to machine accuracy. The flow is modulated in time with period $T$. A circular fluid body of radius $r_0 = 0.15$ and center at (0.5, 0.75) is positioned in this flow at $t = 0$ then it rotates clockwise while stretching and deforming till $t = T/2$, then it goes back to its initial position at $t = T$. The initial maximum CFL number is always equal to 1. For the $T = 2$ test, at the lowest resolution $n = 32$ the reconstruction and advection algorithms cannot follow correctly the interface deformation near the head and the tail of the filament where the local radius of curvture is comparable or even smaller than the grid spacing. A second-order convergence rate with grid resolution is found for this test. The $T = 8$ test is even more demanding, as the thickness of the filament becomes of the order of the grid spacing $h_0$ and the reconstruction algorithm cannot reproduce its shape with just one segment in each cell. The interface is artificially broken while mass is conserved. The algorithm acts as a numerical surface tension by collecting small debris into small object with a characteristic size of the order of the grid spacing (Figs. 3.28 and 3.29).

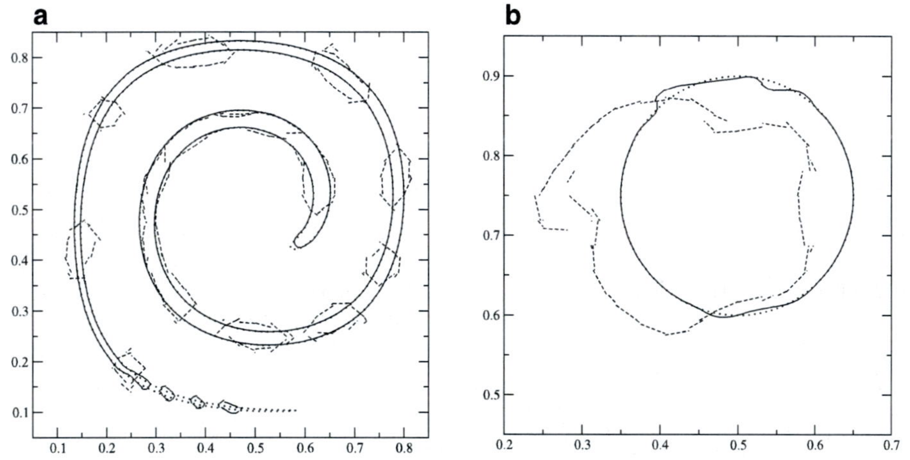


**Fig. 3.29** Vortex-in-a-box test with period $T = 8$: **a** the exact and reconstructed interface at $t = T/2$ (left) and $t = T$ (right), with grid resolutions $n = 32$ (dashed segments) and $n = 128$ (solid segments)

## 3.2 Level Set

### *3.2.1 Level Set Definition*

The level-set approach is mathemtatically based on the concept of implicit function. Indeed, the interface separating two medium can be viewed as a surface representing the zero iso-level of a certain scalar function $\phi(\boldsymbol{x})$. This level-set function is called implicit function for defining the interface location as can be seen in Fig. 3.30a. Here the interface is represented by the two points $\{-1, 1\}$. This definition of the interface by implicit function includes some interesting properties. For example, we can easely determine the side of the interface a point is on by simply looking at the local sign of $\phi(\boldsymbol{x}_0)$. If $\phi(\boldsymbol{x}_0) > 0$, $\boldsymbol{x}_0$ is outside the interface, inside is If $\phi(\boldsymbol{x}_0) < 0$. In order to give more smoothness to the level-set function, we will choose it among the signed distance functions which are a subset of the implicit functions as shown in Fig. 3.30b. Being $\boldsymbol{x}$ a point defined into the space $\Omega$, the value of the Level Set function $\phi(\boldsymbol{x}, t)$ is, at each instant of time $t$, the minimum distance of the point from the interface $\Gamma$, with a sign depending on which side it is contained.

This function $\phi(\boldsymbol{x}, t)$ can be seen as an infinite set of contour lines each of them at a fixed distance $d$ from the interface, as shown in Fig. 3.31. When the distance is zero, the contour line represents the interface itself, $\Gamma = \{\boldsymbol{x} \mid \phi(\boldsymbol{x}) = 0\}$. This is one possible option for describing a Level Set function. It is however the simplest and has the advantage in the adaptive mesh refinement perspective to allow the mesh to "know" how far is the interface, thus making the refinement and de-refinement easier to manage. In order to define the properties on the two regions an arbitrary plus sign is assigned to one of the two sides, the minus being the other one. Any

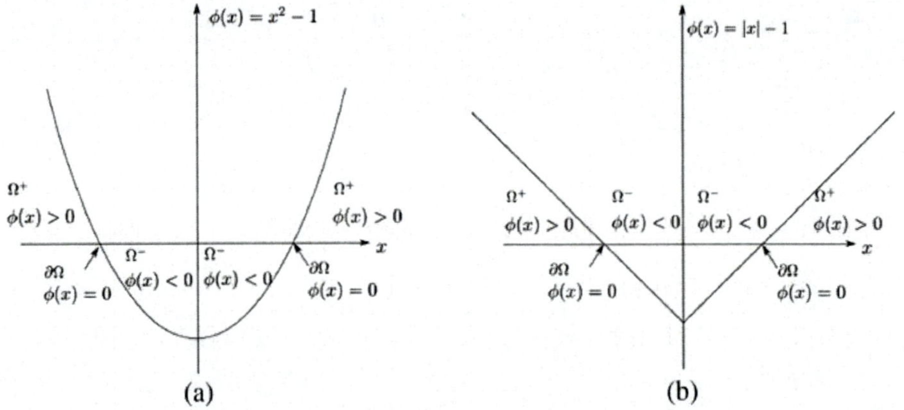


**Fig. 3.30** **a** Implicit level-set function, **b** implicit signed distance function

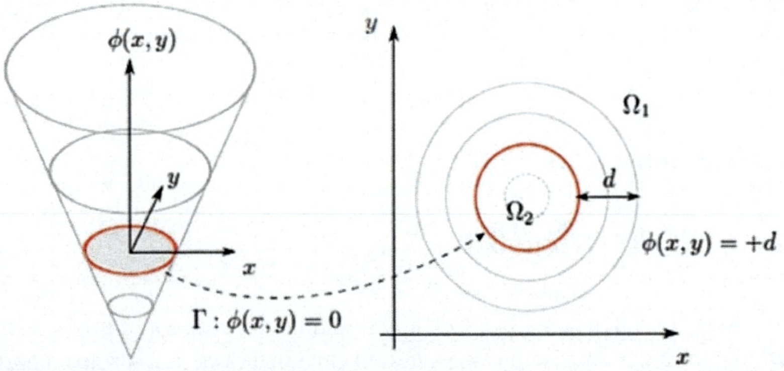


**Fig. 3.31** The level set "point of view": the contour lines of the function $\phi$ are the the locus of points in the plane distant $\phi$ from the interface, which is represented by the $\phi(x, y) = 0$ points

quantity $\alpha$ becomes

$$\alpha(\boldsymbol{x}, t) = \begin{cases} \alpha^{(1)}(\boldsymbol{x}, t) & , \text{if } \phi(\boldsymbol{x}, t) \geq 0 \\ \alpha^{(2)}(\boldsymbol{x}, t) & , \text{if } \phi(\boldsymbol{x}, t) < 0 \end{cases} \tag{3.61}$$

The so formulated level-set has some distinct advantages. The first is that the interface is exactly known at each time step, no reconstruction is needed. It has still to be numerically extracted the zero contour, but this is not necessary to get its geometrical properties. The normal vector and the curvature can be directly obtained from the values of $\phi$:

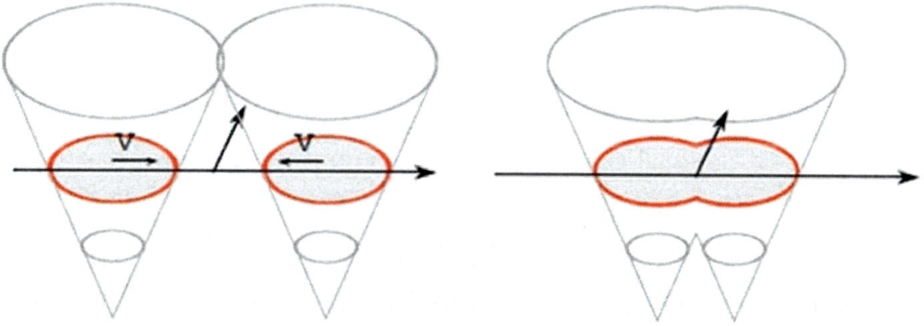


**Fig. 3.32** The level set simulated coalescence: two approaching droplets merge and create naturally a new structure (flow changes not taken in account)

$$\mathbf{n} = \left.\frac{\nabla\phi}{\|\nabla\phi\|}\right|_{\phi=0} \tag{3.62}$$

$$\kappa = \nabla\cdot\left.\left(\frac{\nabla\phi}{\|\nabla\phi\|}\right)\right|_{\phi=0} \tag{3.63}$$

The second is smoothness property $\|\nabla\phi(\boldsymbol{x})\| = 1$ as $\phi$ is a signed distance function. Being a signed function, no discontinuity appear on the interface, with the obvious advantages of an enhanced accuracy, stability and diffusion in the numerical schemes. This property can be locally lost when the function must describe sharp corners or filaments, especially in situations of under resolution. This aspects shows how the Level Set function may naturally ask for a local refinement of the mesh. Another advantage is its capability to naturally handle the topological changes such as the coalescence and ligaments break-up, as shown as example in Fig. 3.32 without the need, differently for the front tracking methods, of heavy dedicated algorithms.

Finally, signed distance functions make boolean operations and more complicated constructive geometry operations easy to apply. For example, if $\phi_1$ and $\phi_2$ are two different signed distance functions, then $\phi(\boldsymbol{x}, t) = \min(\phi_1(\boldsymbol{x}, t), \phi_2(\boldsymbol{x}, t))$ is the signed distance function representing the union of the interior region. The function $\phi(\boldsymbol{x}, t) = \max(\phi_1(\boldsymbol{x}, t), \phi_2(\boldsymbol{x}, t))$ is the signed distance function representing the intersection of the interior regions.

#### 3.2.1.1 Numerical Resolution

The evolution of the Level Set follows the linear hyperbolic advection equation

$$\frac{\partial\phi}{\partial t} + \boldsymbol{u}.\nabla\phi = 0 \tag{3.64}$$

which means that it is passively advected by the velocity field $\boldsymbol{u}$. The numerical schemes should satisfy some requirements in order to correctly treat this problem.

First of all the precision: the advection must be accurate enough to give a precise localization of the interface to allow a correct application of the jump conditions. In [136, 137] it has been demonstrated how the second order schemes are insufficient for the interface tracking problem. A higher order method should be used. The chosen scheme should also limit the numerical issues of diffusion and dispersion, in order to limit the loss of mass and the errors in the advection velocity for all the interfacial scales. Another fundamental property is the robustness. Although the Level Set is a regular function, some local under resolution situations may appear, especially in presence of strong deformations and gradients.

In [8] several high order advection schemes are combined with different interface tracking methods. In particular, the fifth order WENO of [59] in conservative and non conservative form ([58]) and the spectral methods for both finite volumes [138] and finite difference variants [139] are compared. The retained scheme will be the WENO, which is reported to give good accuracy and to have optimized variants which help to improve the mass conservation. The results from [8, 58] suggest a better behaviour in the conservative form of the WENO applied to the Eq. (3.64). The temporal discretization must be realized by a high order scheme as well to couple a good temporal accuracy with the spatial one. The third order Runge-Kutta is used as for the previous cited algorithms.

#### 3.2.1.2 Spatial Discretization

In this section are presented the WENO (Weighted Essentially Non-Oscillatory) schemes based on the work in [59, 140, 141], which are an evolution of the precedent ENO schemes of [142]. They were initially developed for the simulation of discontinuities propagation (shock waves) in the compressible domain. The requirements were mainly the robustness and the ability to conserve the step with as limited as possible numerical smearing. These characteristics make them valuable to spatially discretize the Eq. (3.64). The scheme is suitable to solve both a conservative and a non conservative formulation of the advection equation, as described in [8, 58].

**Non conservative form**

The equation in the "non conservative" form of the Level Set advection is written as

$$\frac{\partial \phi}{\partial t} + \boldsymbol{u}.\nabla\phi = 0 \tag{3.65}$$

Its discrete form in two dimensions looks like

$$\frac{\partial \phi_{ij}}{\partial t} + u_{ij} \left.\frac{\partial \phi}{\partial x}\right|_{ij} + v_{ij} \left.\frac{\partial \phi}{\partial y}\right|_{ij} = 0 \tag{3.66}$$

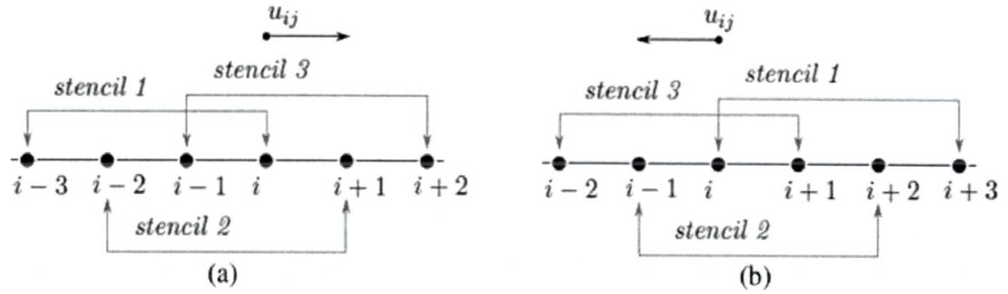


**Fig. 3.33** WENO 5 stencil in the non conservative form: **a** $(\phi_x)^-_{ij}$, **b** $(\phi_x)^+_{ij}$

Leaving the temporal derivative for the next section, the interest is now in the spatial derivatives discretization. A centered discretization for this hyperbolic equation is unstable; the stability can be obtained (under opportune CFL condition $\Delta t < min(\frac{\Delta x}{|u|}, \frac{\Delta y}{|v|})$) with uncentered *upwind* schemes, which respect the direction of propagation of the characteristics by choosing the stencil according to the local direction of information propagation. With a compact notation of $\partial\phi/\partial x = \phi_x$, in the $x$ direction this can be written as

$$\begin{cases} (\phi_x)_{ij} = (\phi_x)^+_{ij} = \dfrac{\phi_{i+1j} - \phi_{i,j}}{\Delta x}\ , \text{if } u_{i,j} < 0 \\ (\phi_x)_{ij} = (\phi_x)^-_{ij} = \dfrac{\phi_{ij} - \phi_{i-1,j}}{\Delta x}\ , \text{if } u_{i,j} > 0 \end{cases} \tag{3.67}$$

which correspond to a first order Euler upwind discretization, the same consideration being applied to the $y$ direction too. The velocity $u_{i,j}$ is in this case the cell centered. The fifth order WENO scheme (WENO5) offers a better approximation of the derivatives $(\phi_x)^+_{ij}$, $(\phi_x)^-_{ij}$, $(\phi_y)^+_{ij}$, $(\phi_y)^-_{ij}$ by exploiting a total of five points organized into an upwind uncentered linear combination of three derivatives based on as many different stencils, as shown in Fig. 3.33. WENO schemes can be viewed as combination of third order ENO interpolants, designed to improve accuracy up to fifth order.

$$(\phi_x)^\pm_{ij} = \sum_{k=1}^{3} \omega_k (\phi_x)^{\pm,k}_{ij} \tag{3.68}$$

where, for the values of $k = 1, 2, 3$, the derivatives are defined as

$$\begin{cases} (\phi_x)^{\pm,1}_{ij} = +\dfrac{1}{3}\, q_1^\pm - \dfrac{7}{6}\, q_2^\pm + \dfrac{11}{6}\, q_3^\pm \\ (\phi_x)^{\pm,2}_{ij} = -\dfrac{1}{6}\, q_2^\pm + \dfrac{5}{6}\, q_3^\pm + \dfrac{1}{3}\, q_4^\pm \\ (\phi_x)^{\pm,3}_{ij} = +\dfrac{1}{3}\, q_3^\pm + \dfrac{5}{6}\, q_4^\pm - \dfrac{1}{6}\, q_5^\pm \end{cases} \tag{3.69}$$

with

$$\begin{cases} q_k^- = \dfrac{\phi_{i-3+k,j} - \phi_{i-4+k,j}}{\Delta x} \\ q_k^+ = \dfrac{\phi_{i+4-k,j} - \phi_{i+3-k,j}}{\Delta x} \end{cases} \tag{3.70}$$

The weighting coefficients $\omega_k$ are a convex combination ($\omega_1 + \omega_2 + \omega_3 = 1$). Of course, if any of the three approximations intrerpolates across a discontinuity, it is given a minimal weight in order to diminish its contribution. The values of the weights $\omega^\pm$ are given by a the following expression of $\alpha^\pm$:

$$\omega_k^\pm = \alpha_k^\pm / \sum_{i=0}^{2} \alpha_i^\pm \tag{3.71}$$

with the $\alpha$ defined as

$$\begin{cases} \alpha_0^\pm = \dfrac{1}{10}\left(\dfrac{1}{\epsilon + IS_0^\pm}\right)^2 \\ \alpha_1^\pm = \dfrac{6}{10}\left(\dfrac{1}{\epsilon + IS_1^\pm}\right)^2 \\ \alpha_2^\pm = \dfrac{3}{10}\left(\dfrac{1}{\epsilon + IS_2^\pm}\right)^2 \end{cases} \tag{3.72}$$

where $\epsilon$ ensures the denominator to be different from zero and the $IS$ are the smoothness indicators:

$$\begin{cases} IS_0^\pm = \dfrac{13}{12}\left(q_1^\pm - 2q_2^\pm + q_3^\pm\right)^2 + \dfrac{1}{4}\left(q_1^\pm - 4q_2^\pm + 3q_3^\pm\right)^2 \\ IS_1^\pm = \dfrac{13}{12}\left(q_2^\pm - 2q_6^\pm + q_4^\pm\right)^2 + \dfrac{1}{4}\left(q_2^\pm - q_4^\pm\right)^2 \\ IS_2^\pm = \dfrac{13}{12}\left(q_3^\pm - 2q_4^\pm + q_5^\pm\right)^2 + \dfrac{1}{4}\left(3q_3^\pm - 4q_4^\pm + q_5^\pm\right)^2 \end{cases} \tag{3.73}$$

As can be seen from (3.73), in smooth regions, the $IS$ are very small and could be small enough compare to $\epsilon$, then (3.72) become $\alpha_1 \approx 0.1\epsilon^2, \alpha_2 \approx 0.6\epsilon^2, \alpha_3 \approx 0.3\epsilon^2$, giving the proper ratio for optimal fifth order accuracy. Normalizing the weighting coefficients as given by (3.71) gives the optimal values for smooth solutions. On the other side, nearly optimal weights are also obtained when the $IS_k^\pm$ are larger than $\epsilon$, as long as the $IS_k^\pm$ are approximatively of the value. For the choice of $\epsilon$, Fedkiw [143] gives the following value:

$$\epsilon = 10^{-6}\max(q_1^2, q_2^2, q_3^2, q_4^2, q_5^2) + 10^{-99} \tag{3.74}$$

where $10^{-99}$ is set to avoid division by zeo in the definition of the $\alpha_k$ and the scaling factor $10^{-6}$ will help in the balance between firth order accurate stencil and the ENO weights.

### Conservative form

Equation (3.64) can be written in the "conservative" form without physically changing its meaning thanks to the incompressibility hypothesis, $\nabla.\boldsymbol{u} = 0$ which is usual when dealing with two-phase interfacial flows and will be assumed in this course.

$$\frac{\partial \phi}{\partial t} + \nabla.(\boldsymbol{u}\phi) = 0 \tag{3.75}$$

The solution $\bar{\phi}$ is, in a finite volume sense, the integral value of the function on the cell of area $\Omega_{ij}$. The the Gauss-Ostrogradsky theorem allows the transformation of the integral of the second term into a surface integral representing the net flux of the conserved quantity across the cell's faces:

$$\int_{\Omega_{ij}} \frac{\partial \phi}{\partial t}\, d\Omega + \int_{\Omega_{ij}} \nabla.(\boldsymbol{u}\phi)\, d(\partial\Omega) = \int_{\Omega_{ij}} \frac{\partial \phi}{\partial t}\, d\Omega + \int_{\partial\Omega_{ij}} (\boldsymbol{u}\phi).\boldsymbol{n}\, dl = 0 \tag{3.76}$$

the numerical fluxes on the faces of the cell $\Omega_{i,j}$, $F$ located on the right cell face and $G$ on the upper one are then defined as

$$\begin{cases} F_{i+\frac{1}{2},j} = \displaystyle\int_{y_{i,j-\frac{1}{2}}}^{y_{i,j+\frac{1}{2}}} u_{i+\frac{1}{2},j}\phi_{i+\frac{1}{2},j}\, dy \\ G_{i,j+\frac{1}{2}} = \displaystyle\int_{x_{i-\frac{1}{2},j}}^{x_{i+\frac{1}{2},j}} v_{i,j+\frac{1}{2}}\phi_{i,j+\frac{1}{2}}\, dx \end{cases} \tag{3.77}$$

The discretization of the Eq. (3.76) then becomes:

$$\left(\frac{\partial \phi}{\partial t}\right)_{i,j} + \frac{F_{i+\frac{1}{2},j} - F_{i-\frac{1}{2},j}}{\Delta x} + \frac{G_{i,j+\frac{1}{2}} - G_{i,j-\frac{1}{2}}}{\Delta y} \approx 0 \tag{3.78}$$

A numerical approximation for the fluxes (3.77) can be written, as for the derivatives in the non conservative form, using the upwind formulation:

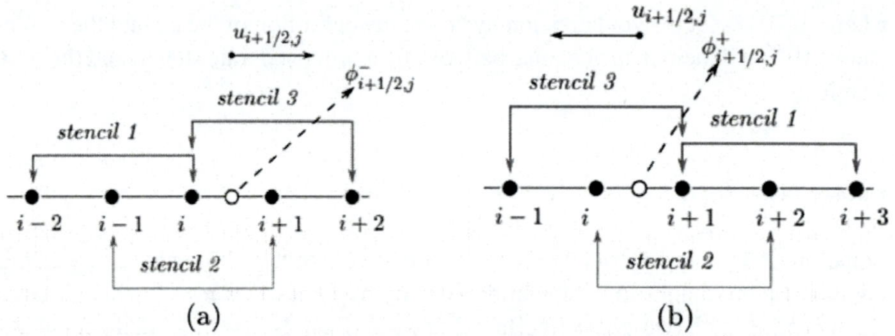


**Fig. 3.34** WENO 5 stencil in the conservative form. **a** $\phi^-_{i+1/2,j}$, **b** $\phi^+_{i+1/2,j}$

$$F_{i+\frac{1}{2},j} \approx \begin{cases} u_{i+\frac{1}{2},j}\bar{\phi}^+_{i-\frac{1}{2},j} & \text{, if } u_{i+\frac{1}{2},j} > 0 \\ u_{i+\frac{1}{2},j}\bar{\phi}^-_{i+\frac{1}{2},j} & \text{, if } u_{i+\frac{1}{2},j} < 0 \end{cases} \tag{3.79}$$

and

$$G_{i,j+\frac{1}{2}} \approx \begin{cases} v_{i,j+\frac{1}{2}}\bar{\phi}^+_{i,j-\frac{1}{2}}, \text{ if } v_{i,j+\frac{1}{2}} > 0 \\ v_{i,j+\frac{1}{2}}\bar{\phi}^-_{i,j+\frac{1}{2}}, \text{ if } v_{i,j+\frac{1}{2}} < 0 \end{cases} \tag{3.80}$$

The value $\bar{\phi}_{i,j}$ can be chosen as the integral mean value $\phi_{ij}$, which would give a first order discretization. A better approximated is obtained by using the same WENO5 discretization as used before: a smoothed ponderation of the quantities $\phi_{i+\frac{1}{2},j}{}^+$ and $\phi_{i+\frac{1}{2},j}{}^-$, as shown in Fig. 3.34.

$$\phi^\pm_{i+1/2,j} = \sum_{k=1}^{3} \omega_k \phi^{\pm,k}_{i+1/2,j} \tag{3.81}$$

with a similar formulation as in Eq. (3.69)

$$\begin{cases} \phi^{\pm,1}_{i+1/2,j} = +\frac{1}{3}\, q_1^\pm - \frac{7}{6}\, q_2^\pm + \frac{11}{6}\, q_3^\pm \\ \phi^{\pm,2}_{i+1/2,j} = -\frac{1}{6}\, q_2^\pm + \frac{5}{6}\, q_3^\pm + \frac{1}{3}\, q_4^\pm \\ \phi^{\pm,3}_{i+1/2,j} = +\frac{1}{3}\, q_3^\pm + \frac{5}{6}\, q_4^\pm - \frac{1}{6}\, q_5^\pm \end{cases} \tag{3.82}$$

and

$$\begin{cases} q_k^- = \bar{\phi}_{i-3+k,j} \\ q_k^+ = \bar{\phi}_{i+4-k,j} \end{cases} \tag{3.83}$$

with the same definition of $\alpha$ and $IS$ as in the Eqs. (3.72) and (3.73).

#### 3.2.1.3 Temporal Discretization

An high order temporal resolution is necessary as well for the correct evolution of the numerical interface. The time derivative $\frac{\partial \phi}{\partial t}$ is discretized by a third order Runge-Kutta scheme from [144]. If the spatial discretization of the advection term is written as $-(\boldsymbol{u}.\nabla)\phi = L(\phi)$, the Eq. (3.64) can be written as

$$\frac{\partial \phi}{\partial t} = L(\phi) \tag{3.84}$$

and an explicit Euler step would look like

$$\phi^{n+1} = \phi^n + \Delta t L(\phi^n) \tag{3.85}$$

stable under the CFL condition $\Delta t_{Euler} < \min\left(\frac{\Delta x}{|u|}, \frac{\Delta y}{|v|}\right)$. The Runge Kutta scheme can be seen as a combination of several Euler like "guesses" of the solution:

$$\begin{cases} \phi^{(i)} = \sum_{k=0}^{i-1} \left(\alpha_{ik}\, \phi^{(k)} + \Delta t\ \beta_{ik}\ L(\phi^{(k)})\right) \\ \phi^{(0)} = \phi^n, \qquad \phi^{(m)} = \phi^{n+1}, \qquad i = 1, ..., m \end{cases} \tag{3.86}$$

This scheme is stable under the following CFL condition:

$$\begin{aligned} &\Delta t < c\Delta t_{Euler} \\ &\text{with } c = \min_{i,k} \frac{\alpha_{i,k}}{\beta_{i,k}} \\ &\text{and } \alpha_{i,k} \geq 0\,,\ \beta_{i,k} \geq 0 \end{aligned} \tag{3.87}$$

The coefficients of the third order TVD-RK can be found in [59], and give the final temporal discretization

$$\begin{cases} \phi^1 = \phi^n + \Delta t L(\phi^n) \\ \phi^2 = \frac{3}{4}\phi^n + \frac{1}{4}\left(\phi^1 + \Delta t L(\phi^1)\right) \\ \phi^{n+1} = \frac{1}{3}\phi^n + \frac{2}{3}\left(\phi^2 + \Delta t L(\phi^2)\right) \end{cases} \tag{3.88}$$

This allows a stability CFL condition of $c = 1$ in Eq. (3.87).

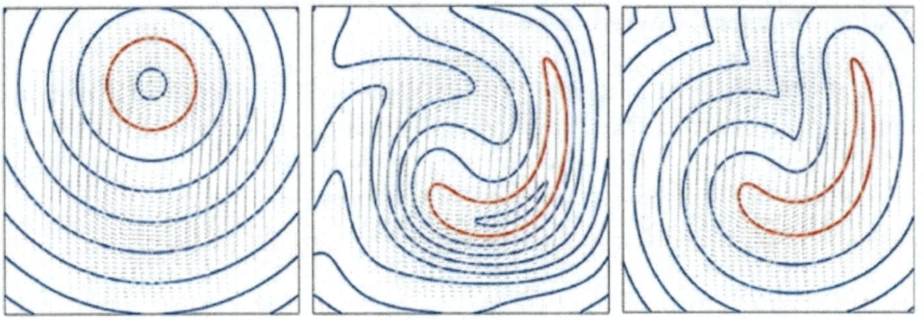


**Fig. 3.35** Example of evolution of a Level Set function. **a** Initial condition. **b** After some iterations, the contour lines are no more iso distance lines. **c** With redistance, the contour lines are parallel again

### 3.2.1.4 The Reinitialization

During the advection of the Level Set, if the velocity field does not impose a rigid translation or rotation, the different contour lines are advected differently, as the local velocity gradient is not uniform: they are no more parallel and they no more represent a constant distance from the interface, $\|\nabla\phi\| \neq 1$ (Fig. 3.35a, b). If the position of the actual interface is not affected by this problem, the characteristic of distance is necessary to the computation of the normal vector and the curvature at the interface. The redistance algorithm has to modify the advected Level Set function, without altering the zero contour, in order to reimpose the $\|\nabla\phi\| = 1$, or the distance property. In [145], Chopp introduced the idea that the level-set could be reinitialised or redistanced, in order to remain a signed distance function. An easely way to do this, is to use a kind of plotting algorithm to extract the $\phi = 0$ iso contour and then geometrically evaluate the distance from each grid point to the interface. Unfortunatly, this procedure is extremely slow as it scales with $N^3$ is $N$ is the number of grid points.

### 3.2.1.5 The Algorithm

[72] developed an algorithm for the Level Set reinitialization problem. The main idea is that the zero contour line is correct, and the others should be corrected in function of it. The steady state solution of the following Hamilton Jacobi equation will give a signed distance function:

$$\begin{cases} \dfrac{\partial\phi}{\partial\tau} = \text{sign}\left(\phi_0(\boldsymbol{x}, t)\right)\left(1 - \|\nabla\phi\|\right) \\ \phi(\boldsymbol{x}, \tau = 0) = \phi_0(\boldsymbol{x}, t) \end{cases} \tag{3.89}$$

as $\|\nabla\phi\| = 1$, and no modifications should be imposed to the interface. Here, $\text{sign}\,(\phi_0(\boldsymbol{x}, t)) = 1$ if $\phi_0(\boldsymbol{x}, t) > 0$, $\text{sign}\,(\phi_0(\boldsymbol{x}, t)) = -1$ if $\phi_0(\boldsymbol{x}, t) < 0$ and $\text{sign}\,(\phi_0(\boldsymbol{x}, t)) = 0$ if $\phi_0(\boldsymbol{x}, t) = 0$. Equation (3.89) can be rewritten into a hyperbolic advection equation:

$$\frac{\partial\phi}{\partial\tau} + \left( \text{sign}\,(\phi_0(\boldsymbol{x}, t)) \frac{\nabla\phi}{\|\nabla\phi\|} \right).\nabla\phi = \text{sign}\,(\phi_0(\boldsymbol{x}, t)) \tag{3.90}$$

Written like this, (3.90) propagates in the normal direction on both sides of the interface (from $\phi_0(\boldsymbol{x}, t) = 0$) away from the interface thanks to the sign function the iso-zero at a celerity of 1.

### 3.2.2 Numerical Method

This equation can be solved by the application of the methods used for the Eq. (3.64), thanks to the extension done in [146] of the WENO to the Hamilton-Jacobi equations: the Eq. (3.90) can be effectively written in this way, as described in [58]. Here we give the detail of the WENO5 Godunov scheme used for the numerical solution of the reinitialization equation. Equation (3.89) in semi-discrete form is written as:

$$\frac{\partial\phi_{i,j}}{\partial\tau} = \hat{H}\left(D_x^+\phi_{i,j}, D_x^-\phi_{i,j}, D_y^+\phi_{i,j}, D_y^-\phi_{i,j}\right) \tag{3.91}$$

where $D_x^+\phi_{i,j}, D_x^-\phi_{i,j}, D_y^+\phi_{i,j}, D_y^-\phi_{i,j}$ are fifth-order accurate WENO approximations of the $\partial_x\phi$ and $\partial_y\phi$. Let us define the folowing quantities:

$$\begin{aligned} \Delta_x^+\phi_{i,j} &= \phi_{i+1,j} - \phi_{i,j} \ , \ \Delta_x^-\phi_{i,j} = \phi_{i,j} - \phi_{i-1,j} \\ \Delta_y^+\phi_{i,j} &= \phi_{i,j+1} - \phi_{i,j} \ , \ \Delta_y^-\phi_{i,j} = \phi_{i,j} - \phi_{i,j-1} \end{aligned} \tag{3.92}$$

The $D^{\pm}$ are defined by:

$$\begin{aligned} D_x^{\pm}\phi_{i,j} &= -\tfrac{1}{12}\left(-\tfrac{\Delta_x^+\phi_{i-2,j}}{\Delta x} + 7\tfrac{\Delta_x^+\phi_{i-1,j}}{\Delta x} + 7\tfrac{\Delta_x^+\phi_{i,j}}{\Delta x} - \tfrac{\Delta_x^+\phi_{i+1,j}}{\Delta x}\right) \\ &\pm \phi^W\left(\tfrac{\Delta_x^-\Delta_x^+\phi_{i\pm2,j}}{\Delta_x}, \tfrac{\Delta_x^-\Delta_x^+\phi_{i\pm1,j}}{\Delta x}, \tfrac{\Delta_x^-\Delta_x^+\phi_{i,j}}{\Delta x}, \tfrac{\Delta_x^-\Delta_x^+\phi_{i\mp1,j}}{\Delta x}\right) \\ D_y^{\pm}\phi_{i,j} &= -\tfrac{1}{12}\left(-\tfrac{\Delta_y^+\phi_{i,j-2}}{\Delta y} + 7\tfrac{\Delta_y^+\phi_{i,j-1}}{\Delta y} + 7\tfrac{\Delta_y^+\phi_{i,j}}{\Delta y} - \tfrac{\Delta_y^+\phi_{i,j+1}}{\Delta y}\right) \\ &\pm \phi^W\left(\tfrac{\Delta_y^-\Delta_y^+\phi_{i,j\pm2}}{\Delta_y}, \tfrac{\Delta_y^-\Delta_y^+\phi_{i,j\pm1}}{\Delta y}, \tfrac{\Delta_y^-\Delta_y^+\phi_{i,j}}{\Delta y}, \tfrac{\Delta_y^-\Delta_y^+\phi_{i,j\mp1}}{\Delta y}\right) \end{aligned} \tag{3.93}$$

where

$$\phi^W(a, b, c, d) = \frac{1}{3}\omega_0(a - 2b + c) + \frac{1}{6}\omega_2(b - 2c + d) \tag{3.94}$$

where $\omega_1 =$ has been substitued by $\omega_1 = 1 - \omega_0 - \omega_2$, and the weights $\omega_i$ are given by:

$$\omega_0 = \frac{\alpha_0}{\alpha_0 + \alpha_1 + \alpha_2} \quad \omega_2 = \frac{\alpha_2}{\alpha_0 + \alpha_1 + \alpha_2}$$

$$\alpha_0^\pm = \frac{1}{10}\left(\frac{1}{\epsilon + IS_0^\pm}\right)^2 \quad \alpha_1^\pm = \frac{6}{10}\left(\frac{1}{\epsilon + IS_1^\pm}\right)^2 \quad \alpha_2^\pm = \frac{3}{10}\left(\frac{1}{\epsilon + IS_2^\pm}\right)^2$$

$$IS_0 = 13(a-b)^2 + 3(a-3b)^2$$

$$IS_1 = 13(b-c)^2 + 3(b+c)^2$$

$$IS_2 = 13(c-d)^2 + 3(3c-d)^2 \tag{3.95}$$

Here the discrete Hamiltonian, based on a Godonuv scheme [146], is given by:

$$\hat{H}(u, v) =$$

$$\begin{cases} S_0\left(1 - \sqrt{\max((u^+)^-, (u^-)^+)^2 + \max((v^+)^-, (v^-)^+)^2}\right) & \text{if } \quad S_0 > 0 \\ S_0\left(1 - \sqrt{\max((u^+)^+, (u^-)^-)^2 +) \max((v^+)^+, (v^-)^-)^2}\right) & \text{otherwise} \end{cases} \tag{3.96}$$

where $(u)^+ = \max(u, 0)$ and $(u)^- = -\min(u, 0)$. Numerical tests [72] show that better results are obtained when $S_0 = \text{sign}(\phi_0(\boldsymbol{x}, t))$ is smeared out over one grid cell:

$$\text{sign}(\phi_0) = \frac{\phi_0}{\sqrt{\phi_0^2 + (\Delta x)^2}} \tag{3.97}$$

Later on, [147] proposed :

$$\text{sign}(\phi) = \frac{\phi}{\sqrt{\phi^2 + \|\nabla\phi\|^2(\Delta x)^2}} \tag{3.98}$$

They shown that (3.98) was a better choice when initial $\phi_0$ is far from a signed distance function. It is worth noting that in this case $\text{sign}(\phi)$ is continually updated as the calculation progresses which is different from (3.91) where $\text{sign}(\phi)$ was kept constant equal to $\text{sign}(\phi_0)$. One can say that smearing the sign function reduces its magnitude and slows the propagation speed of information near the interface. Finally, temporal integration is performed by third order TVD-RK scheme. From a pratical point of view, only 2–3 temporal iteration in pseudo time $\tau$ are needed for level-set reinitialization. Indeed, one need only the level-set to be a distance function in the neighbourhood of the interface. Moreover, for CLF like stability constraint for the reinitialization Eq. (3.91), one can take the following condition $\Delta\tau = \Delta x/2$. Figure 3.35c shows the application of the reinitialization step to the configuration of Fig. 3.35b. One can clearly see the performance of this algorithm to redistance the level-set function of Fig. 3.35b. Nevertheless, even wich such high order scheme, numerical discretization errors lead to small shifts of the interface, with the tendency to thicken the interface and smooth the sharp corners leading to mass conservation errors.

#### 3.2.2.1 Disk in a Deformation Velocity Field

This test is widely used to study the porperties of interface capturing method. The velocity fied convecting the level-set is given by the following stream function:

$$\Psi(x, y, t) = \frac{1}{\pi} \sin(\pi x^2) \sin(\pi y^2) \tag{3.99}$$

The initial condition is a disk of radius $r = 0.15$ centered at $(0.5, 0.75)$ in a unit box. The resulting velocity field will stretch the disk into ever thinner filament. We compare the numerical solution of the level-set approach with and without reinitialization to the exact solution provided by surface markers located on the circle bounding the disk and moved by a high order accurate lagrange approach. We present in Fig. 3.36a, b, the solution at $t = 3s$ on a 128 by 128 grid. We compare the pure advected level-set in Fig. 3.36a and the same with reinitialization. For this case, convective term in the level-set equation are discretized by non conservative WENO5 schemes. Both schemes fail to properly capture the tail of the filament which is clearly under-resolved with 128 by 128 grid. One can observe that reinitialization performs better as more filament is captured. However, we can see that reinitialization seems to truncate the head of the filament (in the central part). Using conservative WENO5 for level-set advection clearly improves the solution as shown in Figs. 3.36c, d.

To get more insgiht into the behaviour of both the effect of the choice of the scheme for level-set advection as well as the reinitialization, we have plotted in Fig. 3.37a, b for different grid sizes from 64 by 64 to 256 by 256. These figures give quantitative informations of mass error problem with level-set method. First of all, whatever is the method, as the initial disk is strecthed and becomes thinner relative to the grid size the error increases with time. One also can see in Fig. 3.37a that WENO5 conservative form gives better results than non conservative one, at least without reinitialization. When reinitialization is activated, see Fig. 3.37b, we observe mass gain when conservative scheme for level-set convection is used and mass loss when non-conservative convection scheme is used. A singular behaviour is also observed when reinitilization is performed, for the non-conservative case, reinitialization decreases a little bit mass error whereas for the conservative case it increases it.

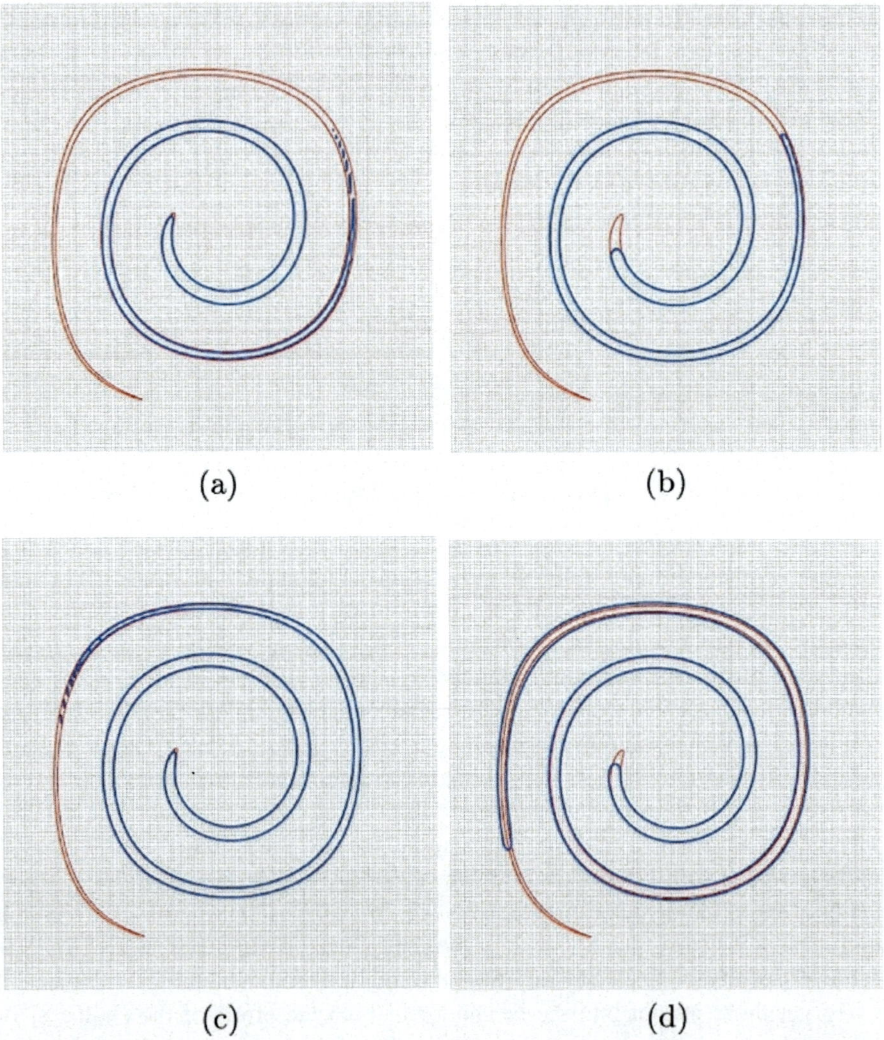


**Fig. 3.36** Deformation problem at t = 3 s evolution of a Level Set function. **a** Non conservative WENO5 scheme for the level-set advection, non reinitilization, **b** non conservative WENO5 scheme for the level-set advection, with reinitilization, **c** conservative WENO5 scheme for the level-set advection, without reinitilization, **d** conservative WENO5 scheme for the level-set advection, with reinitilization reprint from [58]

#### 3.2.2.2 About Mass Conservation

The Level Set method is affected by a mass loss effect, an aspect very disadvantageous if compared to the exact conservation of the VOF. If the use of high order schemes and mesh refinement greatly improves this behaviour, a more definitive solution is still a current topic. Different strategies have been proposed for this problem which can be classified in three family. The first one is based on the coupling of level-set

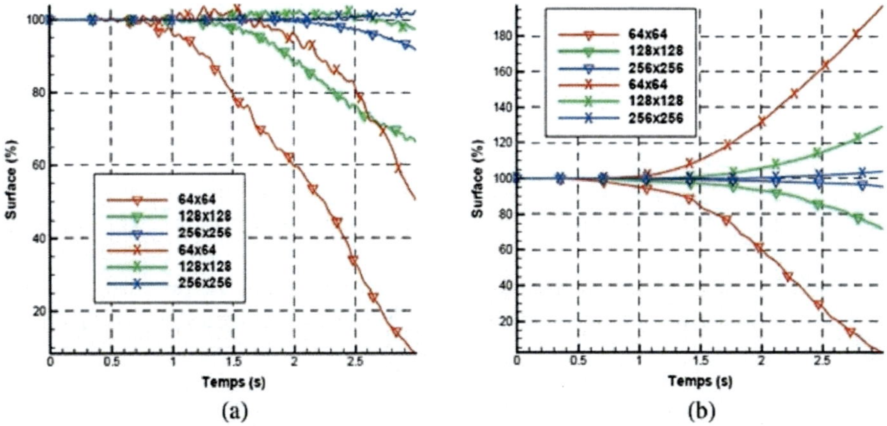


**Fig. 3.37** Deformation problem: surface evolution. **a** Without reinitilization, conservative WENO5 cross, non conservative triangle up, **b** with reinitilization, conservative WENO5 cross, non conservative WENO5 triangle up, reprint from [58]

method with with particle approachs like in [92] with volume markers or in [94] with surfaces markers. The second one mainly developped by [148, 149] and then improved by [150, 151] is based on what is called conservative level-set approach. The last one is related to the coupling of the level-set method with VOF method mainly developped by [89, 152]. We will give in the following section the detail of the CLSVOF as it is one of the most efficient method to get mass conservation with level-set method, particularly when one deals with high density ratio two-phase flow like gas-liquid two-phase flow.

### 3.2.3 *Coupled Level-Set Volume of Fluid*

The coupling between level-set method and volume of fluid method was initially developed by [89, 152] under the acronym CLSVOF (Coupled Level-Set Volume Of Fluid) and later on by [91] under the acronym MCLS (Mass Conserving Level-Set). The main difference of the two approach could be find in the way the volume fraction is build from the level-set and convected. Whereas in [89], this is done geometrically, in [91] this is done algebraicaly. We will now present in more detail the CLSVOF algorithm. The main objective of this coupling is to cure mass error inherent to level-set method by using mass-conservation property of VOF method. In other words, the aim is to locally correct the level-set function by the VOF function in order to preserve mass conservation. From the level-set function $\phi(x, y, t)$ one can define the local liquid volume fraction in a cell $\Omega$ by

$$F(\Omega, t) = \frac{1}{\mid \Omega \mid} \int_{\Omega} H(\phi(x, y, t)) \mathrm{d}\omega \tag{3.100}$$

where $H$ is the Heaviside function defined by:

$$H(\phi) = \begin{cases} 1 & \text{if } x > 0 \\ 0 & \text{otherwise} \end{cases} \tag{3.101}$$

#### 3.2.3.1 Obtaining the Liquid Volume Fraction from the Level-Set

To define the liquid volume fraction from the level-set, as proposed by [89], we will approximate the real interface by straight line and define a local reconstructed $\phi^{n,R}(x, y)$ given by:

$$\phi^{n,R}(x, y) = a_{i,j}(x - x_i) + b_{i,j}(y - y_j) + c_{i,j} \tag{3.102}$$

here $a_{i,j}, b_{i,j}$ are the coordinate of a vector normal to line defined by:

$$a_{i,j}(x - x_i) + b_{i,j}(y - y_j) + c_{i,j} = 0 \tag{3.103}$$

when $a, b$ are normalized by $\sqrt{a^2 + b^2}$, they represent the components of the normal of the reconstructed interface and $\phi i, j^{n,R} = c_{i,j}$ the signed distance to this interface in cell $(i, j)$.

Where the coefficients $a_{i,j}, b_{i,j}, c_{i,j}$ are obtained as the best fit which minizes the following error:

$$E_{i,j} = \int_{y_{i,j-1/2}}^{y_{i,j+1/2}} \int_{x_{i-1/2,j}}^{x_{i+1/2,j}} H'(\phi)\left(\phi - a_{i,j}(x - x_i) - b_{i,j}(y - y_j) - c_{i,j}\right)^2 \tag{3.104}$$

at the discrete level, we minimize the following error:

$$E_{i,j} = \sum_{i'=i-1}^{i'=i+1} \sum_{j'=j-1}^{j'=j+1} \omega_{i'-i,j'-j} H'_\epsilon(\phi_{i',j'}) \left(\phi_{i',j'} - a_{i,j}(x_{i'} - x_i) - b_{i,j}(y_{j'} - y_j) - c_{i,j}\right)^2 \tag{3.105}$$

where the weights $\omega_{r,s} = 16$ if $r = s = 0$ and $\omega r, s = 1$ otherwise. $H'_\epsilon$ is a smoothed delta function with a thickness $\epsilon = \sqrt{2}\Delta x$. Minimization of (3.105) is done by solving a $3 \times 3$ linear system given by:

$$\begin{pmatrix} \sum\sum \omega h X^2 & \sum\sum \omega h XY & \sum\sum \omega h X \\ \sum\sum \omega h XY & \sum\sum \omega h Y^2 & \sum\sum \omega h Y \\ \sum\sum \omega h X & \sum\sum \omega h Y & \sum\sum \omega h X \end{pmatrix} \begin{pmatrix} a_{i,j} \\ b_{i,j} \\ c_{i,j} \end{pmatrix} = \begin{pmatrix} \sum\sum \omega h \phi X \\ \sum\sum \omega h \phi y \\ \sum\sum \omega h \phi \end{pmatrix} \tag{3.106}$$

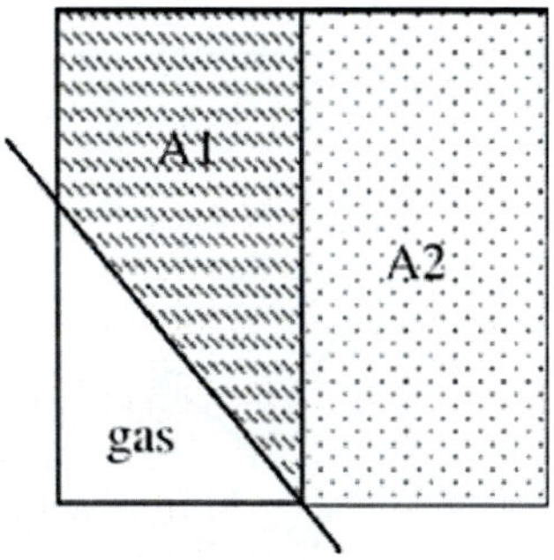


**Fig. 3.38** VOF geometrical scheme from [153]

where:

$$\begin{aligned}\sum\sum &\longrightarrow \sum_{i'=i-1}^{i'=i+1}\sum_{j'=j-1}^{j'=j+1}\\ \omega h &\longrightarrow \omega_{i'-1,j'-j}H'_{\epsilon}(\phi_{i',j'})\\ X &\longrightarrow x_{i'}-x_i\\ Y &\longrightarrow y_{j'}-y_j\\ \phi &\longrightarrow \phi_{i',j'}\end{aligned} \quad (3.107)$$

that can be solved by direct $LU$ method, as well as analytical method. The value of the interceipt $c_{i,j}$ is corrected in order that the line defined by (3.103) cuts out the same volume in cell $i$, $j$ as given by $F^n_{i,j}$

$$F^n_{i,j} = \frac{1}{dxdy}\int_{y_{i,j-1/2}}^{y_{i,j+1/2}}\int_{x_{i-1/2,j}}^{x_{i+1/2,j}} H(a_{i,j}(x-x_i)+b_{i,j}(y-y_j)+c_{i,j})dxdy \quad (3.108)$$

Several methods can be used to get the new value of the interceipt $c_{i,j}$. We present here the method of [153]. As can be seen in Fig. 3.38 calculating $F^n_{i,j}$ from (3.108) is just findind the intersection of the line given (3.103) and the edges of the cell and then calculate the area of the trapeze $A_1$ and the rectangle $A_2$.

In order to reduce the number of possible interface configurations can exist, we make some transformations. We choose the origin of the cell as $(x_{i-1/2,j-1/2}$, $y_{i-1/2,j-1/2}$. We then modify define new coefficients $a'_{i,j}$, $b'_{i,j}$ in order that the normal vector of the interface points towards this new origin and also to be sure that the evaluated volume is the real volume fraction of liquid in the cell by:

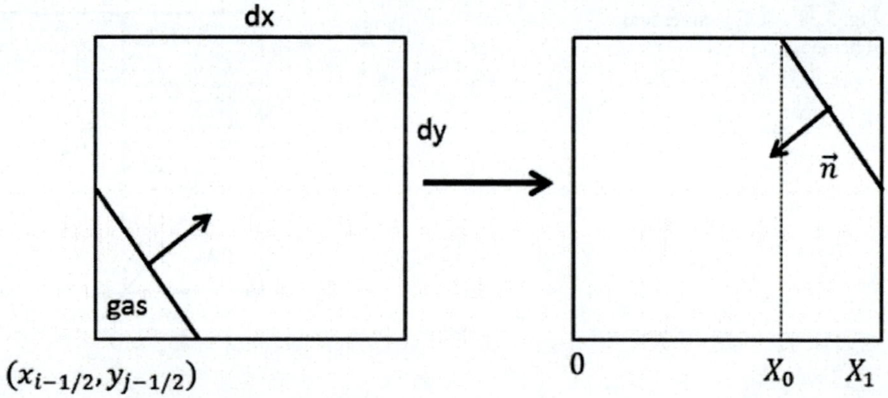


**Fig. 3.39** Coordinate transformation

$$\begin{cases} a'_{i,j}(x - x_{i-1/2,j-1/2}) + b'_{i,j}(y - y_{i-1/2,j-1/2}) + c'_{i,j} = 0 \\ \text{with} \\ a'_{i,j} = -a_{i,j} \quad \text{if} \quad a_{i,j} > 0 \\ b'_{i,j} = -b_{i,j} \quad \text{if} \quad b_{i,j} > 0 \\ c'_{i,j} = c_{i,j} - \frac{1}{2}\left(a'_{i,j}dx + b'_{i,j}dy\right) \end{cases} \tag{3.109}$$

The coordinate transformation is shown in Fig. (3.39).
we can then calculate the liquid volume fraction by:

$$F_{i,j} = \frac{1}{dxdy}\left(\left[AX^2 + BX\right]_{X_0}^{X_1} + X_0 dy\right)$$
$$\text{with} \quad A = \frac{-a'_{i,j}}{b'_{i,j}} \quad \text{with} \quad B = \frac{-c'_{i,j}}{b'_{i,j}}(b'_{i,j} \neq 0) \tag{3.110}$$

When one of the coefficient $a'$ or $b'$ is zero, that means vertical or horizontal interface, in this case the calculation is obvious.

#### 3.2.3.2 Correcting the Level-Set Function from the Volume Fraction

From a given liquid volume fraction $F^n{}_{i,j}$, we have:

$$F^n_{i,j} = \frac{1}{dxdy}\int_{y_{i,j-1/2}}^{y_{i,j+1/2}}\int_{x_{i-1/2,j}}^{x_{i+1/2,j}} H(a_{i,j}(x - x_i) + b_{i,j}(y - y_j) + c_{i,j})dxdy \tag{3.111}$$

coming from the definition of $\phi^{n,r}{}_{i,j}$, if this equality is not verified, a Newton method for $c_{i,j}$ is performed in order to shift the line inside the cell $i$, $j$ to fit the right liquid volume fraction $F^n_{i,j}$:

$$c_{i,j}^{\text{new}} = c_{i,j} - \frac{\dfrac{1}{dxdy}\displaystyle\int_{y_{i,j-1/2}}^{y_{i,j+1/2}}\int_{x_{i-1/2,j}}^{x_{i+1/2,j}} H(a_{i,j}(x-x_i)+b_{i,j}(y-y_j)+c_{i,j})dxdy - F^n_{i,j}}{\displaystyle\int_{y_{i,j-1/2}}^{y_{i,j+1/2}}\int_{x_{i-1/2,j}}^{x_{i+1/2,j}} \delta(a_{i,j}(x-x_i)+b_{i,j}(y-y_j)+c_{i,j})dxdy} dxdy \tag{3.112}$$

### 3.2.4 Advection of the Level-Set Function and the Volume Fraction

As for discrete the level-set function $\phi_{i,j}$, we define the volume fraction $F_{i,j}$of the liquid at the cell center of the grid. The equations governing the interface motion are:

$$\phi_t + \nabla \cdot (\boldsymbol{u}\phi) = 0 \tag{3.113}$$

for the level-set function and

$$F_t + \nabla \cdot (\boldsymbol{u}F) = 0 \tag{3.114}$$

for the volume fraction thanks to $\nabla \cdot \boldsymbol{u} = 0$. We also assume that $\phi^0{}_{i,j}$ is a signed distance function. However in practice one rather solves the following equations:

$$\phi_t + \nabla \cdot (\boldsymbol{u}\phi) = \phi\nabla \cdot \boldsymbol{u} \tag{3.115}$$

for the level-set function and

$$F_t + \nabla \cdot (\boldsymbol{u}F) = F\nabla \cdot \boldsymbol{u} \tag{3.116}$$

Given $\phi^n{}_{i,j}$ and $F^n{}_{i,j}$, we will use a second order conservative Strang operator split [154] to obtain $\phi^{n+1}{}_{i,j}$ and $F^{n+1}{}_{i,j}$. Here are the detail for the 2D operator split algorithm for a scalar $s$

$$\tilde{s}_{i,j} = \frac{s^n_{i,j} + \Delta t/\Delta x\left(G_{i-1/2,j} - G_{i+1/2,j}\right)}{1 - \Delta t/\Delta x\left(u_{i+1/2,j} - u_{i-1/2,j}\right)} \tag{3.117}$$

$$s^{n+1}_{i,j} = \tilde{s}_{i,j} + \frac{\Delta t}{\Delta y}\left(\tilde{G}_{i,j-1/2} - \tilde{G}_{i,j+1/2}\right) + \tilde{s}_{i,j}\left(v_{i,j+1/2} - v_{i,j-1/2}\right) \tag{3.118}$$

where at each iteration the splitting directions are alterned to get second ordre accuracy. Here Eq. (3.117) is semi-implict whereas (3.118) is explicit. One can also used semi-implicit formulation form both direction as mentionned in [153]. We then get the following splitting:

$$\tilde{s}_{i,j} = \frac{s_{i,j}^n + \Delta t/\Delta x \left(G_{i-1/2,j} - G_{i+1/2,j}\right)}{1 - \Delta t/\Delta x \left(u_{i+1/2,j} - u_{i-1/2,j}\right)} \tag{3.119}$$

$$\hat{s}_{i,j} = \frac{\tilde{s}_{i,j} + \Delta t/\Delta y \left(\tilde{G}_{i,j-1/2} - \tilde{G}_{i,j+1/2}\right)}{1 - \Delta t/\Delta y \left(v_{i,j+1/2} - v_{i,j-1/2}\right)} \tag{3.120}$$

$$s_{i,j}^{n+1} = \hat{s}_{i,j} - \Delta t \left(\frac{\tilde{s}_{i,j}}{\Delta x}\left(u_{i+1/2,j} - u_{i-1/2,j}\right) + \frac{\hat{s}_{i,j}}{\Delta y}\left(v_{i,j+1/2} - v_{i,j-1/2}\right)\right) \tag{3.121}$$

where $G_{i+1/2,j} = u_{i+1/2,j}s_{i+1/2,j}$ is the flux of $s$ across the face between cell $i,\ j$ and cell $i+1,\ j$ and $\tilde{G}_{i,j+1/2} = v_{i,j+1/2}\tilde{s}_{i,j+1/2}$ is the flux of $s$ across the face between cell $i,\ j$ and cell $i,\ j+1$. The calculation of $s_{i+1/2,j}$ is depending whether $s$ represents the level-set function $\phi$ or the volume fraction $F$. When $s = \phi$ and $u_{i+1/2,j} > 0$ then:

$$s_{i+1/2,j} = s_{i,j}^n + \frac{\Delta x}{2}\left(1 - u_{i+1/2,j}\frac{\Delta t}{\Delta x}\right)\frac{s_{i+1,j}^n - s_{i-1,j}^n}{\Delta x} \tag{3.122}$$

whereas $u_{i+1/2,j} < 0$

$$s_{i+1/2,j} = s_{i+1,j}^n - \frac{\Delta x}{2}\left(1 + u_{i+1/2,j}\frac{\Delta t}{\Delta x}\right)\frac{s_{i+2,j}^n - s_{i,j}^n}{\Delta x} \tag{3.123}$$

which can be seen as a second order accurate extrapolation in space and time of $s$ at the cell edge $i + 1/2,\ j$ see for detail [108]. When $s$ is the volume fraction $F$, we have the followling equation is $u_{i+1/2,j} > 0$:

$$s_{i+1/2,j} = \frac{\left(\int_{y_{i,j-1/2}}^{y_{i,j+1/2}} \int_{x_{i+1/2,j}-u_{i+1/2,j}\Delta t}^{x_{i+1/2,j}} H(\phi^{n,R}(x,y)\mathrm{dxdy}\right)}{u_{i+1/2,j}\Delta t \Delta y} \tag{3.124}$$

when $u_{i+1/2,j} < 0$

$$s_{i+1/2,j} = \frac{\left(\int_{y_{i,j-1/2}}^{y_{i,j+1/2}} \int_{x_{i+1/2,j}\Delta t}^{x_{i+1/2,j}-u_{i+1/2,j}} H(\phi^{n,R}(x,y)\mathrm{dxdy}\right)}{|u_{i+1/2,j}|\Delta t \Delta y} \tag{3.125}$$

The same formula applied in the transverse direction for calculating $s_{i,j+1/2}$. Here in (3.124) $\phi^{n,R}(x, y)$ is a local linear reconstruction of the interface define in the previous paragraph which will help to calculate $F$ from (3.100) as we cannot directly use (3.100). Detail of implementation can be found for example in [90].

#### 3.2.4.1 Putting All Together: The CLSVOF Algorithm

The overall algorithm is describet below:

1. Initialization: start from $\phi_{i,j}^{n=0}$
   - calculate $a_{i,j}, b_{i,j}, c_{i,j}$ and $\phi_{i,j}^{n,R}$
   - calculate $F_{i,j}^{n=0}$ from (3.108)
2. Coupled advection of $\phi$, $F$
   - get $\tilde{F}$ and $\tilde{\phi}$ from $F^n$ and $\phi^n$ in $x$ or $y$ direction (alternate sweepings)
   - calculate $a_{i,j}, b_{i,j}, c_{i,j}$ for $\tilde{\phi}_{i,j}^R$
   - calculate $c_{i,j}^{\text{new}}$ from (3.110) and correct $\tilde{\phi}_{i,j}^R$ for mass conservation
   - get $\hat{F}$ and $\hat{\phi}$ from $\tilde{F}$ and $\tilde{\phi}$ in $x$ or $y$ direction (alternate sweepings)
   - get $F_{i,j}^{n+1}$ and $\phi_{i,j}^{n+1}$ from (3.120)
   - correct $F_{i,j}^{n+1}$ by

$$F_{i,j}^{n+1} = 0 \quad \text{if} \quad F_{i,j}^{n+1} < 0 \quad \text{or} \quad \phi_{i,j}^{n+1} < -\Delta x$$

$$F_{i,j}^{n+1} = 1 \quad \text{if} \quad F_{i,j}^{n+1} > 1 \quad \text{or} \quad \phi_{i,j}^{n+1} > \Delta x$$

     to avoid VOF flotsam or jetsam
   - calculate $a_{i,j}, b_{i,j}, c_{i,j}$ for correcting $\phi_{i,j}^{n+1}$
   - apply redistancing algorithm on $\phi_{i,j}^{n+1}$ where $F_{i,j}^{n+1} = 0$ or $F_{i,j}^{n+1} = 1$, i.e. away from the interface.

In Fig. 3.40, are shown a comparison of different CLSVOF methods for the problem of a disk in a deformation field, the grid is the same as the results shown in Fig. 3.36.

We can see in this figure the clearl improvement of the CLSVOF method compared to the results of the classical level-set method of Fig. 3.36 as the tail of filament is now well captured even if some cuts appears which are linked to the VOF part of the method.

## 3.3 Front Tracking

The VOF and Level Set methods belong to the class of implicit Eulerian representation of resolved scale interfaces. As for the mass and momentum conservation

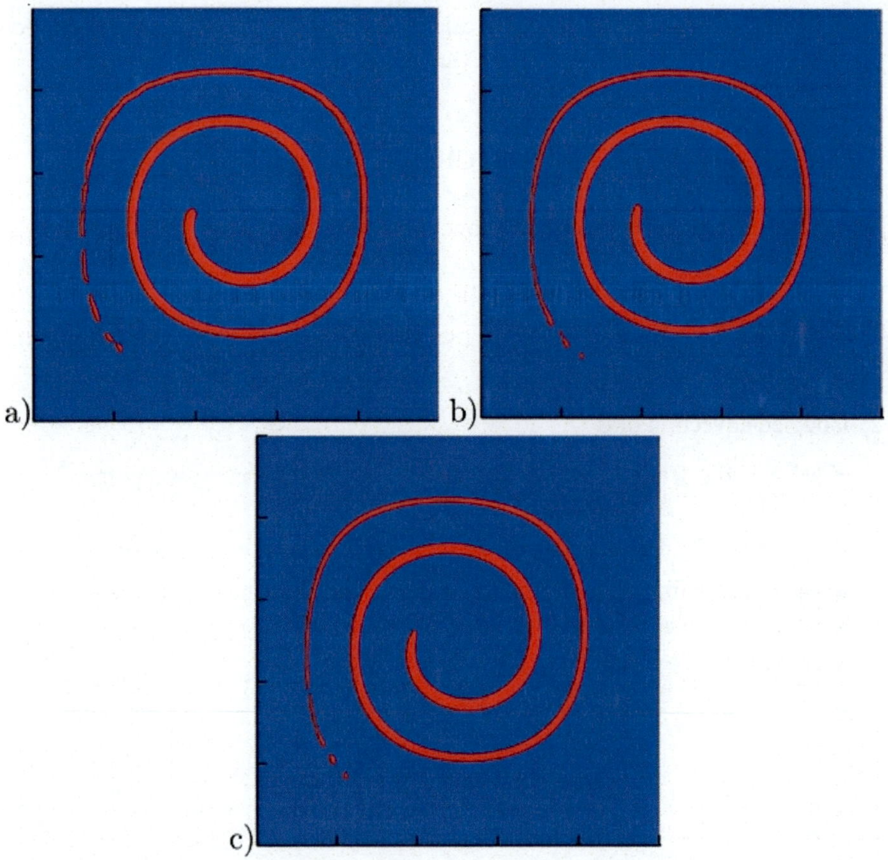

**Fig. 3.40** Solution of the serpentine test for the Sussman (**a**), Ménard (**b**) and Wang (**c**) reconstructions, spatial resolution $128^2$, $t = 3$s, from [155]

equations, they rely on a fixed grid. Equation (2.3) can be interpreted by definition as the transport of a material mass discontinuity interface that satisfies

$$\frac{d\boldsymbol{x}_i}{dt} = \boldsymbol{v}_i \tag{3.126}$$

with $\boldsymbol{x}_i$ the position of marker points on the interface and $\boldsymbol{v}_i$ their corresponding velocity. The Lagrangian tracking of the interface by Eq. (3.126) is called Front Tracking [82, 156]. In the Front Tracking approach, two grids are interacting: the Eulerian grid for the one fluid model (Eqs. (2.1–2.2) for mass and momentum conservation) and the Lagrangian grid of the interface $\Sigma_i$. In fact, on a numerical point of view, this approach is Lagrangian as it consists in describing the trajectories of marker placed on the interface during time. On a physical point of view, as Eq. (3.126) is coupled to the mass and momentum conservation equations (2.1–2.2), the Front Tracking is an Eulerian approach as are the VOF and Level Set methods. Once the marker positions

$\boldsymbol{x}_i$ are known, an Eulerian Heavyside or VOF function $C_i$ is built to regenerate the one-fluid properties $\rho$ and $\mu$ with Eqs. (2.4–2.5) as follows [82]

$$\nabla^2 C_i = \nabla \cdot \int_{\Sigma_i} \boldsymbol{n}_i \delta\left(\boldsymbol{x} - \boldsymbol{x}_i\right) ds \tag{3.127}$$

where $\boldsymbol{x}$ is the position of any of the vertices of the Eulerian mesh, $\boldsymbol{x}_i$ a marker position on the interface, $\delta$ a Dirac function indicating interface, $\boldsymbol{n}_i$ the local normal to the interface and $\Sigma_i$ the interface between phase $i$ and the rest of the considered medium, i.e. the surface located by the $\boldsymbol{x}_i$ points. Instead of solving Laplacian equation (3.127), a ray casting method can be equivalently used [157]. The $C_i$ function is obtained with a Ray-Casting (RC) method. The principle is to cast a ray from each Eulerian point to infinity and to test the number of intersections between the ray and the surface mesh. If the number of intersections is odd, the Eulerian point is inside the object, and outside otherwise. The main interest of the RC approach compared to (3.127) is that it can easily deal with objects that are partly immersed in the simulation domain. For parallel implementations with domain decomposition strategies, this property is very important. The binary $C_i$ function obtained with the RC technique can be extended to obtain a "real" VOF function by using a Level Set function, as described in [157].

Once $\boldsymbol{x}_i$ is known at each time step, the definition of $C_i$ according to the marker position allows to solve the multi-phase flow with the one-fluid model in an equivalent way as with VOF or Level Set methods. The main point now is the time evolution of $\Sigma_i$ and the associated Lagrangian solving of the interface tracking through Eq. (3.126). The immersed Lagrangian surface $\Sigma_i$ is initially defined as a set of linear (2D) or triangle (3D) elements whose vertices are defined thanks to the coordinates of the marker points $\boldsymbol{x}_i$. Examples of $\Sigma_i$ are given in Fig. 3.41. The time evolution of the Lagrangian surface requires the solving of Eq. (3.126). Classically, a Runge-Kutta approximation of second order is used

$$\boldsymbol{x}_i^1 = \boldsymbol{x}_i^n + \boldsymbol{v}_i(\boldsymbol{x}_i^n)\Delta t \tag{3.128}$$

$$\boldsymbol{x}_i^{n+1} = \boldsymbol{x}_i^n + \frac{1}{2}\Delta t \left(\boldsymbol{v}_i(\boldsymbol{x}_i^n) + \boldsymbol{v}_i(\boldsymbol{x}_i^1)\right) \tag{3.129}$$

with $n$ the time index and $\Delta t$ the time step. In the absence of mass transfer or phase change at the interface, the marker interfacial velocity is the fluid velocity. The Lagrangian velocities $\boldsymbol{v}_i$ at the marker position $\boldsymbol{x}_i$ are obtained by using for example a bilinear (trilinear in 3D) interpolation of the Eulerian Navier-Stokes velocities in 2D. Other methods can be considered such as the Peskin discrete Dirac functions [82, 158] or the Parabolic Edge Reconstruction method (PERM) [159]. The interest of the latter approach is that it provides a divergence free interpolation. However, it requires much more calculations than bilinear and Peskin techniques [158].

Instead of considering linear (planar in 3D) shape elements in the Front Tracking, spline shape element can also be investigated, in order to improve the curvature representation of interfaces. In two dimensions, this has been proposed with success

**Fig. 3.41** Example of front tracking surfaces in 2D [106] and 3D [157]

by [106]. The main interest of the Front Tracking approach is that is allows for keeping thin interfacial structures at scale that is smaller than the size of the local Eulerian mesh. The other important feature of this method is that is can directly estimate the surface tension forces on the Lagrangian surface $\Sigma_i$ instead of using volume representation of these forces through gradients of $C_i$. By using the Frenet relation $\kappa \boldsymbol{n}_i = \frac{d\boldsymbol{t}_i}{ds}$ [82], it can be demonstrated that the surface tension force $\boldsymbol{F}_{st}$ is

$$\boldsymbol{F}_{st} = \int_{\Sigma_i} \sigma \kappa \boldsymbol{n}_i \delta_i ds = \sum_{I=1,dim} \int_{A_{I,1}}^{A_{I,2}} \sigma \frac{d\boldsymbol{t}_i}{ds} ds = \sigma \sum_{I=1,dim} (\boldsymbol{t}_{i,A_{I,2}} - \boldsymbol{t}_{i,A_{I,1}}) \quad (3.130)$$

Concerning the drawbacks of the Front Tracking approach, we can cite the mass conservation that is not maintained over time and also the management of coalescence and rupture of interfacial structures that is not straightforward. The first point can be addressed by imposing a dilatation/compression of the global $\sigma_i$ surface so as to keep a constant volume of each phase $i$. Coalescence and rupture are managed either by defining a minimum physical size to cut/connect different elements of $\Sigma_i$ [106] or by projecting the Lagrangian surface elements on the Eulerian grid to define $C_i$ and by generating the new surface $\Sigma_i$ as the 0.5 iso-surface of $C_i$ in a kind of VOF-Front Tracking method [82].

# Chapter 4
# Adaptive Mesh Refinement

This chapter is part of the paper by [160] .

## 4.1 Introduction

The numerical simulation of flows with interfaces is a vast topic, with applications ranging from environment, geophysics to fundamental physics and engineering. In engineering the study of liquid-gas interactions is important in combustion problems: the formation of droplets clouds or sprays that burn in combustion chambers are originated by interfacial instabilities, such as the Kelvin-Helmholtz instability. The numerical simulation can be a powerful tool for predicting the behaviour of these physical phenomena, especially in scales of time or space where experimental visualization is difficult or impossible. A major obstacle in a two fluid representation is the discrete definition of the interface itself. A number of methods have been developed to approximate the fronts, and they can be classified in two groups: Lagrangian and Eulerian methods. The Lagrangian methods, consisting in the passive advection of marker particles, are known for their intrinsic precision, as well as for their computational cost. In the Eulerian approach, a discrete scalar quantity is transported by the velocity field. The Level Set method, as defined by [70], belong to the latter: each cell contains the signed distance from the interface as scalar quantity, so that the fixed zero contour of that function represents the interface itself. The following problem is the imposition of the jump conditions on the cells surrounding the interface. If certain methods consist into a regularization of the discontinuous variables as the Continuum Surface Force [96], the Ghost Fluid of [102] gives a discontinuous discrete solution at the interface without changing the discretization schema, instead adding the variable jump contribution to the right hand side of the equations. In a two fluid configuration the adaptive mesh is naturally advisable, as



S. Vincent et al., *Small Scale Modeling and Simulation of Incompressible Turbulent Multi-Phase Flow*, CISM International Centre for Mechanical Sciences 607,
https://doi.org/10.1007/978-3-031-09265-7_4

if the precision of the interface discretization depends on the size of the cells, it is often necessary to extend the computational domain well beyond the interfacial region of the flow, thus involving a lot of computational effort spent on less interesting zones. Berger and co-workers [161] have pioneered AMR for more structured grids. They use a hierarchy of logically Cartesian grids and sub-grids to cover the computational domain. They allow for logically rectangular sub-grids, which can overlap, have arbitrary shapes and which can be merged with other sub-grids at the same refinement level whenever appropriate. This is a flexible and memory-efficient strategy, but the resulting code is very complex and has proven to be very difficult to parallelize. A simplified variant, where the domain is divided in blocks which can be bisected in each directions when needed, was realized by Zeeuw and Powell [162]. This approach intuitively produces mesh with lower refinement efficiency, that is a ratio between the total number of new cells and the cells which really need refinement: the refinement test is no longer applied to single cells, but instead to a whole block. However, given all the grid blocks have an identical logical structure, the dedicated tree memory storage is lighter, and the cache managing is easier. The projection method for the incompressible Navier-Stokes adds an elliptic equation for the pressure. Poisson solvers on oct-tree meshes generally employ some type of multigrid or sparse linear solver iteration scheme. Sussman [163] has found multigrid alone not able to converge for high density ratio problems, choosing a multigrid preconditioned Conjugate Gradient: in the AMR extension, this solver is applied in a refinement level basis, so that its discretization matrix is symmetrical. Other oct-tree users have instead employed a Bi-Conjugate Gradient method on composite grid. The same Bi-CG stabilized algorithm has been used preconditioned with a composite grid basis multigrid, in order to get a rapid and robust solver. The adaptive mesh represents here an improvement over the algorithm of Couderc [164], and retains the Cartesian structured mesh discretization. The PARAMESH libraries [165] are here employed, designed to provide to an application developer an easy route to extend an existing code which uses a logically Cartesian structured mesh into a code with parallelized AMR.

## 4.2 AMR

Adaptive mesh refinement is a technique for automatically and dynamically refining (or de-refining) certain regions of the physical domain in a finite difference calculation, in order to lessen the computational effort required to perform a simulation with limited loss of precision. The chosen approach is a block-type AMR (sometimes referred as *oct-tree AMR*) like in Fig. 4.1a , preferred to the patch-based AMR like in Fig. 4.1b. In the first algorithm, the computational domain is split recursively into smaller ones (the *blocks*) where needed, until the finest grid is generated; in the second, a certain number of computational cells are tagged for refinement, then clustered to form new fine grids which are superimposed to the coarser ones. In both cases,

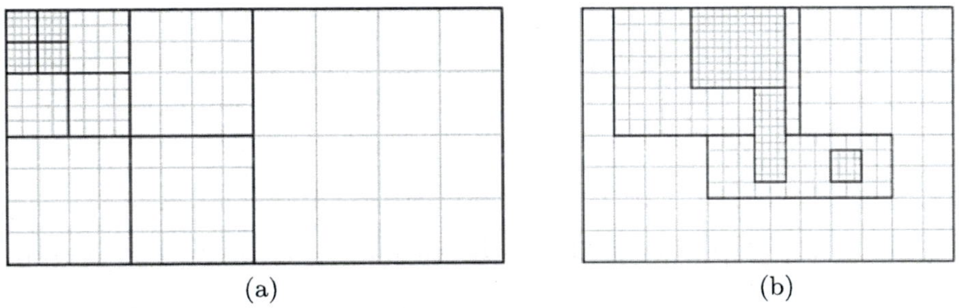

**Fig. 4.1** Representation of the two kind of AMR, **a** block-based, **b** right patch-based

the different grids need guard cells at their boundaries to complete the discretization stencils. Here, some coarse-fine (and vice-versa) interpolation is needed.

In the present work, the PARAMESH package [165] has been chosen to be used with the code. It is an open-source software package of Fortran90 subroutines developed by Peter McNeice and Kevin Olson. The package manages the creation of a block-type AMR: builds and maintains the tree-structure which tracks the spatial relationships between blocks, distributes the blocks amongst the available processors and handles all inter-block and inter-processor communication. It can distribute the blocks so that it can maximize block locality and minimize inter-processor communications. Each block is an uniform mesh with user defined dimensions; typically used blocks are $4 \times 4$ or $8 \times 8$. The refinement ratio is fixed to $r = 2$, that means refined cells are always two times smaller than their parents. PARAMESH keeps also track of physical boundaries on which particular boundary conditions are to be enforced, ensuring that child blocks inherit this information when appropriate. The communication between coarse and fine grids is assured by the prolongation operation, that involves giving values at the fine points overlapping coarser ones, and the reciprocal one, the restriction. The interpolation algorithm can be implemented by the user, or a default Lagrangian polynomial interpolation is provided. For the cell centred variables a symmetrical bi-quadratic nine points stencil interpolation is used. In Fig. 4.2a, the fine point represented by the small black rhombus are needed. A first set of quadratic interpolation in the two directions is performed to find the intermediate points (small squares); then the point can be found with a second set of quadratic interpolations of the intermediate points, as in Fig. 4.2b. Given the nine coarse points, the fixed interpolation coefficients can be calculated only once, thus leading to a direct formulation of the fine point value. For the three other points surrounding the internal cell the matrix of coefficients is rotated. The restriction operation is simpler, as it consists of an average of the four fine points.

The discretization stencils are completed by a user defined set of ghost cells around each block; Dirichlet boundary conditions are exchanged among the blocks every synchronization point. This structure allows the single grid discretization to be applied to each block without modifications. However, at the coarse-fine interface some quantities are evaluated twice. That means numerical fluxes on the shared face are not consistent, as they are calculated from interpolated values: this incongruence

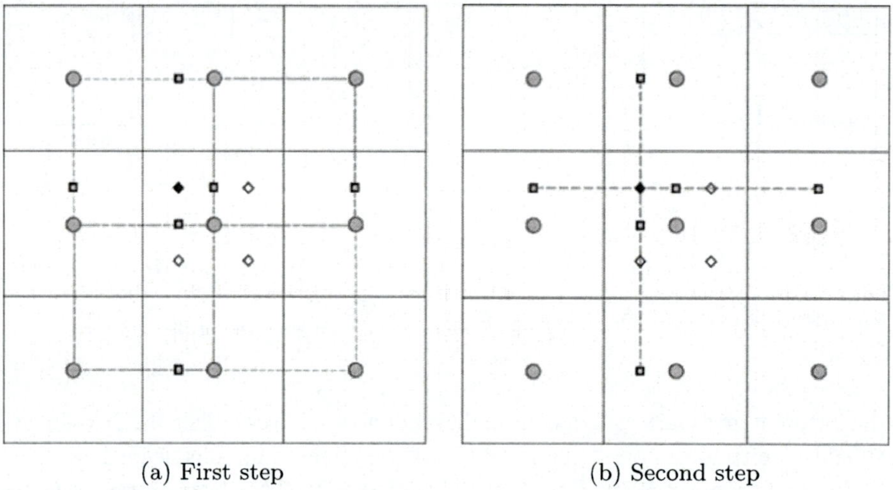


**Fig. 4.2** Bi-quadratic symmetrical interpolation for cell centred variables

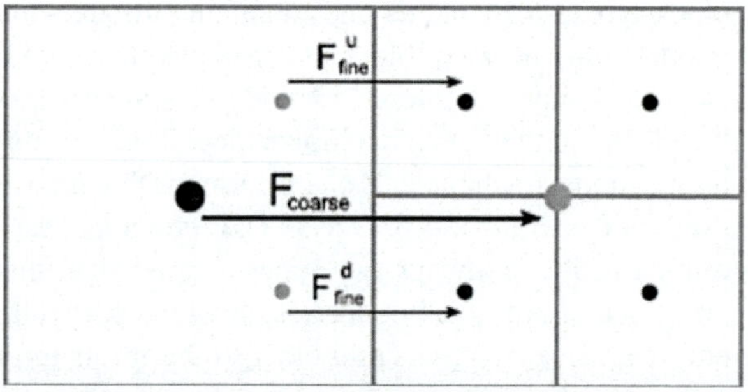


**Fig. 4.3** Representation of the 2D flux matching procedure. In black are represented the mesh point; in grey the interpolated guardcells points. $F_{coarse} = \left(F^u_{fine} + F^d_{fine}\right)/2$

is actually an artificial breaking of conservation. In an incompressible code, this would mean the inability to respect the condition $\nabla \boldsymbol{u} = 0$. An explicit solution for this problem, proposed by Martin and Colella [166], consists in the advection of an constant scalar $\Lambda$: in the refinement jumps, the variation of its initial value is used as free stream correction. However, in order to compute the correction an elliptic equation has to be solved. A simpler operation called *flux fixing* can be employed, so that the additional elliptic equation solving is avoided. When fluxes are evaluated on cell faces near to refinement jumps, they are stored in dedicate vectors; then the coarse cell flux is compared to the average of the fine ones. The difference, that is actually the inconsistency in conservation, is removed from the coarse flux, supposed less precise due to the coarser cell size see Fig. 4.3. The *flux match* operation is applied to both the predicted and corrected velocities, as they are naturally face centered variables, as well as into the Level Set advection equation, in the finite volume fluxes computation.

As a fixed time step strategy has been chosen, the application of the single grid solver on the AMR data structure is oriented more towards a composite grid idea than a level-by-level basis. The finest block covering one region are defined *leaf blocks*; that is what one would see if he looked from above the computational domain. It is on those blocks that the variables are stored and the computation is performed; a level of blocks beneath (the *parent* blocks) is used to provide data for the guard cells filling operation. The actual computation, represented in the algorithm 3, is performed by a loop over the leaf blocks; before and after synchronization points are set in order to refresh boundary data. At the end of the time step, all the leaf and parent blocks are checked for refinement or de-refinement. Then the regridding operation is performed, and data are initialized on the new blocks.

**Algorithm 3** Pseudocode of a time step

```
while t < t_max do
  Begin Integrate (t^n) ⟶ (t^{n+1})
  for lb = 1, nblocks do
    Synchronization point
    Update (u^n_lb, p^n_lb ⟶ u^{n+1}_lb, p^{n+1}_lb)
    Synchronization point
  end for
  Check blocks for refinement/de-refinement
  Update mesh
  Initialize new blocks
end while
```

PARAMESH parallelization is done via MPI. The repartition of the workload is realized by equally distributing the blocks among the processors, as each block requires the same computational effort. An additional criterion is followed: the blocks on the same processor should be physically near, in order to minimize the borders between regions covered by a different processor. This is achieved by drawing a Morton filling curve as shown in Fig. 4.4 through the list of grid blocks; this curve has the property that, in general, points that are close in value are close in physical space. The list of blocks is sorted by arranging them in order of increasing position along the line. By doing this the chance that neighbouring blocks will be located on the same processor is increased.

The refinement criterion is user-defined, and it is imposed by setting the two block flags "refine" or "derefine" to true or false. The PARAMESH is able to rebuild and redistribute the modified tree. The chosen refinement criterion is the near presence of the interface, as this is the region where we desire the best possible precision. This is easily applied into a Level Set formulation, as each cell contains the information of its distance from the zero contour, so that, when a value of $\phi$ falls beneath a certain threshold (usually 4-8 cell diagonals), the block "senses" the near interface, it asks for refinement. The same applies in the opposite sense, the derefinement. This way the interface never overlaps a refinement jump, and the implementation of the jump conditions does not change.

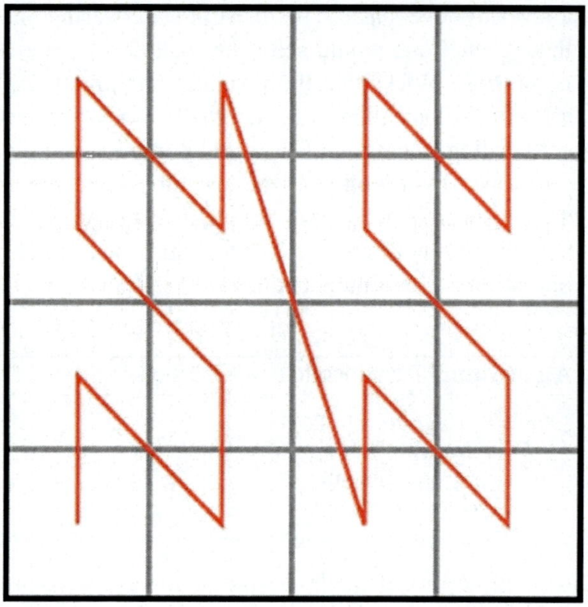

**Fig. 4.4** Morton filling curve

## 4.3 Poisson Solver

### Multigrid with AMR

In the second step of the Chorin projection method the pressure field is computed in order to project the velocity into a divergence free field, by solving the linear elliptic equation (2.18).

In the single phase solver a geometrical Multigrid strategy [62] has been chosen to inverse the associated linear system's matrix, due to its optimal convergence rate (up to an $O(N)$ number of operations, with $N$ representing the number of unknowns). However, performing Multigrid with adaptive mesh refinement requires some modifications on the "regular" algorithm, as the idea of the grid hierarchy has to be redefined. In this context, the mesh is not evolving during the Poisson solver iterations, as the grid is allowed to evolve at the end of each timestep only: it would be more precise to say that it consists of a fixed composite grid. As stated by Brown and Lowe [167] and in the book *Multigrid* [62], there are two natural approaches. In the first, called *Multilevel* (MLAT, Multi Level Adaptive Technique, as defined by Brandt [168]), the grids in the multigrid hierarchy coincide with the AMR resolution levels, so that each level consists of an uniform mesh or, better, of an union of grids with same mesh spacing. This method is well suited for the patch based AMR, where each patch can be solved as an independent grid (with multigrid, for example, as done by Sussman [169]), and offers the advantage of a symmetrical associated system matrix. The second method is known as FAC (Fast Adaptive Composite grid solver, as defined by McCormick [170]). In this algorithm the relaxation sweeps are performed across the entire computational domain, and one defines the succession of multigrid levels by restricting the highest resolution sub-grids to the newt lower resolution, building at each level a composite grid, that is defined as an union of blocks not all at the same refinement level. In this configuration more work is done in the relaxation sweep, but a stricter coupling between levels can be maintained.

If the two are told to work well, the latter has been chosen for some reasons. The first is for parallel efficiency purpose: in the PARAMESH framework the repartition of the blocks among the processors is done for the leaf blocks first, because the most of the time is spent here. So, if a relaxation sweep is performed on a fixed level of refinement, it will likely make only some processors work. Instead, if it is applied on the entire domain at each level, it can benefit from a more efficiently parallelized relaxation. The second reason is that PARAMESH multigrid routines are meant to work in a global sense, so that it would not be possible to perform "sub-multigrid" relaxation over a few blocks only, like in a patch based Multigrid [169], and the FAC is meant to work in a more global sense. Moreover, it is more straightforward to treat boundary conditions over the coarse/fine interface in each Multigrid level, making the prolongation and restriction operations easier. The relaxation operation consists in the application of a conservative Jacobi smoother, applied into a loop over all the blocks, after a synchronization step, as shown in algorithm 4. Actually, results show that updating a block values inside the loop instead of doing at the end improves greatly the convergence of the solver (it is something like a Gauss Seidel use of blocks instead of single grid points). The discretization matrix A is actually a block local matrix, so that each block is treated as an independent Laplacian problem, until convergence of their boundary conditions; as boundary conditions elimination is applied, even at the borders of the domain the discretization four point stencil never changes.

**Algorithm 4** Relaxation routine pseudocode

*Subroutine* RELAX
**while** $(iter < iter_max)$ **do**
  *Synchronization point*
  **for** $(l = 1, nbl)$ **do**
    $x(l) \leftarrow Jacobi\ (A(l), f(l), x(l))$
  **end for**
  $r(:) = f(:) - A(:)x(:)$
  $iter = iter + 1$
**end while**

In order to build the composite meshes, following the example of an unofficial version of Multigrid for PARAMESH written by Kevin Olson, some "ghost blocks" are created. They are "clones" of the leaf blocks not at maximum refinement level, and are placed side by side to the finer blocks, so that together they cover the entire computational domain. In Fig. 4.5 the active blocks are showed in gray for both the original and modified mesh.

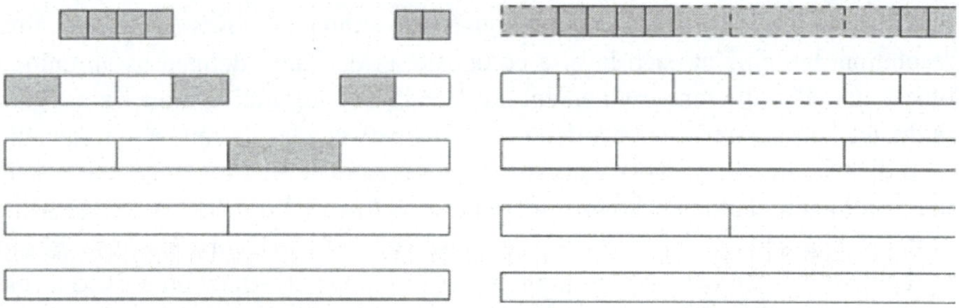

(a) AMR data structure (b) Composite grid Multigrid structure

**Fig. 4.5** Modifications to the AMR data structure for the Multigrid solver

In a two grid cycle with AMR the relaxation is performed on all the leaf blocks, and the residuals computed. Then, those residuals are restricted from the finest grid to the underlying coarse one; all the other leaf blocks copy their residuals to their underlying "ghost", completing the right hand side of the lower multigrid level. Then relaxation is performed on this level, and the correction is passed to the finer level by prolongation, where a finer grid exists, and copy elsewhere. It has to be kept in mind that, here, the term level does not indicate a refinement level, but the maximum level contained in that composite grid.

In a full Vcycle, shown in algorithm 5, the operation continues until reaching the coarsest grid, where a more precise correction is computed by a unpreconditioned CGM (at a certain level, the mesh becomes always uniform). The same user made routine used for AMR prolongation is used here. The stencil and coefficients do not change at the block boundaries because, previous to the prolongation operation, all the blocks receive guardcell data. As stated by Martin and Cartwright [171], quadratic interpolation is the minimum necessary to maintain second order accuracy.

Both the Vcycle and Full Multigrid cycle have been tested: even if in terms of iterations the FMG is faster to converge, the Vcycles has proved to be more effective in terms of computational time. Concerning the number of relaxation sweeps, there are no theoretical basis to predict the optimum value. After some trial-and-error tests, a number of sweeps increasing toward the coarser grids seems to give the best convergence speed.

Special care has to be taken in the coarse/fine interface, as stated by Martin and Cartwright [171, 172], in a similar way to the flux matching problem in the conservative advection equation. Since the way of communicating data between blocks is an exchange of Dirichlet boundary conditions via ghost cells (following restriction and prolongation where needed), the operators working on each single block are not supposed to change near the boundaries: they can be applied inside a loop that covers all the concerned blocks, without taking care of the actual position or refinement level of the block itself (Fig. 4.6).

**Algorithm 5** FAC Multigrid algorithm for $Ax = f$, recursive formulation with V-cycle

*Procedure* MG
$r_{lmax} = f - A_{lmax} x_0$
**while** ($r_{lmax} > toll$) **do**
  MGLEVEL(*lmax*)
  $x = e_{lmax}$
  $r_{lmax} = f - A_{lmax} x$
**end while**

*Procedure* MGLEVEL($l$)

**if** ($l > 0$) **then**
  **if** ($l = l_{max}$) **then**
    $e_{lmax} = x$
  **else**
    $e_l = 0$
  **end if**

  $e_l \leftarrow RELAX(A_l, f_l)$
  $r_l = f_l - A_l e_l$
  $f_{l-1} \leftarrow Restrict\,(r_l)$
  MGLEVEL($l - 1$)
  $\delta e_l \leftarrow Prolong\,(e_{l-1})$
  $e_l = e_l + \delta e_l$
  $e_l \leftarrow RELAX(A_l, f_l)$

**else**
  $e_l = 0$
  $e_0 \leftarrow RELAX(A_0, f_0)$
**end if**

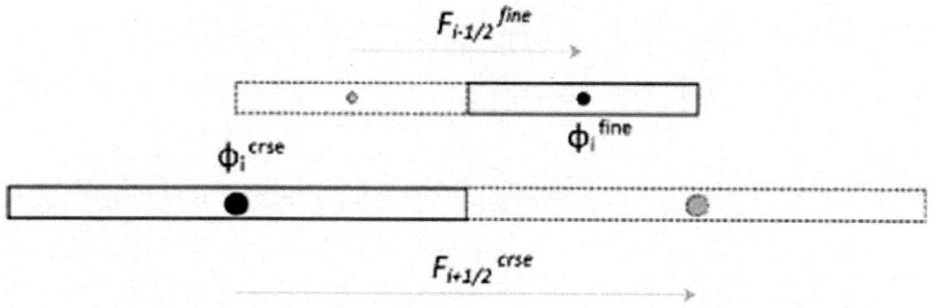


**Fig. 4.6** Computation of fluxes at refinement jump

In that way, the coarse and fine solution are not sufficiently linked, because the numerical fluxes, in this case the pressure gradient, evaluated at a coarse/fine interface are computed twice, once by the coarse mesh and once by the fine. This causes a discontinuity in the first derivative which is $O(h_c)$, that acts like a singular charge proportional to the first derivative mismatch, and corrupts the solution preventing the attainment the fine grid precision or even convergence

$$\nabla\phi + C\delta\left(x_{cf}\right)\left(\frac{\partial\phi^{fine}}{\partial n} - \frac{\partial\phi^{crse}}{\partial n}\right) = f$$

In order to eliminate this artificial charge, one need to enforce the continuity of both the variable and his first derivative over the interface, condition called *elliptic matching*. As for the advection equation, a correction on the fluxes is done exploiting the dedicate PARAMESH routines: the pressure gradient are stored on the two sides of the interface, and, across refinement jumps, the coarser one (the less precise) is corrected with the computed difference between the two. Actually, it is like a "counter-charge" on the right hand side with the same value as the initial mismatch. If the standard Laplacian operator in one dimension is

$$\left(\frac{\partial^2\phi}{\partial x^2}\right)_i = \frac{1}{\Delta x^2}\left(\phi_{i+1} - 2\phi_i + \phi_{i-1}\right) = \frac{1}{\Delta x}\left(F_{i+\frac{1}{2}} - F_{i-\frac{1}{2}}\right) \tag{4.1}$$

with the fluxes $F$ defined as

$$\begin{aligned} F_{i+\frac{1}{2}} &= \tfrac{1}{\Delta x}\left(\phi_{i+1} - \phi_i\right) \\ F_{i-\frac{1}{2}} &= \tfrac{1}{\Delta x}\left(\phi_i - \phi_{i-1}\right) \end{aligned} ll \tag{4.2}$$

The expressions for the two sides of the refinement jump, considering that interpolated guard cells are used, are

$$\begin{aligned} \left(\tfrac{\partial^2\phi}{\partial x^2}\right)_i^{crse} &= \tfrac{1}{\Delta x^{crse}}\left(F_{i+\frac{1}{2}}^{crse} - F_{i-\frac{1}{2}}^{crse}\right) \\ \left(\tfrac{\partial^2\phi}{\partial x^2}\right)_i^{fine} &= \tfrac{1}{\Delta x^{fine}}\left(F_{i+\frac{1}{2}}^{fine} - F_{i-\frac{1}{2}}^{fine}\right) \end{aligned} \tag{4.3}$$

But the fine flux value must be imposed to the coarse grid to maintain the continuity of the $\frac{\partial\phi}{\partial x}$, so

$$\begin{aligned} \left(\tfrac{\partial^2\phi}{\partial x^2}\right)_i^{crse,match} &= \tfrac{1}{\Delta x^{crse}}\left(\tfrac{\Delta x^{crse}}{\Delta x^{fine}} F_{i-\frac{1}{2}}^{fine} - F_{i-\frac{1}{2}}^{crse}\right) \\ \left(\tfrac{\partial^2\phi}{\partial x^2}\right)_i^{fine} &= \tfrac{1}{\Delta x^{fine}}\left(F_{i+\frac{1}{2}}^{fine} - F_{i-\frac{1}{2}}^{fine}\right) \end{aligned} \tag{4.4}$$

Finally the expression for the coarse grid can be written as the standard formulation plus a correction

$$\begin{aligned}\left(\frac{\partial^2\phi}{\partial x^2}\right)_i^{crse,match} &= \frac{1}{\Delta x^{crse}}\left(F^{crse}_{i+\frac{1}{2}} - F^{crse}_{i-\frac{1}{2}}\right) + \\ &\left(\frac{1}{\Delta x^{crse}}F^{crse}_{i+\frac{1}{2}} - \frac{1}{\Delta x^{fine}}F^{fine}_{i-\frac{1}{2}}\right) \\ &= \left(\frac{\partial^2\phi}{\partial x^2}\right)_i^{crse} + \delta F^{c/f}\end{aligned} \tag{4.5}$$

## The Preconditioned Bi-CGMStab Solver

In a preliminary development, the Multigrid solver has been tested on the two phase Poisson equation. However, for problems subject to high density ratio it usually fails to converge, or needs too much time to invert the matrix. As stated by Sussman [173], a density ratio of about 1:10 seems to be the uppermost limit. Some variants of the standard Multigrid seem to be able to overcome this problem: in particular the *semicoarsening* method [174], in which the coarsening of the grid is performed differently in the two (or three) directions. Another method is the *Algebraic Multigrid* [175]: it was developed as a more robust version of a standard multigrid, able to work with strong discontinuous and anisotropic coefficients. Differently from the geometrical Multigrid, it does not set the coarser grids by a fixed coarsening strategy; instead, it dynamically makes the best choice of coarse point in order to reduce efficiently the error. For the strong anisotropy problem, it actually performs a sort of "dynamical semicoarsening". However, this means that, at every call to the solver, some important time has to be spent in the restriction and prolongation operators setup. In the case of repeated calls of the solver with the same matrix coefficients this would not be a problem; in our case the interface motion makes the matrix values change at every time step, so that an AMG algorithm should continuously perform this set-up. On the contrary, a geometrical multigrid the coarser grids and the interpolation procedures are fixed and never change, as they always work in the same way. For this reason those kind of algorithms were not found attractive for this particular problem.

Unluckily, the discretization of the laplacian on the composite grid leads to an unsymmetrical matrix, as the interpolation coefficients make their appear, thus preventing the use of Conjugate Gradient Method. To overcome this problem, one of the most reliable unsymmetrical linear system solvers of the Krilov family, the Bi-Conjugate Gradient Stabilized Method (Bi-CGMStab [176]) has been chosen. Its most interesting features are the quite regular convergence, no need to build the global matrix transpose (as for the Conjugate Gradient Squared), and the simplicity of the preconditioning. As for the PCGM, there is no need to multiply explicitly the system by the preconditioning matrix $M$, but just to compute the effect of applying $M^{-1}$ to a vector. On the other side, within the Bi-CGMStab algorithm this operation is performed twice, so that the preconditioner is called two times per iteration. Finally, the Multigrid FAC algorithm can be applied as preconditioner, as CG does with standard Multigrid. Again, some attention has to be paid to the coarse/fine interface. Te procedure is different from the *Jacobi* iterations *flux-match* correction: there is no explicit flux computation in the Bi-CGMStab. As suggested by Teigland and Eliassen [177], the elliptic matching condition can be imposed by a Dirichlet and Neumann boundary conditions coupling. Across the interface a preliminary Dirichlet conditions exchange between blocks is allowed; then the coarse grid ghost cells values are corrected to make the first derivative match the fine grid value, so that the coarse grid actually receives an imposed Neumann boundary values. The smoothing routine is changed, being now a *Red-Black Ordered Gauss Seidel*, that shows better smoothing properties.

The main algorithm inefficiency source lies in the communication. At every relaxation step, as well as at every Bi-CGMStab matrix-vector product, guardcell data have to be exchanged. In a block-based AMR, this is the most part of the additional work imposed by the adaptive mesh because, if in a single mesh solver boundary conditions are only needed at the edges of the computational domain, here they are needed at each block edge (and one would like to have more smaller blocks, so that he can easier isolate refined regions and reduce the total number of points). In particular, in the strong density ratio problems the benefit of an increased number of relaxation sweeps would raise excessively the call to the guardcell filling routine. A number of relaxation proportional to the Multigrid level seems to reduce the total number of iteration as well as the time spent on the coarser levels, where the parallelization in less effective.

## 4.4 Numerical Results

### Static Bubble

The simulation of a round static bubble is the first interesting test for the numerical method. Although very simple in theory, it can give informations about the precision of the interface treatment, in particular of the surface tension. A round bubble of a dense fluid is centred in a square domain, surrounded by a lighter one; there are no initial velocities and no gravity. In this case the pressure jump is imposed by the interface curvature and the surface tension product, as given by the Laplace equation

$$\begin{aligned} p_{int} - p_{ext} &= \sigma/R, \quad \text{if in 2D} \\ p_{int} - p_{ext} &= 2\sigma/R, \quad \text{if in 3D} \end{aligned} \tag{4.6}$$

Any error resulting from the discrete computation of the level set curvature leads to an erroneous value of the pressure gradient between two adjacent cells: this non zero pressure gradient then acts as a wrong correction for the predicted velocity field. This the cause of the so called spurious currents, non zero components of the velocity that arise from the first iteration, and may in some cases destroy the interface. A comparison between two different test cases is proposed: in the first mesh convergence results from some uniform meshes are shown; in the second one the same computation is performed with an adaptive mesh localized on the interface. The parameters

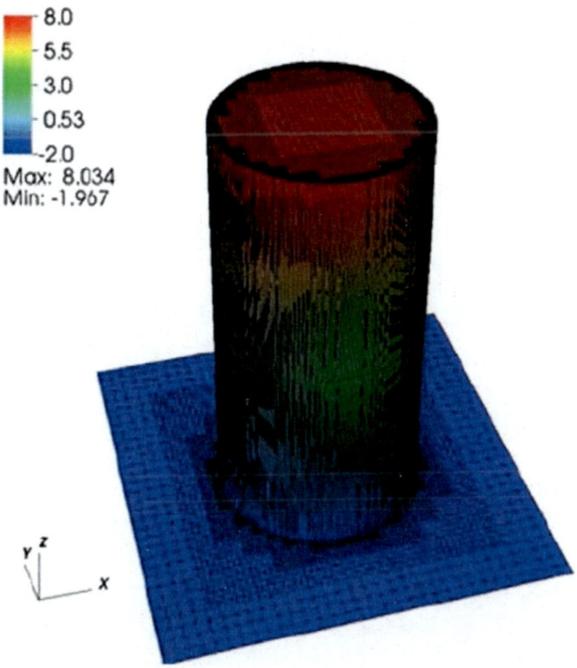


**Fig. 4.7** Pressure field of a static round bubble after one timestep captured with four grid levels

of the simulation are taken from [58]: $L = 4$ cm, $R = 1$ cm, $\rho_{int} = 1000$ kg m$^{-3}$, $\rho_{ext} = 1$ kg m$^{-3}$, $\mu_{int} = 10^{-3}$ Pa.s, $\mu_{ext} = 10^{-5}$ Pa.s, $\sigma = 0.1$ N m$^{-1}$. Only one iteration of level set reinitialization is performed in order to reduce any artificial displacement (Fig. 4.7).

In the first Table 4.1 the norm of the velocity is computed after one time step alone: here the predicted velocities are equal to zero, so that the spurious currents are determined by the curvature computation alone.

$$\|\boldsymbol{u}\|_{L2} = \left( \frac{1}{nx\, ny} \sum_{i,j} \left( u_{i,j}^2 + v_{i,j}^2 \right) \right)^{1/2}$$

If the analytical curvature is used as input, then the solution is always precise to the roundoff error. As shown in Table 4.1, the convergence rate, given by $q = \frac{ln(err_{2h}) - ln(err_h)}{ln(2)}$ are, for the uniform mesh computation, close to 3: this behaviour, justified by Oevermann and Berger [178], comes from the zero gradient solution of the pressure on both the sides that nullify the effects of the interface position relative to the grid. In the adaptive mesh test a slight decrease of the code performance, mainly for the finer resolutions, is visible. It has to be kept in mind that there is no lower limit to the refinement level, so that the lower resolution are kept far from the interface as the resolution increases. The effects of the coarser grids and the interpolation are visible in the small spurious current increase. Still, the convergence rate never drops under 2.5.

**Table 4.1** Spurious currents norm, first iteration

| Mesh | Grid | $\|U\|_2$ | q |
|---|---|---|---|
| Uniform | $16^2$ | 1.933e−06 | – |
| | $32^2$ | 1.910e−07 | 3.34 |
| | $64^2$ | 2.485e−08 | 2.94 |
| | $128^2$ | 3.277e−09 | 2.92 |
| | $256^2$ | 3.893e−10 | 3.07 |
| AMR | $16^2$ | 1.933e−06 | – |
| | $32^2$ | 2.125e−07 | 3.19 |
| | $64^2$ | 3.259e−08 | 2.70 |
| | $128^2$ | 5.717e−09 | 2.51 |
| | $256^2$ | 9.451e−10 | 2.60 |

AMR grid correspond to the finest level

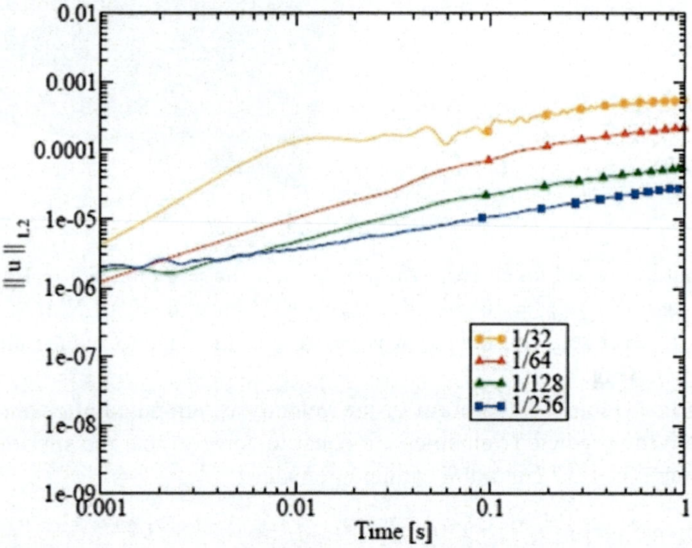


**Fig. 4.8** Evolution of the spurious currents for a long time simulation, $t = 1$ s

The growth of the spurious currents are here checked for a prolonged time of simulation as shown in Fig. 4.8. In this case the code tries to achieve an equilibrium for those non-physical velocities, with the generation of non zero vorticity near the interface. An initial steep growth followed is followed by a stabilization in time. In the AMR computation the initial error grows, as for the first iteration, a little more than in the uniform mesh; still it does not rise dangerously, but it stabilizes as for the previous test. The convergence rates in Table 4.2 are now near to 1.5.

## Rising Bubble

In this section the dynamics of a low density 2D bubble immersed into a heavier fluid and subjected to gravity are presented. Figure 4.9 shows the numerical configuration and the initial conditions. Lacking any analytical solution to compare, extensive

**Table 4.2** Spurious currents norm, $t = 1$ s

| Mesh | Grid | $\|U\|_2$ | q |
|---|---|---|---|
| Uniform | $16^2$ | 1.885E−03 | – |
| | $32^2$ | 5.828E−04 | 1.69 |
| | $64^2$ | 1.778E−04 | 1.71 |
| | $128^2$ | 3.879E−05 | 2.20 |
| | $256^2$ | 1.060E−05 | 1.87 |
| AMR | $16^2$ | 1.885E−03 | – |
| | $32^2$ | 5.676E−04 | 1.73 |
| | $64^2$ | 2.181E−04 | 1.38 |
| | $128^2$ | 5.545E−05 | 1.98 |
| | $256^2$ | 2.953E−05 | 0.91 |

AMR grid correspond to the finest level

research work has been done in creating *benchmarks* for this problem, mainly from Sussman [173] and Hysing [179]. Some comparison with the Hysing's simulations are performed in this section.

In the initial configuration there is a column of fluid with no slip conditions imposed on the horizontal surfaces and slip conditions on the vertical ones. In the lower half of the $[1 \times 2]$ domain, centered at $[0.5, 0.5]$ is the circular bubble of radius $r = 0.25$ composed of a lighter fluid, all the velocities initialized to zero. As the gravity applies its effect, the bubble starts rising, and at the same time begins to change its shape. As in the Hysing's *benchmarks*, computations are performed for two configurations described in Table 4.3.

Two dimensionless numbers can describe the regime of the bubble rise. Given $L = 2r_0$ and $U_g = \sqrt{g2r_0}$, these are the Reynolds and the Eötvös number:

$$Re = \frac{\rho_1 U_g L}{\mu_1}, \qquad Eo = \frac{\rho_1 U_g^2 L}{\sigma}$$

In the first case the density and viscosity ratio are equal to 10. The surface tension should be enough to hold the bubble together, and the final shape should be of an ellipsoidal regime. The second test is more challenging, as the density ratio reaches 1000 and the viscosity ratio 100; this bubble lies somewhere between the skirted and dimpled ellipsoidal-cap regimes indicating that break up can possibly occur [180].

The evolution of the two bubbles is tracked for three time units and shown in Fig. 4.10a, b, during which three quantities are measured. The first is the centroid of mass $(x_c, y_c)$, of which, given the symmetry of the problem, the $y$ coordinate only is considered:

$$y_c = \frac{\int_{\Omega_2} y \, dy}{\int_{\Omega_2} dy}$$

**Fig. 4.9** Initial condition for the rising bubble test

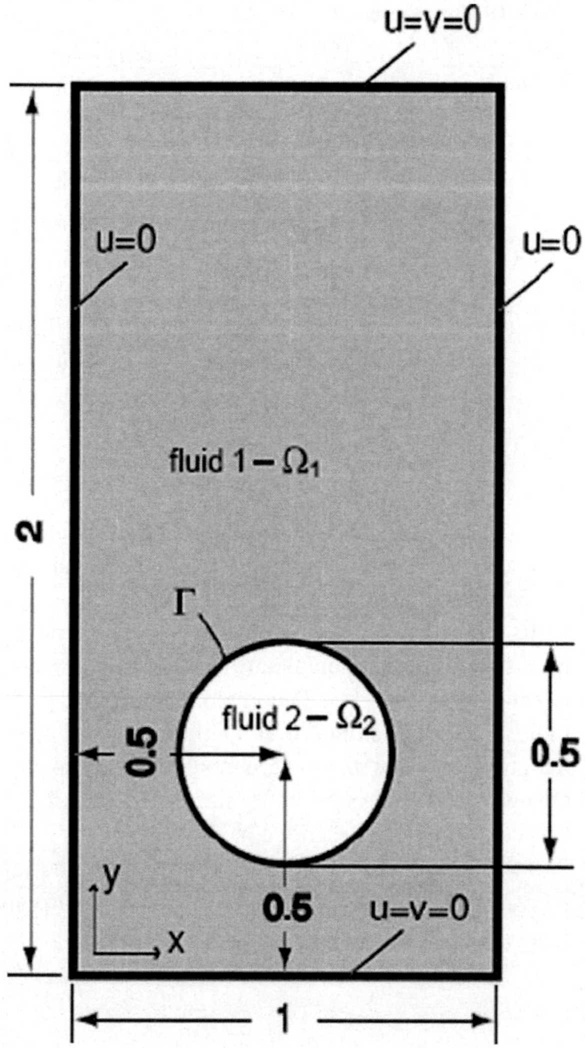


the instantaneous rise velocity $(u, v)$, again $y$ component only

$$v_{rise} = \frac{\int_{\Omega_2} v \, dy}{\int_{\Omega_2} dy}$$

This is an interesting quantity to compute, because it does not only measure how the interface tracking algorithm behaves, but also it gives an idea of the quality of the overall solution; this means a check to the reduction of precision of the AMR grid far from the interface, and its influence on the interface itself. The last quantity is the *circularity*, an index of how much the shape of the bubble differs from the circle

**Table 4.3** Physical parameters for the two test cases

| Test case | $\rho_1$ | $\rho_2$ | $\mu_1$ | $\mu_2$ | $g$ | $\sigma$ | $Re$ | $Eo$ | $\frac{\rho_1}{\rho_2}$ | $\frac{\mu_1}{\mu_2}$ |
|---|---|---|---|---|---|---|---|---|---|---|
| 1 | 1000 | 100 | 10 | 1 | 0.98 | 24.5 | 35 | 10 | 10 | 10 |
| 2 | 1000 | 1 | 10 | 0.1 | 0.98 | 1.96 | 35 | 125 | 1000 | 100 |

$$c = \frac{\pi d_a}{P_b}$$

Here the numerator is the perimeter or circumference of a circle with diameter $d_a$, which has an area equal to that of a bubble with perimeter $P_b$: it always starts from the maximum value of one, an then decreases as the bubble begin to be deformed. Figure 4.10a, b show the interface locations for both cases $t = 3$ s. Some qualitative comparison between our results and Hysing's are proposed in Fig. 4.11, where Fig. 4.11a, c are respectively Hysing results and our result for case 1 and Fig. 4.11b, d are respectively Hysing results and our result for case 2.

A comparison is done with a reference computation made with an uniform fine mesh of [256 × 1024]. The difference between the two is calculated at the final timestep: err $= |q_{t=3} - q_{ref,t=3}|$ for both case and are summarized in Tables 4.4 and 4.5.

The temporal evolutions of the bubble values shown in Figs. 4.12, 4.13 and 4.14 follow the *benchmarks* for both the configurations; however he code could not track the circularity for the second test case after $t = 1.5$ s, approximately when the bubble forms the two sharp corners. In this location Hysing could find, with a finite elements code, a small ligament that subsequently gives birth to two small drops.

In the first simulation set the convergence towards the reference solution is very good. The position of the center of mass is almost the same for all the meshes, and grows almost linearly in time; the convergence rate is quadratic for this value. The circularity computation is more difficult: the lowest mesh computed value seems quite far from the good solution, mainly in the last half of the simulation; still there are similar values of $q$. The rise velocity seems a little bit underestimated, for the lesser mesh, in the region around the maximum; the terminal velocity is however well estimated for all the grids, and $q$ keeps quadratic.

### Damped Surface Wave

In this section the computational performances of the code are measured in the simulation of the damped wave of a surface dividing two viscous fluids. Without gravitational effects, an initial sinusoidal perturbation starts to oscillate under the effect of the surface tension; the amplitude is damped by the viscous effects. A square computational domain of side L with periodic boundary conditions on the vertical faces and slip conditions on the horizontal ones is given. Each half of the domain is occupied by a different fluid. The surface height is

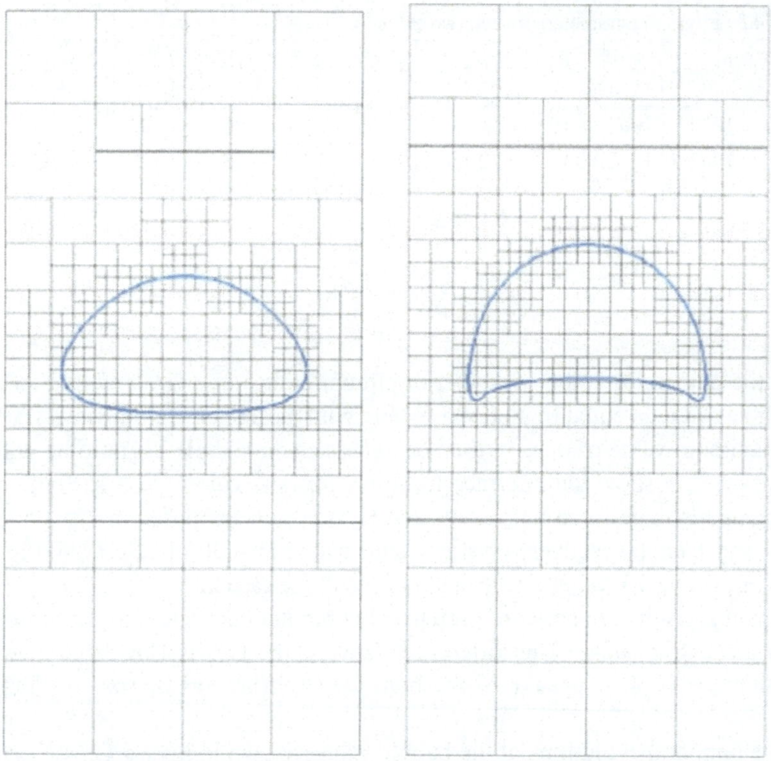

(a) Solution test 1 (b) Solution test 2

**Fig. 4.10** Solution after $t = 3$ s of the two configurations of bubble, four levels of mesh up to $128 \times 256$

$$a(t) = a_0 cos(kx), \quad \text{with} \quad k = \frac{2\pi}{L} \quad \text{and} \quad a(0) = \frac{0.04L}{2}$$

and $L = 2\pi$. In the case where the two fluids have the same viscosity, an analytical solution of the temporal evolution of the amplitude $a(t)$ is given by Prosperetti [181]:

$$\begin{aligned} a(t) = \tfrac{4(1-\beta)\nu^2 k^2}{8(1-4\beta)}\, a(0)\text{erfc}\left(\sqrt{\nu k^2 t}\right) + \\ \textstyle\sum_{i=1,4}^{4} \frac{z_i}{Z_i} \left(\frac{\omega^2 a(0)}{z_i^2 - \nu k}\right) \exp\left((\omega^2 a(0) z_i^2) t\right)\ \text{erfc}\left(z_i t^{\frac{1}{i}}\right) \end{aligned} \tag{4.7}$$

where the terms $z_i$ are the four roots of the algebraic equation

$$\begin{aligned} z^4 - 4\beta(k^2\nu)^{1/2} z^3 + 2(1-6\beta)k^2\nu z^2 + \\ 4(1-3\beta)(k^2\nu)^{3/2} z + (1-4\beta)k^4\nu^2 + \omega^2 = 0 \end{aligned}$$

The $Z_i$ came from the cyclical index permutations of

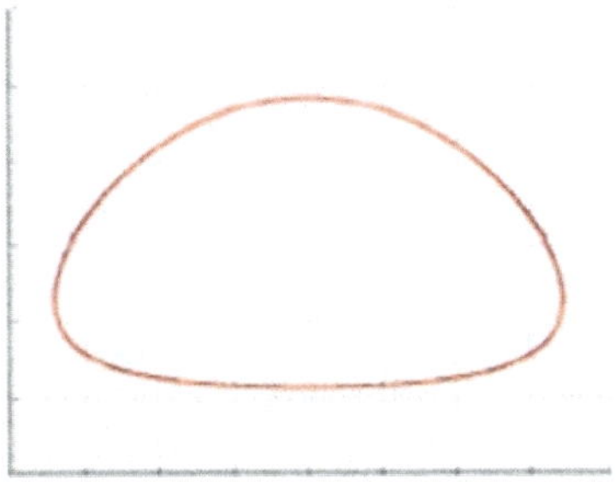

(a) Hysing, $Re = 35$, $Eo = 10$

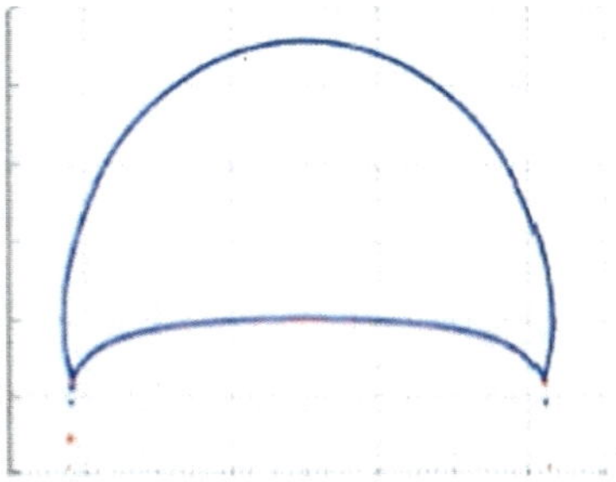

(b) Hysing, $Re = 35$, $Eo = 125$

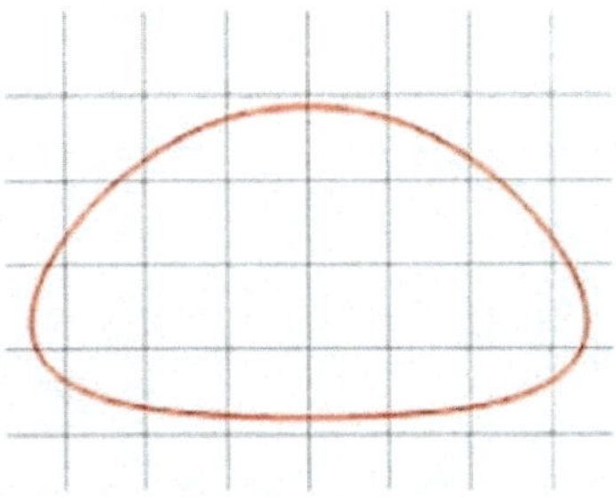

(c) AMR solution, $Re = 35$, $Eo = 10$

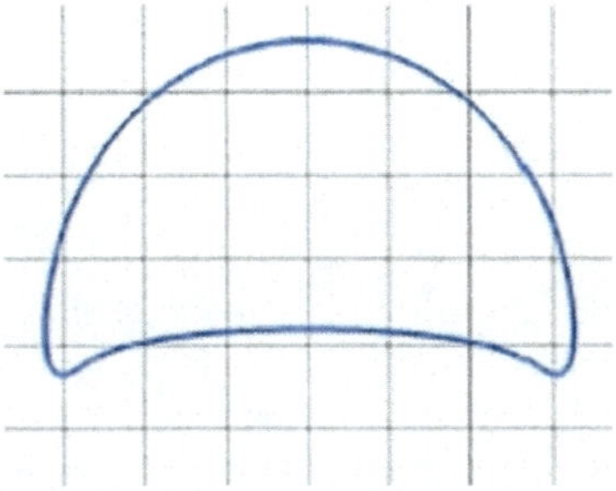

(d) AMR solution, $Re = 35$, $Eo = 125$

**Fig. 4.11** Zoom over the bubble shape, comparison with Hysing

**Table 4.4** Computed values for the rising bubble, test 1

| | Mesh | Value | Err | q |
|---|---|---|---|---|
| $y_c$ | 32 | 1.0894 | 4.035E−3 | – |
| | 64 | 1.0863 | 9.399E−4 | 2.11 |
| | 128 | 1.0856 | 2.399E−4 | 1.97 |
| | Ref | 1.0854 | – | – |
| $c$ | 32 | 0.9313 | 1.018E−2 | – |
| | 64 | 0.9233 | 2.161E−3 | 2.23 |
| | 128 | 0.9206 | 5.175E−4 | 2.06 |
| | Ref | 0.9212 | – | – |
| $v_{rise}$ | 32 | 0.1986 | 2.861E−3 | – |
| | 64 | 0.1964 | 6.984E−4 | 2.03 |
| | 128 | 0.1955 | 1.571E−4 | 2.15 |
| | Ref | 0.1956 | – | – |

**Table 4.5** Computed values for the rising bubble, test 2

| | Mesh | Value | Err | q |
|---|---|---|---|---|
| $y_c$ | 32 | 1.1619 | 1.86E−2 | – |
| | 64 | 1.1525 | 9.21E−3 | 1.02 |
| | 128 | 1.1477 | 4.42E−3 | 1.06 |
| | Ref | 1.1434 | – | – |
| $c$ | 32 | 0.8630 | 2.41E−2 | – |
| | 64 | 0.8532 | 1.43E−2 | 0.75 |
| | 128 | 0.8443 | 5.38E−3 | 1.11 |
| | Ref | 0.8389 | – | – |
| $v_{rise}$ | 32 | 0.2105 | 5.61E−3 | - |
| | 64 | 0.2125 | 3.59E−3 | 0.65 |
| | 128 | 0.2144 | 1.66e−3 | 1.11 |
| | Ref | 0.2161 | – | – |

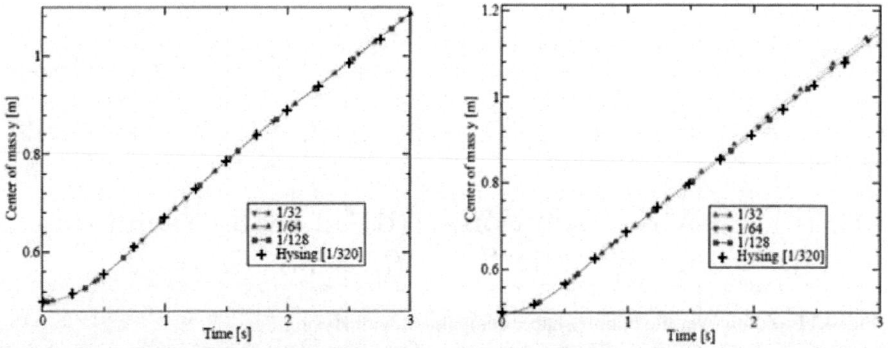


(a) Center of mass, test 1 (b) Center of mass, test 2

**Fig. 4.12** Comparison of the temporal evolution of three quantities

$$Z_i = (z_2 - z_1)(z_3 - z_1)(z_4 - z_1)...$$

The inviscid oscillation frequency $\omega$ is

$$\omega^2 = \frac{\sigma k^3}{\rho_l + \rho_g}$$

and the parameter $\beta$ is given by the fluid densities as

$$\beta = \frac{\rho_l \, \rho_g}{(\rho_l + \rho_g)^2}$$

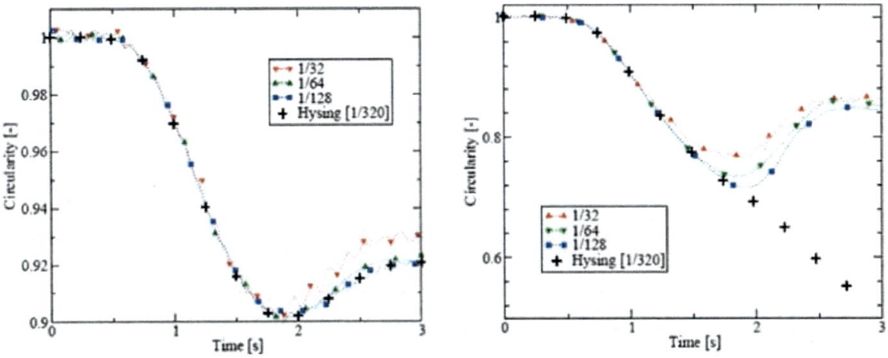


(a) Circularity , test 1 (b) Circularity , test 2

**Fig. 4.13** Comparison of the temporal evolution of three quantities

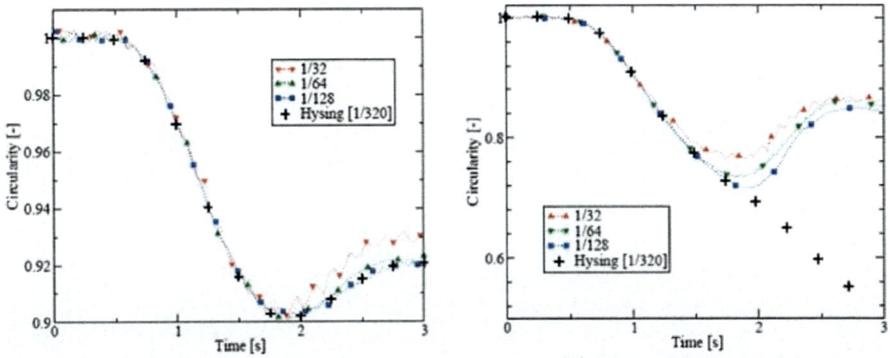


(a) Rise velocity , test 1 (b) Rise velocity , test 2

**Fig. 4.14** Comparison of the temporal evolution of three quantities

Two set of simulation are presented. In the first the density is set to 1 kg m$^{-3}$ for both fluids; in the second one the density of the lower fluid is raised to 1000 kg m$^{-3}$. In both cases the surface tension is $\sigma = 0.1$ N m$^{-1}$ and the viscosity is $\nu = 10^{-3}$ m$^2$.$s^{-1}$. One Level Set redistance iteration is performed.

In the graph Fig. 4.15 a very good agreement between the analytical and the computed solution is shown. The first case is clearly the easier one, as the convergence is quite faster; it has to be pointed out that in the second case the CFL condition was reduced up to 0.15 to preserve the stability of the code. In both graphs, even the coarsest meshes give quite good results.

For all the test cases the refined mesh is used, but it should be pointed out that for these simulation the mesh never changes in time: given the oscillation amplitude, there is not simply room enough for refinement, as the resolutions should have been risen excessively in order to follow the interface with very small blocks. The influence of the connection between different grid resolutions can still be tested, leaving behind for a moment the effect of the creation and removal of blocks. This will be tested

**Fig. 4.15** Analytical and computed amplitude time histories for the damped surface wave test

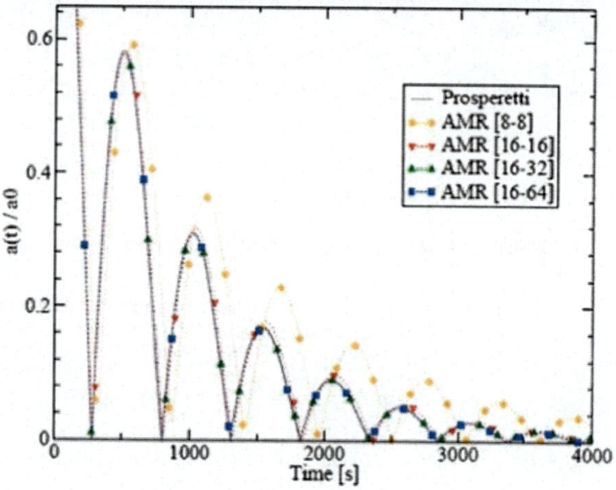


(a) Density ratio $\Delta\rho = 1$

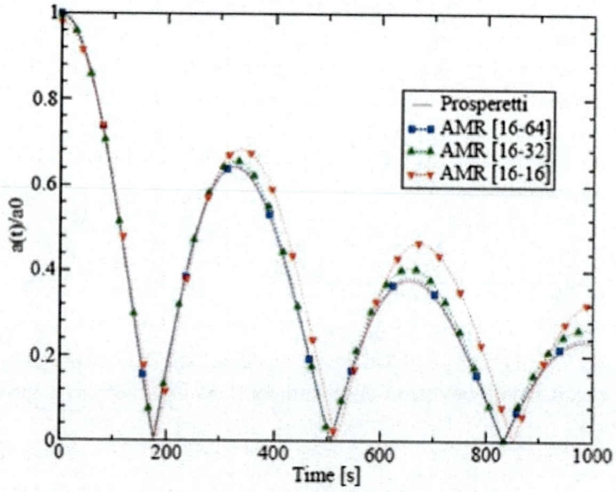


(b) Density ratio $\Delta\rho = 1000$

more extensively in the next section. In order to quantify the error the following norm is built:

$$\|e\| = \frac{1}{N}\sum_{i=1}^{N} |a_{num}(t_i) - a_{theo}(t_i)|$$

that is the mean value of the distance of our computed value from the Prosperetti's theoretical one. The simulation is stopped at time $t = 4000$ s for the first case and $t = 1000$ s for the second. The convergence rate in the first test falls below the unity for the finest mesh: this sublinear convergence has been reported by Oevermann [178] as well as from Couderc [58], and could be caused by the approximation of the infinity depth domain with our slip conditions. Nevertheless, it seems that an

**Table 4.6** Results of damped wave simulation, error integrated over time

| | Grid | $\|e\|$ | q |
|---|---|---|---|
| $\Delta\rho = 1$ | $8^2$ | 0.0689 | – |
| | $16^2$ | 0.0140 | 2.30 |
| | $32^2$ | 0.0053 | 1.41 |
| | $64^2$ | 0.0029 | 0.85 |
| $\Delta\rho = 1000$ | $8^2$ | 0.1492 | – |
| | $16^2$ | 0.0512 | 1.54 |
| | $32^2$ | 0.0140 | 1.87 |
| | $64^2$ | 0.0041 | 1.77 |

extension of the vertical dimension of the domain does not improve the results. Quite surprisingly, the convergence rates for the strong density ratio are well superlinear. In any case the intrinsic precision of our code can be appreciated, as well as the ability to achieve almost the fine grid resolution: the AMR computation error is greater than the corresponding fine mesh but always smaller than the immediately coarser grid one (Table 4.6).

### Rayleigh-Taylor Instability

A common test problem for a two phases flow numerical method is the Rayleigh-Taylor type instability. This phenomenon can be observed by superimposing a heavy fluid over a lighter one; without any perturbation they are in equilibrium. An initial sinusoidal perturbation of a given wavelength is subject to the destabilizing effect of gravity and the stabilizing effect of surface tension: if the destabilizing effect prevails, then the perturbation grows, as the heavier fluid moves down and deplaces the lighter one. With this kind of simulation the aptitude of the code to respect the solution predicted from the linear stability theory can be checked. By considering a long time simulation, the instability's amplitude is allowed to exceed the limit of the linear theory approximation. The subsequent large deformations of the interface and the consequent formation of the typical thin ligaments are shown in this simulation.

### Small amplitude, linear theory hypothesis

For short time simulations the initially sinusoidal disturbance $a = a(0)\cos(kx)$ grows following an exponential law $a(t) = a(0)\exp(nt)$, with $n$ predicted by the linear theory relation

$$n^2 = kg\left(\frac{\rho_h - \rho_l}{\rho_h + \rho_l} - \frac{k^2\sigma}{g\,(\rho_h + \rho_l)}\right) \tag{4.8}$$

given the value of surface tension and the wavelength small enough to let the instability to take place. Given $k = 1$, a limit value $\sigma_{crit}$ that describes a marginal stability state (where $n = 0$) can be found. For this simulation the physical parameters are the following: $L_x = 2\,\pi$, $L_x = 4\,\pi$, $\rho_h = 3\,\mathrm{kg\,m^{-3}}$, $\rho_l = 1\,\mathrm{kg\,m^{-3}}$, $\sigma_{crit} = 20\,\mathrm{N\,m^{-1}}$. The interface is placed in the middle of the computational domain; periodic boundary

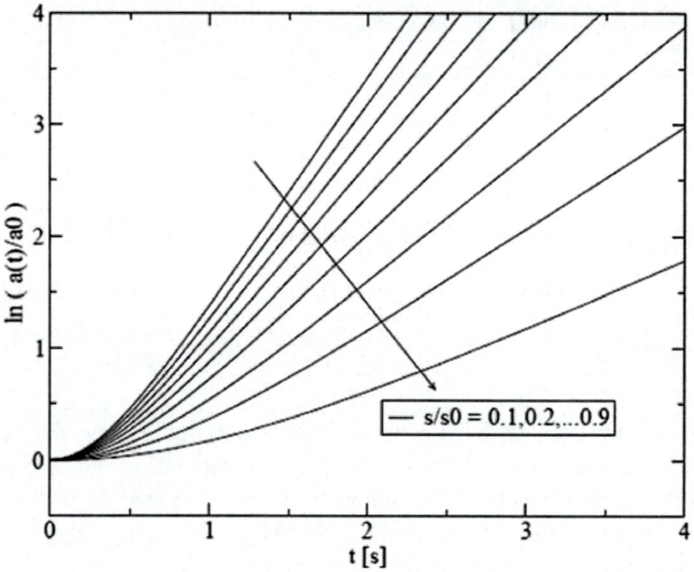


**Fig. 4.16** Computed time histories of the initial deformation amplitude

conditions are placed on the vertical walls, slip conditions on the horizontal ones. The initial amplitude of the perturbation is $a_0 = 1.E - 8$. Nine different values of $\sigma$ are treated, in the range $\sigma/\sigma_{crit} = [0.1, 0.2...0.9]$. In Fig. 4.16, time history of the initial interface deformation is plotted for 9 values of the surface tension coefficient. One clearly can observe the damping effect of the surface tension force on the amplitude evolution. In Fig. 4.17, the perturbation growth rate for the 9 values of the surface tension is plotted and compared to the linear stability theory. The mesh is adaptive in all the simulations, as the results can immediately be compared with the analytical solution; for the value $\sigma/\sigma_{crit} = 0.5$ a convergence study has been realized, for both the uniform and adaptive mesh (Tables 4.7 and 4.8).

In general, found values of $n$ are very close to the predicted ones, in particular for high values of surface tension; however, in this region the computation is more difficult, as some stability issues appeared: the CFL condition had to be reduced closer to the critical value of $\sigma$. As in other works, linear convergence is found for the uniform mesh test; with the adaptive mesh something in both errors and convergence rates is lost, remaining however close to the linear behaviour. The error never falls near or under the immediately coarser level.

#### Large amplitude, non-linear growth

In this section the evolution of the Rayleigh-Taylor instability is well outside the linear theory approximation limits. The simulation is performed in an extended domain $L_x = 1$ m, $L_y = 4$ m. The interface is placed, as before, in the middle of the domain. The new computational parameters are: $\rho_h = 1.225\,\text{kg m}^{-3}$, $\rho_l = 0.1694\,\text{kg m}^{-3}$, $\mu_h = \mu_l = 3.13E - 3\,\text{kg m}^{-1}.s^{-1}$, $\sigma = 0\,\text{N m}^{-1}$. The new initial perturbation is

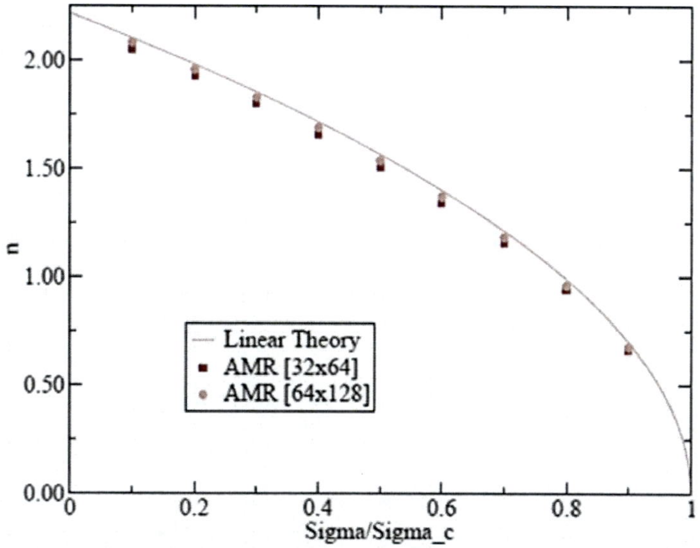


**Fig. 4.17** Computed growth rate for different values of $\sigma$ and for two (adaptive) meshes

**Table 4.7** Numerical computation of the growth rate for $\sigma/\sigma_{crit} = 0.5$: $n = 1.5508$, uniform mesh

| Grid | n | Err | q |
|---|---|---|---|
| 32 | 1.498 | 0.053 | – |
| 64 | 1.525 | 0.026 | 1.033 |
| 128 | 1.541 | 0.010 | 1.397 |

$y = 0.05\cos(kx)$ with $k = 2\pi$. Those parameters are the same used by Popinet [182]: into his work a comparison is made between an eulerian VOF method and a lagrangian type interface tracking that make use of markers. In this simulation the interface sustains strong non linear deformations, the initial perturbation grows into a "mushroom" shape until 0.8 s. At this time the extremities begin to form two thin ligaments that quickly shrink to the grid cell size. With insufficient resolution the ligaments are broken and, subsequently, two drops are detached, so that the computed velocity field deviates quickly from the good solution. The precise lagrangian solution in [128] shows us that until $t = 0.9$ s the ligaments are kept. The first computation is performed with the same base resolution, $64 \times 256$ with four levels of AMR. At the last time step the ligaments are already somewhat fragmented, even if the shape remains as can be seen in Fig. 4.18. The resolution has been subsequently improved by adding an AMR level, reaching a fine resolution of $128 \times 512$. On the results of Fig. 4.19, the ligaments are clearly maintained, showing the improvement on the two-phase flows code by a local refinement over the interface.

**Table 4.8** Numerical computation of the growth rate for $\sigma/\sigma_{crit} = 0.5$: $n = 1.5508$, adaptive mesh

| Grid | n | Err | q |
|---|---|---|---|
| 32 | 1.496 | 0.055 | – |
| 64 | 1.522 | 0.029 | 0.928 |
| 128 | 1.536 | 0.015 | 0.960 |

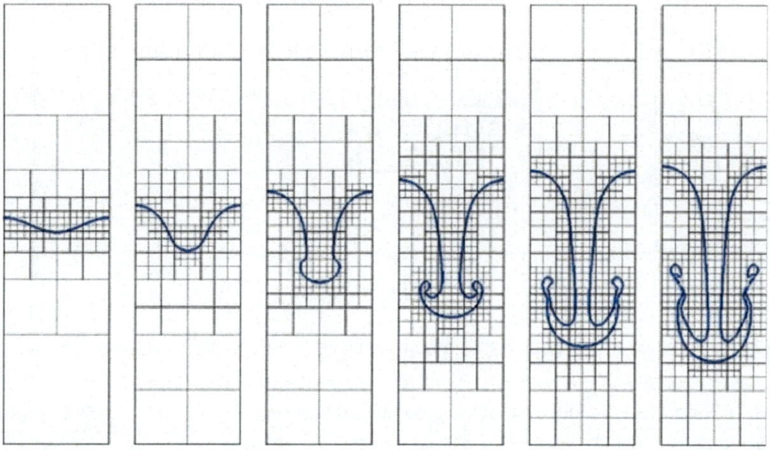

**Fig. 4.18** Results from the Rayleigh-Taylor instability computations, five levels of mesh, finest grid equivalent to $64 \times 256$, $t = 0.9$ s

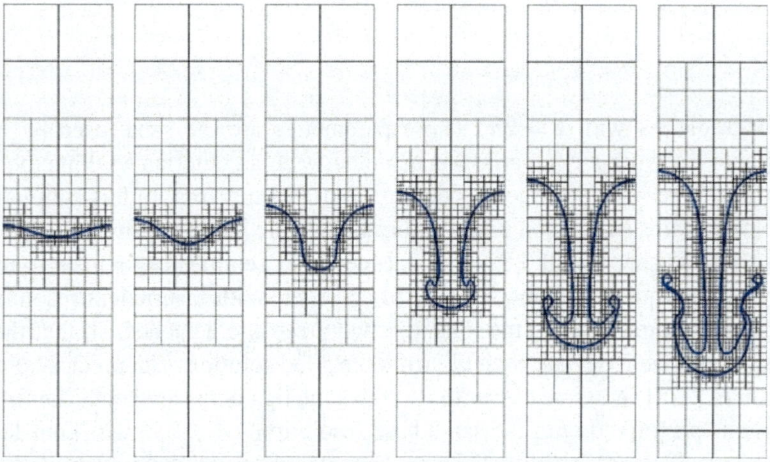

**Fig. 4.19** Results from the Rayleigh-Taylor instability computations, four levels of mesh, finest grid equivalent to $128 \times 512$, $t = 0.9$ s

### Parallel Performances

In the subsequent section the parallel performances of the code are tested. The two types of parallel study are realized, the strong and the weak scaling. In the first the domain is fixed, so that the maximum number of points a single processor can handle are given; then some computations are performed with an increasing number of processors. The ideal reduction of time, that would be achieved if the communication time were zero, is equal to the number of processors, as the domain become more partitioned among the CPUs. In the second case the domain addressed to a single processor is fixed; the dimension of the problem grows together with the number of cores. The expected computation time should be almost constant.

### Strong scaling

For the strong scaling test a square domain has been reconstructed with an adaptive mesh composed by 12 levels of blocks (the chosen test case is the damped surface wave oscillation). Each block is composed by $8 \times 8 = 64$ internal points plus 132 ghost points (three cells on each direction, corners included). When run on a single processor, this mesh requires an allocation of at least 40,000 blocks in order to build the whole tree (actually not all the blocks are active, but PARAMESH needs some room for moving the data). The resulting fine resolution is $16{,}384^2$. The computational domain is then distributed among an increasing number of processors, from two up to 256, each time their number doubled. The computational time is measured for a fixed number of iterations; the mean iteration time is kept. For each processor number increase, the maximum allocated block are accordingly reduced (approximately divided by two). The test has been repeated with a maximum level of 8 (resolution 1024, 3000 initial blocks) in order to compare the limit value of points per processor which stops the speed-up.

The results in Fig. 4.20a show a good constant decrement in the computational time, even if not the ideal one. The speed-up limit seems reached with 256 processors; the time still decreases from the 128 test, but quite far from the ideal division by two. This result can be expected, as in the AMR code the communication is more complicated than an uniform mesh code alone, as the quantity of exchanged data may vary in function of the mesh configuration. Moreover, a particular “idle processors” problem arises from the multigrid preconditioning: a more precise evaluation of this issue is given afterwards. As expected, in the 8 levels test the limit is reached sooner, at 32 processors. The two results agree reasonably well in the limit number of blocks per processor: if the last point of the first test and the next-to-last of the second one are kept as limits, the number of blocks reach its limit at around 100 blocks, corresponding to 6400 points. If the next-to-last point is kept for the fist case, this values jumps to around 300 blocks or 19,200 points. This is an interesting value because in the experience of [58] the limits of the sub domains is reached with a $128 \times 128$ mesh per processor, corresponding to 16,384 points.

### Weak scaling

In this section the domain assigned to a processor is kept constant at 200 blocks, the number retained from the previous tests. The resolution is each time increased as

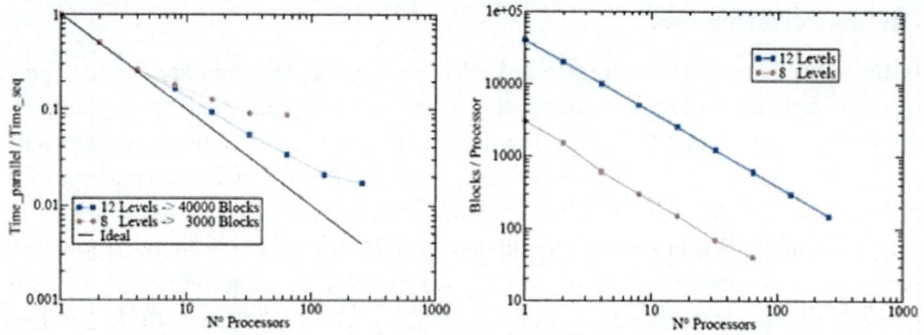


(a) Speed-up ratio for the two test cases. (b) Corresponding blocks per processor allocated. Each block contains 64 active points.

**Fig. 4.20** Strong scaling speed-up test

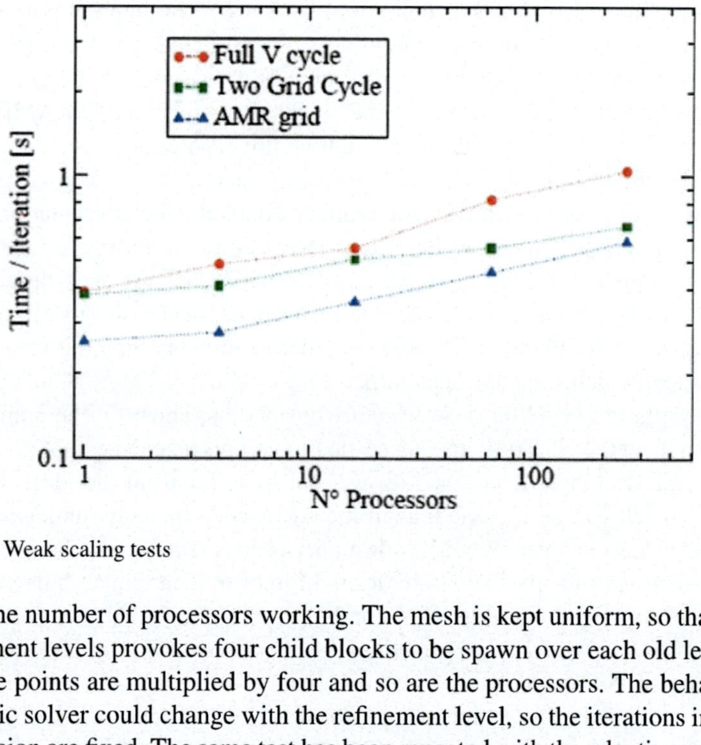


**Fig. 4.21** Weak scaling tests

well as the number of processors working. The mesh is kept uniform, so that adding a refinement levels provokes four child blocks to be spawn over each old leaf block: the active points are multiplied by four and so are the processors. The behaviour of the elliptic solver could change with the refinement level, so the iterations instead of the precision are fixed. The same test has been repeated with the adaptive mesh, thus losing the correspondence between the number of points and processors increase. Results from this test are shown in Fig. 4.21. The time per iteration is not really constant: the time per iteration rises slightly, from about 0.4–1 s approximately for respectively 1 and 256 processors. It has to be pointed out, however, that the multigrid algorithm utilizes more levels (so more points) at the finer resolutions. The use of the adaptive mesh does not change the general behaviour, but the computational time obviously decreases.

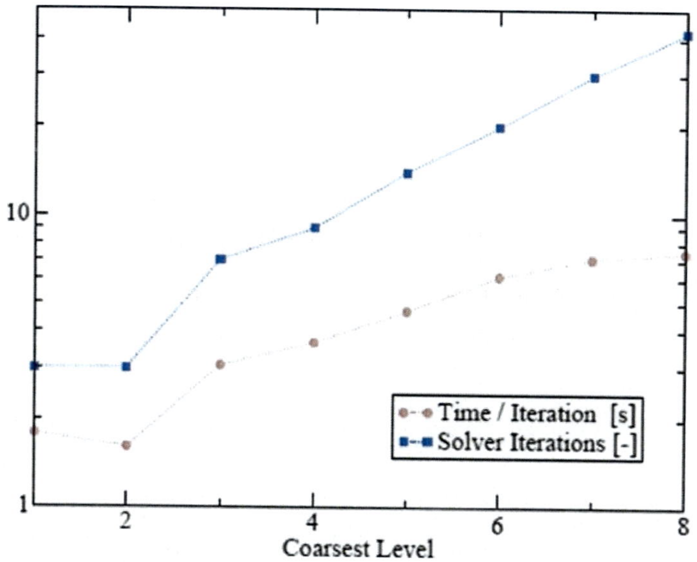


**Fig. 4.22** Multigrid coarsest level test: augmentation of iterations and computational time due to the decreasing of total multigrid levels

Finally, the multigrid full V-cycle has been replaced by a two grid cycle, in order to check the loss of speed-up caused by the preconditioning. This is caused by the PARAMESH peculiar utilization of the AMR tree for the multigrid: for each descent step the hidden parent block are defined as active blocks and used for the relaxation; however, at the coarser levels the number of active blocks approaches and surpass the number of processors, making an increasing number of those to wait. As can be seen in the graph, the computational time is now much more constant. This result confirms the limits of the preconditioner scalability. One obvious solution would be to stop the descent to a finer multigrid level. However, this would reduce the multigrid efficiency, as it depends on the number of levels employed. A simple test has been done to check the influence of the coarsest level on the convergence speed: a computation with a fixed precision is performed with an increasing coarsest level (one block, four, sixteen, and so on) that corresponds to a decrease of the total number of multigrid levels. The results in Fig. 4.22 show how the loss of even a single level may raise the number of solver iteration, ad consequently the computational time. This block based multigrid characteristic may even be an advantage, because usual fixed mesh codes are limited in their coarsest grid by the number of processors used.

# Chapter 5
# Numerical Treatment of Constraints with Fictitious Domains

## 5.1 Augmented Lagrangian Methods

In the following subsections, the existing augmented Lagrangian approaches and its theoretical bases are first presented briefly in order to justify the introduction of an algebraic method for managing augmented Lagrangian techniques when multi-phase flows are dealt with [53–55].

**Theoretical formulation**

For the sake of simplicity, let us consider the stationary incompressible Stokes equations on a domain $\Omega \in \mathbb{R}^d$, where $d$ is the number of space dimensions:

$$-\nabla p + \mu \Delta \boldsymbol{u} = \boldsymbol{f} \text{ in } \Omega \tag{5.1}$$

$$\nabla \cdot \boldsymbol{u} = 0 \text{ in } \Omega \tag{5.2}$$

$$\boldsymbol{u} = 0 \text{ on } \partial\Omega \tag{5.3}$$

Contrary to scalar projection methods, the augmented Lagrangian (AL) [52] proposes to satisfy the two equations (2.1) and (2.2) at the same time, resulting a divergence free flow $\boldsymbol{u}$ with the resolution of one equation only. The method uses at the same time a minimization under constraint and a penalty term to accelerate the convergence. The pressure $p$ is here a Lagrange multiplier which allows the constraint to be ensured.

The functional $J(\boldsymbol{v})$ defined for $\boldsymbol{f} \in \mathrm{L}^2(\Omega)^d$ and $\boldsymbol{v} \in (H_0^1(\Omega))^d$ is built from the weak formulation of the original problem:

$$J(\boldsymbol{v}) = \frac{\mu}{2}\boldsymbol{a}(\boldsymbol{v},\boldsymbol{v}) - \int_\Omega \boldsymbol{f} \cdot \boldsymbol{v} \, \mathrm{d}x = \frac{\mu}{2}\boldsymbol{a}(\boldsymbol{v},\boldsymbol{v}) - (\boldsymbol{f},\boldsymbol{v}) \tag{5.4}$$

with

$$\boldsymbol{a}(\boldsymbol{u},\boldsymbol{v}) = \sum_{i,j=1}^{d} \int_\Omega \frac{\partial u_i}{\partial x_j}\frac{\partial v_i}{\partial x_j} \, \mathrm{d}x. \tag{5.5}$$



S. Vincent et al., *Small Scale Modeling and Simulation of Incompressible Turbulent Multi-Phase Flow*, CISM International Centre for Mechanical Sciences 607,
https://doi.org/10.1007/978-3-031-09265-7_5

and

$$(\boldsymbol{u}, \boldsymbol{v}) = \int_{\Omega} \boldsymbol{u}\boldsymbol{v}\, \mathrm{d}x. \tag{5.6}$$

This functional has to be minimized under the constraint $\boldsymbol{u}, \boldsymbol{v} \in M = \{\boldsymbol{v} \in (H_0^1(\Omega))^d, \nabla \cdot \boldsymbol{v} = 0\}$. The problem is now:

$$\begin{cases} J(\boldsymbol{u}) \leq J(\boldsymbol{v}), \quad \forall \boldsymbol{v} \in M \\ \boldsymbol{u} \in M \end{cases} \tag{5.7}$$

Practically, a solution in a constrained space such as $M$ cannot be easily computed. This problem of minimization under constraint is transformed into a problem of minimization without constraint thanks to a Lagrange multiplier. We define the following Lagrangian:

$$L(\boldsymbol{v}, q) = J(\boldsymbol{v}) - \int_{\Omega} q \nabla \cdot \boldsymbol{v}\, \mathrm{d}\Omega \tag{5.8}$$

with $q \in L^2(\Omega)$. The minimization problem (5.7) consists in finding a saddle-point $(\boldsymbol{u}, p) \in (H_0^1(\Omega))^d \times L^2(\Omega)$ of the Lagrangian (5.8):

$$L(\boldsymbol{u}, q) \leq L(\boldsymbol{u}, p) \leq L(\boldsymbol{v}, p) \quad \forall \boldsymbol{v} \in (H_0^1(\Omega))^d, \ \forall q \in L^2(\Omega), \tag{5.9}$$

which implies

$$\begin{aligned} L(\boldsymbol{u}, p) &= \min_{\boldsymbol{v} \in (H_0^1(\Omega))^d} \max_{q \in L^2(\Omega)} L(\boldsymbol{v}, q) \\ &= \max_{q \in L^2(\Omega)} \min_{\boldsymbol{v} \in (H_0^1(\Omega))^d} L(\boldsymbol{v}, q) \end{aligned} \tag{5.10}$$

In order to increase the convergence rate [52], the constraint is used to build a penalty term $\frac{1}{2} r |\nabla \cdot \boldsymbol{v}|^2, \ r \in \mathbb{R}$ [183]. The augmented Lagrangian is denoted as:

$$L_r(\boldsymbol{v}, q) = J(\boldsymbol{v}) - \int_{\Omega} q \nabla \cdot \boldsymbol{v}\, \mathrm{d}x + \int_{\Omega} \frac{r}{2} |\nabla \cdot \boldsymbol{v}|^2\, \mathrm{d}x. \tag{5.11}$$

The saddle-point of $L_r$ is the same as for $L$ as the penalty term vanishes when the constraint $\nabla \cdot \boldsymbol{v} = 0$ is satisfied. We admit that the solution $(\boldsymbol{u}, p)$ of the saddle-point problem for the weak formulation of the initial equations is the solution of the strong formulation (5.1), (5.2) of the problem [52]. The saddle-point can be calculated using the Uzawa algorithm [184]:

$$p^0 \in \mathbb{R} \text{ arbitrarily given} \tag{5.12}$$

$p^n$ being known, we compute $\boldsymbol{u}^{n+1}$, then $p^{n+1}$ with:

$$L_r(\boldsymbol{u}^{n+1}, p^n) \leq L_r(\boldsymbol{v}, p^n) \quad \forall \boldsymbol{v} \in \mathbb{R}^d, \boldsymbol{u}^{n+1} \in \mathbb{R}^d \tag{5.13}$$

$$p^{n+1} = p^n + r \nabla \cdot \boldsymbol{u}^{n+1} \tag{5.14}$$

Satisfying (5.13) is equivalent to solve

$$- \nabla p^n + \mu \Delta \boldsymbol{u}^{n+1} + \frac{1}{2} r \nabla \cdot \boldsymbol{u}^{n+1} = \boldsymbol{f}^n \tag{5.15}$$

which is the form used to build a finite-volume approximation in our approach. At this point, the value of $r$ has not been explicitly given. For Stokes equations, with constant fluid properties, Fortin and Glowinski [52] have demonstrated that the optimal value of $r$ is the average between the smaller and the larger eigenvalues of the discretization matrix. For the Navier-Stokes equations, no demonstration of optimal value are available. Without loss of generality, and for the sake of simplicity, $\frac{r}{2}$ is replaced by $r$.

### Standard augmented Lagrangian (SAL)

Following a quite similar walkthrough [52], the augmented Lagrangian can be applied to the unsteady Navier-Stokes equations. The standard augmented Lagrangian (SAL) method was first introduced by Fortin and Glowinski [52]. Starting with $\boldsymbol{u}^{*,0} = \boldsymbol{u}^n$ and $p^{*,0} = p^n$, the solution predictor reads: while $||\nabla \cdot \boldsymbol{u}^{*,m}|| > \epsilon$ , solve

$$\begin{aligned}
(\boldsymbol{u}^{*,0}, p^{*,0}) = (\boldsymbol{u}^n, p^n)\rho \left( \frac{\boldsymbol{u}^{*,m} - \boldsymbol{u}^{*,0}}{\Delta t} + \boldsymbol{u}^{*,m-1} \cdot \nabla \boldsymbol{u}^{*,m} \right) - r \nabla (\nabla \cdot \boldsymbol{u}^{*,m}) \\
= -\nabla p^{*,m-1} + \rho \boldsymbol{g} + \nabla \cdot [\mu (\nabla \boldsymbol{u}^{*,m} + \nabla^T \boldsymbol{u}^{*,m})] + \sigma k \boldsymbol{n}_i \delta_i \\
p^{*,m} = p^{*,m-1} - r \nabla \cdot \boldsymbol{u}^{*,m}
\end{aligned} \tag{5.16}$$

where $r$ is the augmented Lagrangian parameter used to impose the incompressibility constraint, $m$ is an iterative convergence index and $\epsilon$ a numerical threshold controlling the constraint. Usually, a constant value of $r$ is used. From numerical experiments, optimal values are found to be of the order of $\rho_i$ and $\mu_i$ to accurately solve the motion equations in the related zone [54]. The momentum, as well as the continuity equations are accurately described by the predictor solution $(\boldsymbol{u}^*, p^*)$ coming from (5.16) in the medium, where the value of $r$ is adapted. However, high values of $r$ in the other zones act as penalty terms inducing the numerical solution to satisfy the divergence-free property only. Indeed, if we consider for example $\rho_1/\rho_0 = 1000$ (characteristic of water and air problems) and a constant $r = \rho_1$ to impose the divergence-free property in the denser fluid, the asymptotic equation system solved in the predictor step is:

$$\begin{aligned}
&\rho \left( \frac{\boldsymbol{u}^* - \boldsymbol{u}^n}{\Delta t} + (\boldsymbol{u}^n \cdot \nabla) \boldsymbol{u}^* \right) - r \nabla (\nabla \cdot \boldsymbol{u}^*) \\
&= \rho \boldsymbol{g} - \nabla p^n + \nabla \cdot [\mu (\nabla \boldsymbol{u}^* + \nabla^T \boldsymbol{u}^*)] + \sigma k \boldsymbol{n}_i \delta_i \quad \text{in } \Omega_1 \\
&\frac{\boldsymbol{u}^* - \boldsymbol{u}^n}{\Delta t} - r \nabla (\nabla \cdot \boldsymbol{u}^*) = 0 \quad \text{in } \Omega_0
\end{aligned} \tag{5.17}$$

Our idea is to locally estimate the augmented Lagrangian parameter in order to obtain satisfactory equivalent models and solutions in all the media.

### Adaptive Augmented Lagrangian (2AL)

Instead of choosing an empirical constant value of $r$ fixed at the beginning of the simulations, we propose at each time step to locally estimate the augmented Lagrangian parameter $r$. Then, $r(t, M)$ becomes a function of time $t$ and space position $M$. It must be two to three orders of magnitude higher than the most important term in the conservation equations. Let $L_0$, $t_0$, $u_0$ and $p_0$ be reference space length, time, velocity and pressure respectively. If we consider one iterative step of the augmented Lagrangian procedure (5.16), the non-dimensional form of the momentum equations can be rewritten as

$$\begin{aligned} &\rho\frac{u_0}{t_0}\frac{\boldsymbol{u}^{*,m}-\boldsymbol{u}^n}{\Delta t}+\rho\frac{u_0^2}{L_0}(\boldsymbol{u}^{*,m-1}\cdot\nabla)\boldsymbol{u}^{*,m}-\frac{u_0}{L_0^2}\nabla(r\nabla\cdot\boldsymbol{u}^{*,m}) \\ &=\rho\boldsymbol{g}-\frac{p_0}{L_0}\nabla p^{*,m-1}+\frac{u_0}{L_0^2}\nabla\cdot[\mu(\nabla\boldsymbol{u}^{*,m}+\nabla^T\boldsymbol{u}^{*,m})]+\frac{\sigma}{L_0^2}k\boldsymbol{n}_i\delta_i \end{aligned} \tag{5.18}$$

Multiplying the right and left parts of equation (5.18) by $L_0^2/u_0$, we can compare the augmented Lagrangian parameter $r$ to all the contributions of the flow (inertia, gravity, pressure and viscosity). We obtain

$$\begin{aligned} &\rho\frac{L_0^2}{t_0}\frac{\boldsymbol{u}^{*,m}-\boldsymbol{u}^n}{\Delta t}+\rho u_0L_0(\boldsymbol{u}^{*,m-1}\cdot\nabla)\boldsymbol{u}^{*,m}-\nabla(r\nabla\cdot\boldsymbol{u}^{*,m}) \\ &=\rho\frac{L_0^2}{u_0}\boldsymbol{g}-\frac{p_0L_0}{u_0}\nabla p^{*,m-1}+\nabla\cdot[\mu(\nabla\boldsymbol{u}^{*,m}+\nabla^T\boldsymbol{u}^{*,m})]+\frac{\sigma}{u_0}k\boldsymbol{n}_i\delta_i \end{aligned} \tag{5.19}$$

It can be noticed that $r$ is comparable to a viscosity coefficient. It is then defined as

$$r(t,M)=\mathcal{K}\ \max\left(\rho(t,M)\frac{L_0^2}{t_0},\rho(t,M)u_0L_0,\rho(t,M)\frac{L_0^2}{u_0}g,\frac{p_0L_0}{u_0},\mu(t,M),\frac{\sigma}{u_0}\right) \tag{5.20}$$

If, for example, $\frac{\rho_1}{\rho_0}=1000$ and $\mu_0<\mu_1<<\rho_0<<\rho_1$, the semi-discrete form of the momentum equations resulting from the new values of $r(t, M)$ given by (5.20) then becomes

$$\begin{aligned} &\rho_1\left(\frac{\boldsymbol{u}^{*,m}-\boldsymbol{u}^n}{\Delta t}+(\boldsymbol{u}^{*,m-1}\cdot\nabla)\boldsymbol{u}^{*,m}\right)-\nabla(r\nabla\cdot\boldsymbol{u}^{*,m}) \\ &=\rho_1\boldsymbol{g}-\nabla p^{*,m-1}+\nabla\cdot[\mu_1(\nabla\boldsymbol{u}^{*,m}+\nabla^T\boldsymbol{u}^{*,m})]+\sigma k\boldsymbol{n}_i\delta_i \\ &\qquad\text{in }\Omega_1\text{ with }r=\mathcal{K}_1\rho_1L_0^2u_0 \\ &\rho_0\left(\frac{\boldsymbol{u}^{*,m}-\boldsymbol{u}^n}{\Delta t}+(\boldsymbol{u}^{*,m-1}\cdot\nabla)\boldsymbol{u}^{*,m}\right)-\nabla(r\nabla\cdot\boldsymbol{u}^{*,m}) \\ &=\rho_0\boldsymbol{g}-\nabla p^{*,m-1}+\nabla\cdot[\mu_0(\nabla\boldsymbol{u}^{*,m}+\nabla^T\boldsymbol{u}^{*,m})]+\sigma k\boldsymbol{n}_i\delta_i \\ &\qquad\text{in }\Omega_0\text{ with }r=\mathcal{K}_0\rho_0L_0^2u_0 \end{aligned} \tag{5.21}$$

where $\mathcal{K}_0$ and $\mathcal{K}_1$ are in between 10 and 1000. In this way, thanks to expression (5.19), the adaptive Lagrangian parameter is assigned a value that exceeds from 10 to 1000 times the order of magnitude of the most important term between inertia, viscosity, pressure or gravity in both $\Omega_1$ and $\Omega_0$ domains. Compared to the SAL approach (5.17), the 2AL is consistent with the Navier-Stokes equations in each phase [54]. The new method (5.20) can be easily extended to other forces such as Coriolis effects or specific source terms. Comparisons between standard and adaptive augmented Lagrangian (2AL) methods are presented in the next section. In particular, the influence of the penalty parameter on the convergence speed of the BiCG-Stab solver and time and space variations of $r(M)$ are discussed.

As a summary, the complete time-marching procedure of the predictor-corrector algorithm including the 2AL method is the following:
⊗ **Step 1**: initial values $\boldsymbol{u}^0$ and $p^0$ and boundary conditions on $\Gamma$ are defined,
⊗ **Step 2**: knowing $\boldsymbol{u}^n$, $p^n$ and a divergence threshold $\epsilon$, the predictor values $\boldsymbol{u}^*$ and $p^*$ are estimated with the Uzawa algorithm (5.16) associated to the local estimate of $r(t, M)$ defined in expression (5.20), so that $\boldsymbol{u}^* = \boldsymbol{u}^{*,m}$ and $p^* = p^{*,m}$ when $m$ verifies $||\nabla \cdot \boldsymbol{u}^{*,m}|| < \epsilon$,
⊗ **Step 3**: the solution $(\boldsymbol{u}^*, p^*)$ is projected on a divergence-free subspace thanks, for example, to projection approaches [185, 186] to get the correction solution $(\boldsymbol{u}', p')$. Then, the numerical solution at time $(n+1)\Delta t$ is $(\boldsymbol{u}^{n+1}, p^{n+1}) = (\boldsymbol{u}^* + \boldsymbol{u}', p^* + p')$,
⊗ **Step 4**: $n$ is iterated in steps 3 and 4 until the physical time is reached.

**Algebraic adaptive augmented Lagrangian (3AL)**

It has been demonstrated that estimating a local and adapted augmented Lagrangian parameter is crucial for simulating multi-phase flows [54]. The main remaining drawback of the 2AL method is linked to the a priori definition of dimensionless parameters for defining $r(t, M)$. The augmented Lagrangian approach is based on the concept of a penalty method. As a consequence, the augmented Lagrangian parameter acts as an algebraic parameter which increases the magnitude of specific coefficients in the linear system in order to verify a specific constraint, while solving at same time the conservation equations. In this section, an estimate of $r(t, M)$ is proposed which is based on a scanning of the matrix coefficients. The main interests of the algebraic adaptive augmented Lagrangian method (3AL) are the following: it does not require any a priori physical information, it applies to any kind of geometry and grid and it takes into account the residual of the linear solver and the fulfilment of incompressible and solid constraints.

At each time step and in two-dimensions (the three-dimensional algorithm is straightforward), the 3AL method determines $r(t, M)$ as follows:
⊗ **Step 1**: two matrices $A$ and $A^*$ are built corresponding respectively to the discretization of the momentum equations (5.21) with $r(t, M) = 0$ and $r(t, M) = 1$. In order to optimize computer memory, a compressed storage row (CSR) structure is chosen to store only the non-null coefficients of each matrix,
⊗ **Step 2**: on the fixed staggered Cartesian grid, $r(t)_{i,j}$ is evaluated according to the discretization coefficients of the surrounding velocity $u_{x,i-\frac{1}{2},j}$, $u_{x,i+\frac{1}{2},j}$, $u_{y,i,j-\frac{1}{2}}$ and $u_{y,i,j+\frac{1}{2}}$ components, as presented in Fig. 5.1. The discretization of each velocity

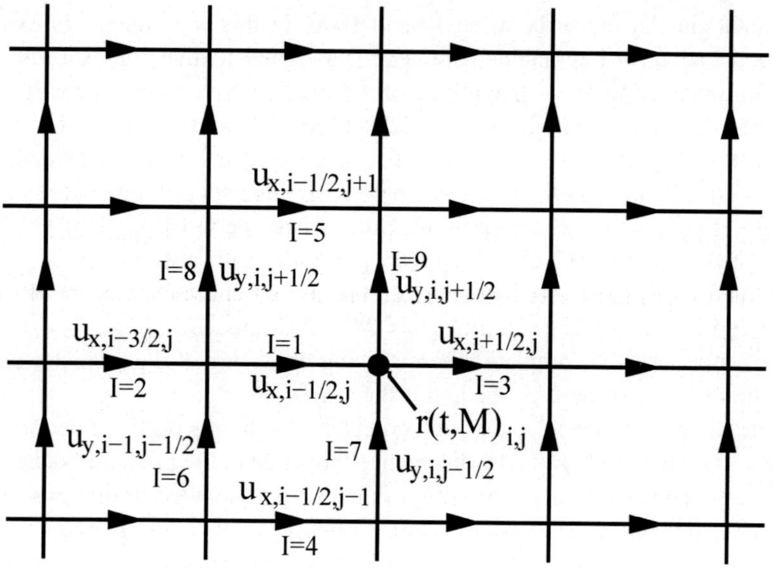


**Fig. 5.1** Typical distribution of discretization variables on a 2D fixed staggered Cartesian grid

component, $u_{x,i-\frac{1}{2},j}$ for example, requires the use of 9 neighboring velocity nodes, i.e. $u_{x,i-\frac{1}{2},j}$, $u_{x,i+\frac{1}{2},j}$,$u_{x,i-\frac{3}{2},j}$, $u_{x,i-\frac{1}{2},j+1}$ and $u_{x,i-\frac{1}{2},j-1}$ for the discretization of the inertial and viscous terms and $u_{y,i-1,j-\frac{1}{2}}$, $u_{y,i-1,j+\frac{1}{2}}$, $u_{y,i,j-\frac{1}{2}}$ and $u_{y,i,j+\frac{1}{2}}$ for the viscous and augmented Lagrangian terms. In this way, we estimate the maximum values of the discretization coefficients $A_I(u)$, $1 \leqslant I \leqslant 9$, associated to velocities $u_{x,i-\frac{1}{2},j}$, $u_{x,i+\frac{1}{2},j}$, $u_{y,i,j-\frac{1}{2}}$ and $u_{y,i,j+\frac{1}{2}}$. We define:

$$\begin{cases} C_{i-\frac{1}{2},j} = \max_{I=1...9} \left|A_I(u_{x,i-\frac{1}{2},j})\right| \\ C_{i+\frac{1}{2},j} = \max_{I=1...9} \left|A_I(u_{x,i+\frac{1}{2},j})\right| \\ C_{i,j-\frac{1}{2}} = \max_{I=1...9} \left|A_I(u_{y,i,j-\frac{1}{2}})\right| \\ C_{i,j+\frac{1}{2}} = \max_{I=1...9} \left|A_I(u_{y,i,j+\frac{1}{2}})\right| \end{cases}$$

⊗ **Step 3**: in the same way, the coefficient $C^*_{i-\frac{1}{2},j}$, $C^*_{i+\frac{1}{2},j}$, $C^*_{i,j-\frac{1}{2}}$ and $C^*_{i,j+\frac{1}{2}}$, corresponding to $A^*$, are estimated,

⊗ **Step 4**: the minimum and maximum discrete coefficient $\text{Rmin}_{i,j}$ and $\text{Rmax}_{i,j}$ at the scalar position of $r(t)_{i,j}$ are then defined as:

$$\begin{cases} \text{Rmin}_{i,j} = min(|\frac{C_{i-\frac{1}{2},j}}{C^*_{i-\frac{1}{2},j}}|, |\frac{C_{i+\frac{1}{2},j}}{C^*_{i+\frac{1}{2},j}}|, |\frac{C_{i,j-\frac{1}{2}}}{C^*_{i,j-\frac{1}{2}}}|, |\frac{C_{i,j+\frac{1}{2}}}{C^*_{i,j+\frac{1}{2}}}|) \\ \text{Rmax}_{i,j} = \max(|\frac{C_{i-\frac{1}{2},j}}{C^*_{i-\frac{1}{2},j}}|, |\frac{C_{i+\frac{1}{2},j}}{C^*_{i+\frac{1}{2},j}}|, |\frac{C_{i,j-\frac{1}{2}}}{C^*_{i,j-\frac{1}{2}}}|, |\frac{C_{i,j+\frac{1}{2}}}{C^*_{i,j+\frac{1}{2}}}|) \end{cases}$$

⊗ **Step 5**: as can be seen in Fig. 5.1, the parameters $r(t)_{i,j}$ and $r^*(t)_{i,j}$ are located at pressure nodes while the $C$ and $C^*$ are located at velocity nodes (in the staggered grid approach). In the calculation of Rmin and Rmax, strong variations of $C$ and $C^*$ can occur in the vicinity of liquid-gas or fluid-solid interfaces. Without modification, the penalty of the incompressible or solid constraint is not ensured due to grid irregularity or strong local variation of physical or penalty parameters. In this way, a threshold is used for $r^*(t)_{i,j}$: if $\frac{\mathrm{Rmax}_{i,j}}{\mathrm{Rmin}_{i,j}} \leqslant 10^{10}$ and $\frac{\max(\Delta_{x,i,j},\Delta_{y,i,j})}{\min(\Delta_{x,i,j},\Delta_{y,i,j})} \leqslant 1000$, $r^*(t)_{i,j} = \mathrm{Rmin}_{i,j}$. Else, $r^*(t)_{i,j} = \mathrm{Rmax}_{i,j}$,

⊗ **Step 6**: once $r^*(t)_{i,j}$ has been estimated for all $i$ and $j$, $1 \leqslant i \leqslant n_x$ and $1 \leqslant j \leqslant n_y$, the local values of the algebraic penalty parameters are normalized as

$$r(t)_{i,j} = \frac{\mathcal{K} r^*(t)_{i,j}}{r^*_{min} + 10^{-40}} \tag{5.22}$$

where $r^*_{min} = min_{i=1..n_x, j=1..n_y}(r^*(t)_{i,j})$. As for the SAL and the 2AL, $\mathcal{K}$ is a numerical parameter used to tune the minimization of the divergence according to the residual of the solver. The higher $\mathcal{K}$ is, the lower is the divergence and higher the residual is. In the 3AL, the order of magnitude of $r(t)_{i,j}$ is naturally balanced so the constant $\mathcal{K}$ is equal to 1 except at the first calculation step where $\mathcal{K} = 100$ if $\frac{\mathrm{Rmax}_{i,j}}{\mathrm{Rmin}_{i,j}} \leqslant 10^{10}$ or if during the five first calculation steps, the norm of the residual of the linear solver divided by the norm of the divergence, obtained at the previous calculation step, is greater than $10^6$. This last particular case can be obtained when a flow simulation is initialized without any imposed velocity field.

Steps 1 to 6 are repeated at each calculation step corresponding to a physical time incremented of $\Delta t$ seconds.

## 5.2 Penalty Methods

### Low order penalty methods

First-order methods are first presented. They are used to penalize control volumes in solid media approximating the solid/fluid interface with stair-steps. They can be applied to both scalar and Navier.Stokes equations. They can also be used to imposed boundary conditions, the physical boundaries of a given problem being immersed in the calculation domain (a staggered grid) for which the components of velocities are not located at the same point. The penalty methods consist in adding specific terms in the conservation equations to play with the order of magnitude of existing physical contributions so as to solve at the same time and with the same set of equations the fluid and solid mechanics of fluid flow and boundary conditions (see [34, 54] for example). Among the existing methods, we can cite the Darcy penalty method (DPM) that can be used to treat fixed obstacles by adding a Darcy term [34] in the momentum equations. The volume penalty method (VPM) (see [34] and the references therein) consists in the addition of a penalty term $\beta_u(\boldsymbol{u} - \boldsymbol{u}_\infty)$ in conservation equations, such that

$$\rho \frac{d\boldsymbol{u}}{dt} = -\nabla p + \rho \boldsymbol{g} + \nabla \cdot \big((\mu + \mu_t)(\nabla \boldsymbol{u} + \nabla^t \boldsymbol{u})\big) + \boldsymbol{f}_{st} + \beta_u(\boldsymbol{u} - \boldsymbol{u}_\infty) \qquad (5.23)$$

where $\boldsymbol{u}$is a unknown vector (it could be similarly a scalar), $\beta_u$ the penalty coefficient and $\boldsymbol{u}_\infty$ the value of a Dirichlet condition to be imposed in the given media. According to the values of $\beta_u$ and $\boldsymbol{u}_\infty$, we have

$$\rho \frac{d\boldsymbol{u}}{dt} = -\nabla p + \rho \boldsymbol{g} + \nabla \cdot \big((\mu + \mu_t)(\nabla \boldsymbol{u} + \nabla^t \boldsymbol{u})\big) + \boldsymbol{f}_{st} \text{ if } \beta_u \to 0 \qquad (5.24)$$

$$\boldsymbol{u} = \boldsymbol{u}_\infty \text{ if } \beta_u \to \infty \qquad (5.25)$$

The values of $\beta_u$ are defined according to the characteristic functions $C_i$ that allow to locate the fictitious domains (liquid, gaz, solid) or if a point belongs to the boundary of the numerical domain. Concerning obstacles moving under flow action, the derivatives of the velocity are penalized through a new formulation of the viscous stress tensor [187] in order to impose no deformation with a velocity resulting from the effect of the surrounding fluid. This method is called the implicit tensorial penalty method (ITPM) which is described in Sect. 5.2. The space convergence of these VPMs is only first order since they consider the projected shape of the fluid/solid interface on the Eulerian grid to define the penalty parameters (see [188, 189]).

**High order penalty methods**

The order of the standard VPMs is limited since they approach at first order the topology of the real interface by fitting its shape to the control volumes. Thus, these penalty methods have to be improved near the fluid-solid interfaces or immersed boundary conditions. Instead of strongly constraining the mean value of the solution in Eulerian control volumes, we decide to impose the local value of the Dirichlet condition at the Lagrangian interface location and so to define implicit penalty functions that are no longer constant into the control volumes. The key points of the method lie on the following concepts. First, the interface $\Sigma_i$ is accurately piecewise reconstructed using a Lagrangian mesh coupled with front-tracking methods. This discrete interface $\Sigma_{i,h}$ is used to build the penalty terms. As the interface is described at a smaller scale than the Eulerian grid scale, the method is called the sub-mesh penalty (SMP) method. The method is fully implicit and the same model, like (5.23–5.25) applies in all domains by adding a penalty term in the conservation equations. The discrete operators are not modified, contrary to ghost fluid [45] or Immersed Interface Methods (IIM) [191]. The method uses interpolation functions $f_p$ of various support $\varsigma_p$ (typically $P1$ or $Q1$) according to the local interface topology.

In two dimensions, let $\boldsymbol{u}$ be the solution of (5.23) in a domain $\Omega = \Omega_0 \cup \Omega_1$, $\Omega_1$ being an embedded domain (solid media for example) of frontier $\Sigma_i$ on which the Dirichlet condition $\boldsymbol{u} = \boldsymbol{u}_\infty$ must be satisfied. We discretize $\boldsymbol{u}$ over the computational domain by a set of unknowns $\boldsymbol{u}_i$ located at the nodes $\boldsymbol{x}_i$ of the staggered grid, with $i \in \{1, 2, ..., Nx \times Ny\}$. We first explain the non-symmetric version used to impose Dirichlet boundary conditions on $\Omega_0$ only, the solution in $\Omega_1$ being known or without interest. That is the case if $\Omega_0$ is a fluid domain and $\Omega_1$ a fixed obstacle of complex shape. Then, we generalize the method for Dirichlet conditions on immersed interfaces.

To locate the zone to be penalized, an Eulerian volume fraction $C_i$ is built with a front-tracking Lagrangian/Eulerian projection method [157] that either solves a Poisson equation and uses Peskin discrete Dirac functions (see equation (3.127) presented before) or uses a ray casting technique. The color function $C_i$ admits the value 0 in $\Omega_0$ and 1 in $\Omega_1$. The interface is easily located on the Eulerian grid by stating $C_i = 0.5$. In all the control volumes $V_i$, where $C_i = 1$ and $\nabla C_i \neq 0$, a standard VPM is used to impose $\boldsymbol{u} = \boldsymbol{u}_\infty$. In our case, we use the VPM and state $\beta_{u_i} = 1/\eta$ with $\eta \approx O(10^{-40})$. The SMP functions $f_p$ are built in all $V_i$ where $\nabla C_i \geq 1/h_i$, with $h_i$ the local space step. We define $\wp$ the set containing each node $x_j \in \Omega_1$ having at least one neighbor in $\Omega_0$, and $\imath$ the set containing each node in $\Omega_1$ but not in $\wp$. Each node of $\wp$ is associated with a specific penalty function $f_p$. If $x_j$ ($x_{j1}$ in Fig. 5.2a) has one neighbor in $\Omega_0$ ($x_{s11}$ in Fig. 5.2a), the support of $f_p$ is the segment composed of the two nodes ($x_{j1}$ and $x_{s11}$) and the interpolation is linear in one direction. If $x_j$ ($x_{j2}$ in Fig. 5.2a) has two neighbors in $\Omega_0$, the support of $f_p$ is a $Q1$ or $P1$ cell containing $x_j$ and its two neighbors in $\Omega_0$. The interpolation is then linear/bilinear. If $x_j$ ($x_{j3}$ in Fig. 5.2a) has three neighbors in $\Omega_0$, the support of $f_p$ is again a segment formed by $x_j$ and the median point of the three neighbors ($x_{s32}$ in Fig. 5.2a).

Each penalty function is associated with a Lagrangian point $\boldsymbol{x}_l$ of $\Sigma_h$ included in the support of $f_p$. Once the support $\varsigma_p$ and the corresponding interpolation type are defined, $f_p$ is built. Each $f_p$ is a polynomial function that satisfies $f_p(\boldsymbol{x}_s) = \boldsymbol{u}_s$, with $\boldsymbol{u}_s$ the solution at the nodes $\boldsymbol{x}_s$ of $\varsigma_p$. In this way, $f_p$ can be expressed as a linear combination of unknowns $\boldsymbol{u}_s$, such as $f_p(\boldsymbol{x}) = \sum_{k/x_k \in \varsigma_p} \alpha_k(\boldsymbol{x})\boldsymbol{u}_k$. Once $f_p$ is defined, the constraint $f_p(\boldsymbol{x}_l) = \boldsymbol{u}_\infty$ defines the penalty constraint. Hence, the penalty term $\beta_u(\boldsymbol{u} - \boldsymbol{u}_\infty)$ from the VPM is replaced by $\beta_u(f_p(\boldsymbol{u}) - \boldsymbol{u}_\infty)$. The support $\varsigma_p$ of each $f_p$ contains some nodes in $\Omega_0$ but only one node in $\Omega_1$. The penalty term is acting only for the discretization of equation (5.23) at this node in $\Omega_1$. The penalty term locally crushes the physical equation in $\Omega_1$ where we are not interested by the solution. Equation (5.23) is then equivalent to a penalty constraint acting on all the nodes

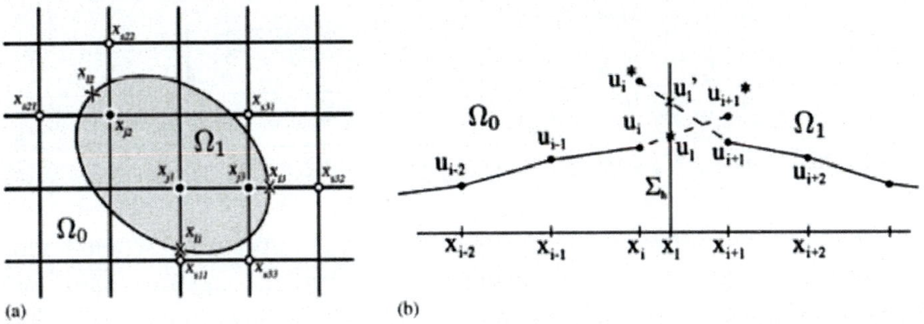


**Fig. 5.2** **a** Location of the kind of nodes used for the SMP method and **b** illustration of the symmetric SMP method [190]

of $\wp$. Hence, the computation of the physical solution in $\Omega_0$ does not require the unknowns located at the nodes of $\imath$. This version of the method is then adapted to take into account complex immersed boundaries of $\Omega_0$. As we deal with a vectorial problem, each component is treated as described before.

To allow an immersed Dirichlet condition from both sides of the interface (that is to say to compute the solution both in $\Omega_0$ and $\Omega_1$, i.e. an immersed interface problem), we symmetrize the procedure previously described. Now we have to consider two unknowns at each node near the interface: a "physical unknown" is used for the physical equation and is seen only by the node of the same sub-domain and an "auxiliary unknown" that supports the penalty constraint and is seen only by unknowns in the other sub-domain (Fig. 5.2b). The auxiliary unknowns are marked with $*$. The creation of auxiliary unknowns near the interface increases the size of the linear system. A way to avoid the creation of auxiliary unknowns is to use nodes only from the same side of the interface for a given penalty constraint. The interface is no more included in the support of the penalty function. The Dirichlet value at the interface becomes an implicit extrapolation of the solution in the support. Instead of being activated for nodes in $\Omega_1$, penalty terms are now active for the discretization of equation (5.23) at nodes in $\Omega_0$, where the physical solution is required. An implementation of this method for scalar problems has showed that convergence order remains the same, but the error grows two orders in magnitude.

**Implicit Tensorial Penalty Method**

The aim of this section is to present an original method for treating in a unified solver and with the same set of equations multi-phase flows in interaction with solids. The idea is to split the stress tensor in the Navier-Stokes equations into four parts, associated respectively to incompressibility, elongation, pure shearing and rotation. This formulation was previously presented in [187, 192]. In Cartesian coordinates and for the general framework of compressible fluids, the stress tensor $\bar{\bar{\gamma}}$ reads

$$
\begin{aligned}
\bar{\bar{\gamma}} &= \left[-p + \lambda \nabla \cdot \boldsymbol{u}\right] \overline{\overline{Id}} + \mu \left(\nabla \boldsymbol{u} + \nabla^t \boldsymbol{u}\right) \\
&= \begin{bmatrix} -p+\lambda\,\nabla\cdot\mathbf{u} & 0 & 0 \\ 0 & -p+\lambda\,\nabla\cdot\mathbf{u} & 0 \\ 0 & 0 & -p+\lambda\,\nabla\cdot\mathbf{u} \end{bmatrix} + \kappa \begin{bmatrix} \frac{\partial u}{\partial x} & 0 & 0 \\ 0 & \frac{\partial v}{\partial y} & 0 \\ 0 & 0 & \frac{\partial w}{\partial z} \end{bmatrix} \\
&\quad + \zeta \begin{bmatrix} 0 & \frac{\partial u}{\partial y} & \frac{\partial u}{\partial z} \\ \frac{\partial v}{\partial x} & 0 & \frac{\partial v}{\partial z} \\ \frac{\partial w}{\partial x} & \frac{\partial w}{\partial y} & 0 \end{bmatrix} - \eta \begin{bmatrix} 0 & \frac{\partial u}{\partial y} - \frac{\partial v}{\partial x} & \frac{\partial u}{\partial z} - \frac{\partial w}{\partial x} \\ \frac{\partial v}{\partial x} - \frac{\partial u}{\partial y} & 0 & \frac{\partial v}{\partial z} - \frac{\partial w}{\partial y} \\ \frac{\partial w}{\partial x} - \frac{\partial u}{\partial z} & \frac{\partial w}{\partial y} - \frac{\partial v}{\partial z} & 0 \end{bmatrix} \\
&= (-p + \lambda\,\nabla\cdot\mathbf{u})\overline{\overline{Id}} + \kappa\overline{\overline{\Gamma}}(\boldsymbol{u}) + \zeta\overline{\overline{\Theta}}(\boldsymbol{u}) - \eta\overline{\overline{\Omega}}(\boldsymbol{u})
\end{aligned}
\tag{5.26}
$$

where $\overline{\overline{Id}}$ is the identity tensor of $R^3$ and $\overline{\overline{\Gamma}}$, $\overline{\overline{\Theta}}$ and $\overline{\overline{\Omega}}$ respectively pseudo-tensors of elongation, pure shearing and rotation. The standard viscous stress tensor for a Newtonian fluid is recovered by stating $\lambda = -\frac{2}{3}\mu$, $\kappa = 2\mu$, $\zeta = 2\mu$ and $\eta = \mu$. First proposed by Fortin and Glowinski [52] for simulating incompressible single phase flows, the formulation (5.26) allows to build a penalty method based on augmented Lagrangian techniques. The idea is to replace the solving of Navier-Stokes equation system by the minimization of a new equation under the constraint of incompressibility in fluid zones and deformation in solid parts of the flow. Practically, an Uzawa algorithm is implemented [192], in which the following equations are solved iteratively to obtain velocity and pressure

$$
\begin{aligned}
&\rho\left(\frac{\boldsymbol{u}^{n+1}-\boldsymbol{u}^{n}}{\Delta t}+\boldsymbol{u}^{n}\cdot\nabla\boldsymbol{u}^{n+1}\right)=-\nabla p^{n}\\
&+\nabla\left[\lambda\left(\nabla\cdot\boldsymbol{u}^{n+1}\right)\right]+\nabla\cdot\left[\kappa\overline{\overline{\Gamma}}(\boldsymbol{u}^{n+1})+\zeta\overline{\overline{\Theta}}(\boldsymbol{u}^{n+1})-\eta\overline{\overline{\Omega}}(\boldsymbol{u}^{n+1})\right]+\rho\boldsymbol{g}+\boldsymbol{f}_{st}\\
&p^{n+1}=p^{n}-\lambda\nabla\cdot\boldsymbol{u}^{n+1}
\end{aligned}
\tag{5.27}
$$

until $\|\nabla\cdot\boldsymbol{u}^{n+1}\|<\epsilon_1$ is satisfied in the fluid zones and at the same time $\|\nabla\cdot\left[\kappa\overline{\overline{\Gamma}}(\boldsymbol{u}^{n+1})+\zeta\overline{\overline{\Theta}}(\boldsymbol{u}^{n+1})-\eta\overline{\overline{\Omega}}(\boldsymbol{u}^{n+1})\right]\|<\epsilon_2$ in the solid zones. As a definition, $\epsilon_1$ and $\epsilon_2$ are small thresholds controlling the minimization of the constraints. Equations (5.27) are presented in semi-discrete form with $n$ the time index corresponding to time $n\Delta t$ and $\Delta t$ the time step.

In the fluid phases, a Standard Augmented Lagrangian (SAL) method of Fortin and Glowinski [52] is obtained by making $\lambda\rightarrow+\infty$ or other adaptive augmented Lagrangian techniques can also be used. An incompressible constraint is thus directly imposed during the solving of the momentum conservation thanks to term $\nabla\left[\lambda\left(\nabla\cdot\boldsymbol{u}^{n+1}\right)\right]$. Non-deformable property can be ensured in solid media by a similar penalty method, acting on the elongation, shearing and rotation viscosities in a kind of Implicit Tensorial Penalty Method (ITPM) that plays on the magnitude of viscosity and associated viscous constraints. It is assumed that $\kappa$, as well as $\zeta$ and $\eta$, tends to infinity. The main difficulty of the penalty numerical method is to choose the numerical value of $\lambda$, $\kappa$, $\zeta$ and $\eta$. These numerical parameters must be several orders of magnitude larger than the larger term in the Navier-Stokes equations. However, they must keep a sufficiently small value in order to maintain an as good as possible conditioning of the linear system.

In SAL approach, $\lambda$ is chosen constant in the range $1<\lambda<10{,}000$. With such a method, multi-phase free surface or interfacial flows have been successfully simulated [193] and [53] with Reynolds number less than 10 and large interfacial structures compared to the size of the grid as well as the one of the computational domain. Concerning particulate flows, penalty methods called Standard Solid Penalty (SSP) approach, which uses constant $\mu$ between 1000 and 100,000 in the solid zones, allows to accurately solve Stokes flow interacting with moving particles [109]. However, as soon as the unsteadiness of the flow increases as well as the complexity of the fluid-solid interactions, numerical failures are noticed in the flow solutions provided by SAL and SSP approaches [54]. In the previous Sect. 5.1, automatic criteria have been proposed for detecting the magnitude of the penalty coefficients $\lambda$ (and equivalently $\kappa$, $\zeta$ and $\eta$) according to the physical characteristics of the problem.

## 5.3 Remarks on Time Splitting Approaches

The class of time splitting methods concern the use of Poisson equation solving to ensure the pressure-velocity coupling and divergence free conditions according to the mass and momentum conservation equations. Pioneer works on this topics are [48, 49, 185]. Instead of trying to solve at the same time and in a coupled way the pressure and velocity, as in augmented Lagrangian approaches (see Sect. 5.1), the time splitting methods, or projection methods, uses a decomposition of pressure and velocity to solve the problem (2.1 and 2.2) in two steps. For example, the Goda method [185] reads

$$\boldsymbol{u}^{n+1} = \boldsymbol{u}^* + \boldsymbol{u}' \tag{5.28}$$

$$p^{n+1} = p^n + p^* \tag{5.29}$$

$$\rho\left(\frac{\boldsymbol{u}^* - \boldsymbol{u}^n}{\Delta t} + \boldsymbol{u}^n \cdot \nabla \boldsymbol{u}^*\right) = -\nabla p^n + \nabla \cdot \mu\left(\nabla \boldsymbol{u}^* + \nabla^t \boldsymbol{u}^*\right) + \rho \boldsymbol{g} + \boldsymbol{f}_{st} \tag{5.30}$$

$$\nabla \cdot \left(\frac{\Delta t}{\rho} \nabla p^*\right) = \nabla \cdot \boldsymbol{u}^* \tag{5.31}$$

$$\boldsymbol{u}' = -\frac{\Delta t}{\rho} \nabla p^* \tag{5.32}$$

with $n$ the time index, Details on projection methods will be given in the course of Jean-Luc Estivalezes. As was previously demonstrated by Sarthou [194], the integration of penalty terms in equation (5.30) requires the modification of equation (5.31) in order to keep the penalty values unchanged after the projection of the predicted velocity field $\boldsymbol{u}^*$. For example, if a VPM is considered we have the following new form for equations (5.30 and 5.31):

$$\rho\left(\frac{\boldsymbol{u}^* - \boldsymbol{u}^n}{\Delta t} + \boldsymbol{u}^n \cdot \nabla \boldsymbol{u}^*\right) = -\nabla p^n + \nabla \cdot \mu\left(\nabla \boldsymbol{u}^* + \nabla^t \boldsymbol{u}^*\right) + \rho \boldsymbol{g}$$
$$+ \boldsymbol{f}_{st} + \beta_u(\boldsymbol{u}^* - \boldsymbol{u}_\infty) \tag{5.33}$$

$$\nabla \cdot \left(\frac{\Delta t}{\rho(1+\beta_u)} \nabla p^*\right) = \nabla \cdot \boldsymbol{u}^* \tag{5.34}$$

Similar modifications are required whatever the penalty method used (VPM, SPM or ITPM).

When multi-phase flows and one-fluid model are investigated with time splitting methods, the density jump at the interface make the solving of the projection equation difficult as soon as the density jump increases. Indeed, when iterative solvers are used to tackle with the Poisson projection equation, the conditioning of the linear system is even worse than the density ratio is large. To circumvent this drawback, other king of projection methods have been proposed. Among then, we can cite the vectorial projection method [186] and the kinematics scalar projection approach [195]. In these last time splitting methods, the density is not present in the projection step (this is achieved by considering the projection as a part of the Hodge-Helmholtz decomposition of the velocity unknown). This way, the numerical complexity induced by the density ratio is avoided. In these methods, the "physical pressure" is different from the potential of the velocity field. The pressure, which is not required to solve the time splitting formulation of the one-fluid model, is built by integrating the gradient of the potential of the velocity along the edges of the computational grid [195].

## 5.4 Validation of Penalty Techniques

Many validations have been considered concerning penalty methods. Among them, we can cite the treatment of incompressibility with augmented Lagrangian methods [53, 54, 192], the management of obstacles [56, 190, 194, 196], the treatment of boundary conditions [54, 197] and particulate flows [23, 192].

**Augmented Lagrangian methods for incompressibility**

- 3D settling of a particle in a rectangular tank [56]

The sedimentation of a spherical particle in a square section tank is considered. This test case corresponds to the experiments of Mordant and Pinton [198] where a rigid steel sphere of diameter $8 \times 10^{-4}$ m and density 7710 kg m$^{-3}$ is initially released without initial velocity. The particle is released into a liquid whose density and viscosity are 1000 kg m$^{-3}$ and $8.9 \times 10^{-4}$ Pa s respectively. The calculation domain is considered unbounded as the radial containment is less than $10^{-3}$. The Reynolds number, based on the particle diameter and its settling velocity, is 280 and the corresponding Stokes number is 240.

In this section, several simulations are leaded on a $50 \times 50 \times 800$ grid associated to dimensions 0.004 m $\times$ 0.004 m $\times$ 0.064 m in Cartesian coordinates. The numerical domain is smaller than the experimental one. However, the magnitude of the Reynolds and Stokes numbers allows a small effect of choosing a numerical containment of 0.2 artificially higher than the experimental one to be evaluated.

The vertical component of the particle velocity is first investigated with the SAL, 2AL and 3AL methods (see Fig. 5.3). It is observed that the 2AL and 3AL approaches accurately predict the experimental measurements. On the contrary, the SAL method involves strong discrepancies with the measurements [198]. As it has been demonstrated in the work of Vincent et al. [56], a constant $r(t, M)$ and the resulting augmented Lagrangian penalty term alter dramatically the conditioning of the linear system in one of the fluids. The resulting simulations are not physically pertinent. Examples of velocity magnitudes and vectors, obtained with the SAL and 3AL methods at $t = 0.15$ s, are shown in Fig. 5.4 in a vertical slice intersecting the particle center. The flow behind and in the rear of the particle is different. The toroidal velocity field, which is experimentally observed near the particle [198] and with the 3AL approach, is not recovered with the SAL method. The magnitude of the resulting velocities is 50% underestimated with the SAL method and the flow is not relevant

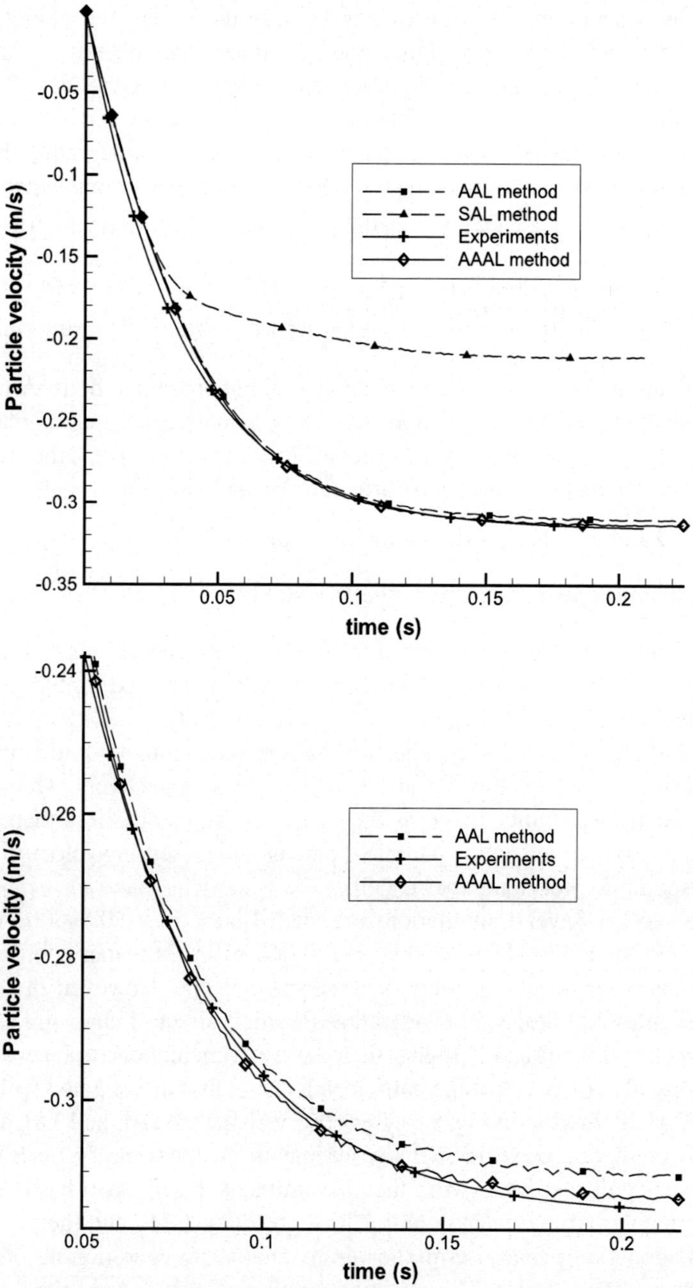


**Fig. 5.3** Simulation of the settling of a spherical particle at $Re = 280$—Evolution of the particle velocity during time—Comparison between SAL, 2AL, 3AL and experimental results (top) and zoom for the same comparison between $t = 0.05$ s and $t = 0.025$ s (bottom)

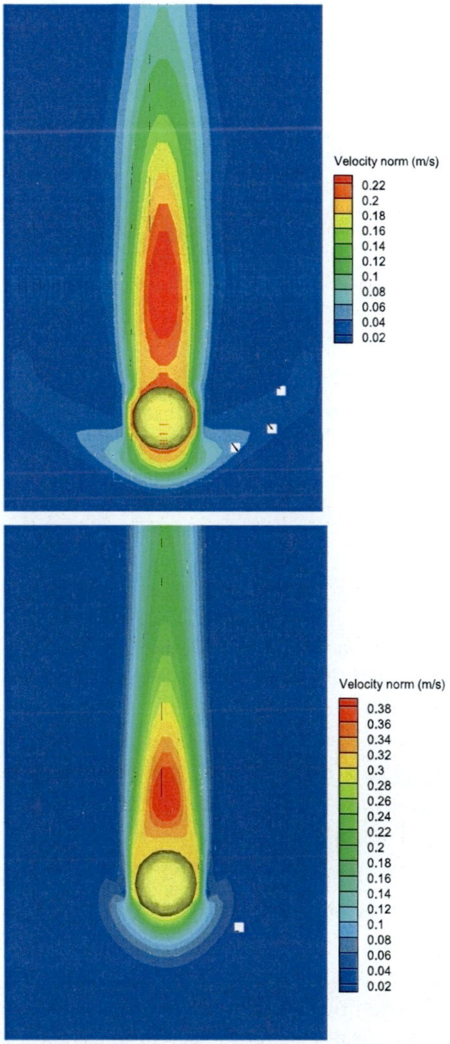


**Fig. 5.4** Simulation of the settling of a spherical particle at $Re = 280$—Isolines of vertical velocity norms in a vertical slice cutting the particle barycenter—Comparison between SAL (top) and 3AL (bottom) results

in this case. The numerical overcost induced by the 3AL method is less than 1% of the time required to solve the Navier-Stokes equations and the extra memory requirements are negligible.

- Free fall of a dense cylinder [56]

The purpose of this case is to demonstrate the ability of the 3AL method to deal with extreme cases and to compare this method with a classic time-splitting approach [185]. A cylindrical object of diameter $d = 0.0125$ m, centered at $(0.05, 0.175)$, is initially dropped in a rectangular cavity full of air whose dimensions are $[-0.05\,\text{m}; 0.05\,\text{m}] \times [-0.1\,\text{m}; 0.1\,\text{m}]$. The physical properties of air and solid media are the following: $\rho_{air} = 1.1768$ kg m$^{-3}$, $\rho_{solid} = 10^5$ kg m$^{-3}$, $\mu_{air} = 1.85 \times 10^{-5}$ Pa s,

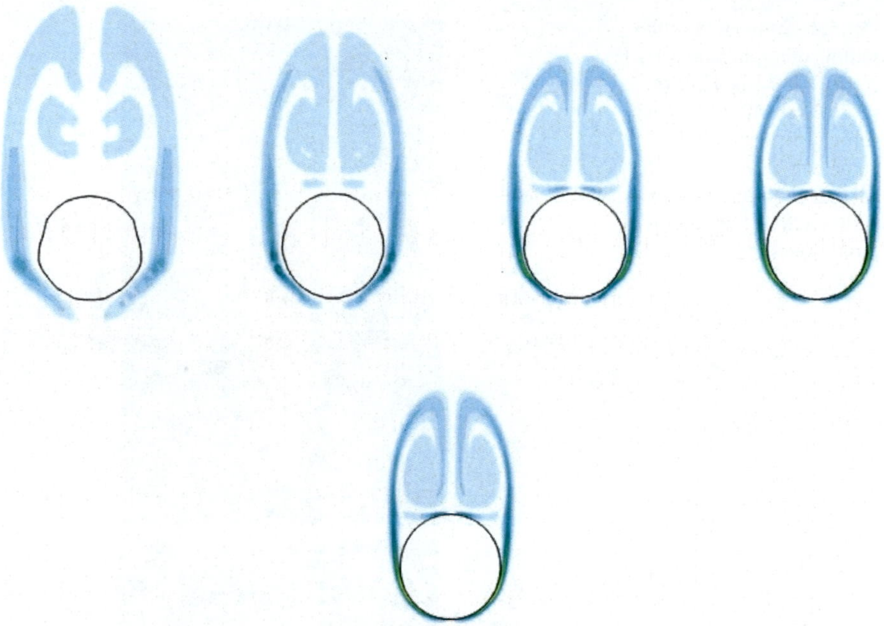

**Fig. 5.5** Vorticity magnitude and iso $C = 0.5$ for the free fall of a dense cylinder with $50 \times 100$, $100 \times 200$, $200 \times 400$, $400 \times 800$ and $800 \times 1600$ meshes

$\mu_{solid} = 10^5$ Pa s. The interest of this case comes from consideration of high density and viscosity ratio. If it is assumed that the air has no effect on the fall of the solid, an analytical solution of the solid velocity $V$ can be easily found: $V = -gt$.

The behavior of the 3AL method is first investigated. The case is simulated on 5 meshes ($50 \times 100$, $100 \times 200$ $200 \times 400$, $400 \times 800$ and $800 \times 1600$ points). The time step is constant and equal to $6.25 \times 10^{-5}$ s. The phase function $C$, which locates the dense cylinder, is advected with a VOF-PLIC method [199] and can induce deformations of cylinder shape. Another interest of this test case is to evaluate the ability of the 3AL method, coupled with an Eulerian interface tracking method and ITPM method, to handle the motion of a rigid object. Figure 5.5 shows, after 2500 time iterations, the isosurface $C = 0.5$ describing the shape of the cylinder and the vorticity magnitude. The space convergence is observed while the cylindrical aspect of the solid is very well kept since the $100 \times 200$ mesh. On a quantitative point of view, the asymptotic velocity of the cylinder after 2500 iterations is estimated thanks to the Richardson extrapolation [200]. A limit value of $-1.52819$ m s$^{-1}$ is obtained and used as reference value. Compared to the analytical inviscid solution equal to 1.5328125 m s$^{-1}$ for $t = 0.15625$ s, the asymptotic numerical solution is in good agreement, even if the two solutions cannot be directly compared as the air medium is taken into account in our simulations. Figure 5.6 shows the relative error of the velocity. A regular convergence order of 0.8 is obtained. The second-order of the 3AL method is overridden by the error of the VOF-PLIC method and the estimate of the fall velocity by the discrete integral of the vertical velocities such

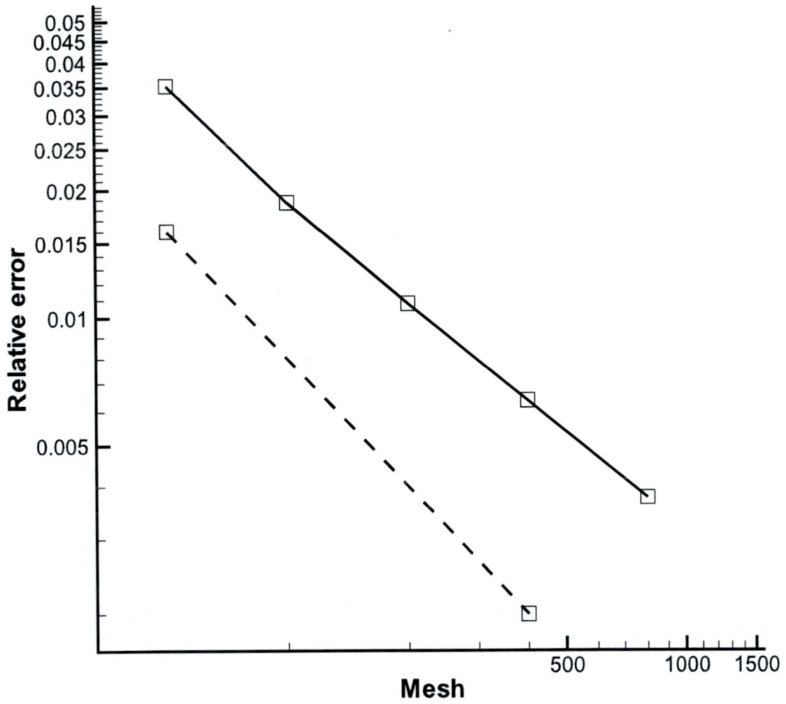


**Fig. 5.6** Convergence of the relative error on the fall velocity for the 3AL method—straight line (numerical results) and dashed line (order 1)

that $C \geq 0.5$. Concerning the time convergence order, a $200 \times 400$ grid and seven time steps, i.e. from $4 \times 10^{-4}$ s to $0.625 \times 10^{-5}$ s by successive divisions by 2, have been considered. The solutions of the velocity are taken at $t = 0.78s$. A Richardson extrapolation have provided an asymptotic value of the velocity of $-0.766807$ m s$^{-1}$. The evolution of the relative error according to the time step is described in table 5.1. The convergence order is irregular, with a mean value of 1.21. The simulation on a $200 \times 400$ grid has been also performed with a time-splitting approach [185]. It has been observed that except for the smaller time step, i.e. $0.625 \times 10^{-5}$ s, the simulations failed after several hundreds of time iterations due to the large divergence level of the prediction step. This step was discretized implicitly as explained in chapter 2. The correction step was not able to recover a correct divergence level after a certain time. For the smallest time step, the fall velocity admits a value of $-0.7704$ m s$^{-1}$ with the projection approach, which is close to the value of $-0.7668$ m s$^{-1}$ obtained with the 3AL method. The main interest of the augmented Lagrangian techniques is to allow the use of much larger time steps than with the time-splitting approaches, even for difficult cases such as the present case.

**Table 5.1** Relative errors and corresponding convergence orders on the fall velocities of a dense cylinder for various time steps—values obtained with the 3AL method

| Time (s) | Relative error | Order |
|---|---|---|
| $4 \times 10^{-4}$ | $6.35 \times 10^{-4}$ | |
| $2 \times 10^{-4}$ | $2.20 \times 10^{-4}$ | 1.53 |
| $1 \times 10^{-4}$ | $1.38 \times 10^{-4}$ | 0.67 |
| $0.5 \times 10^{-4}$ | $4.34 \times 10^{-5}$ | 1.67 |
| $2.50 \times 10^{-5}$ | $3.48 \times 10^{-5}$ | 0.32 |
| $1.25 \times 10^{-5}$ | $1.0 \times 10^{-5}$ | 1.66 |
| $6.25 \times 10^{-6}$ | $2.6 \times 10^{-6}$ | 1.89 |

## VPM and SPM for obstacles

- Couette flow [194]

We consider a Couette flow between two cylinders of radius $R_1 = 0.5$ m and $R_2 = 3$ m. Their angular velocities are $\omega_1 = 0$ rad s$^{-1}$ and $\omega_2 = 2$ rad s$^{-1}$. The analytical solution for velocity is

$$u_\theta(r) = \frac{\omega_2 R_2^2 - \omega_1 R_1^2}{R_2^2 - R_1^2} r + \frac{(\omega_1 - \omega_2) R_2^2 R_1^2}{R_2^2 - R_1^2} \frac{1}{r} \tag{5.35}$$

with $r$ the radial coordinate and $\theta$ the angular coordinate in a polar coordinate system. The one-fluid equations are solved in a domain $\Omega = [0.15, 0.15] \times [0.15, 0.15]$. The analytical solution is imposed on the boundaries of $\Omega$. A penalty method is used to impose a Dirichlet boundary conditions on the inner circle. The solutions for the SMP and VPM method are compared in Fig. 5.7 with the same case simulated without immersed boundary method in polar coordinates. As expected, the SMP method reaches a second order convergence in space while the VPM approach is of first order. The simulation in polar coordinates reaches also a second order, in agreement with the order of the discretization schemes. This later method is 20 times more accurate than the Cartesian case with the SMP method. Indeed, the polar mesh is body-fitted in this configuration and the mesh lines are collinear with the flow.

- Flow past a cylinder (Re = 100) [57]

The main goal of the present study is consequently to compute the drag coefficient $C_D$ and visualize the laminar and turbulent unsteady flow field for different Reynolds numbers. The drag coefficient defined as a global average of a normalized drag force on the cylinder from the flow

$$C_D = \frac{D}{0.5 \rho U_\infty A}$$

where $U_\infty$ is the inflow velocity and $A$ is the projected area of the cylinder.

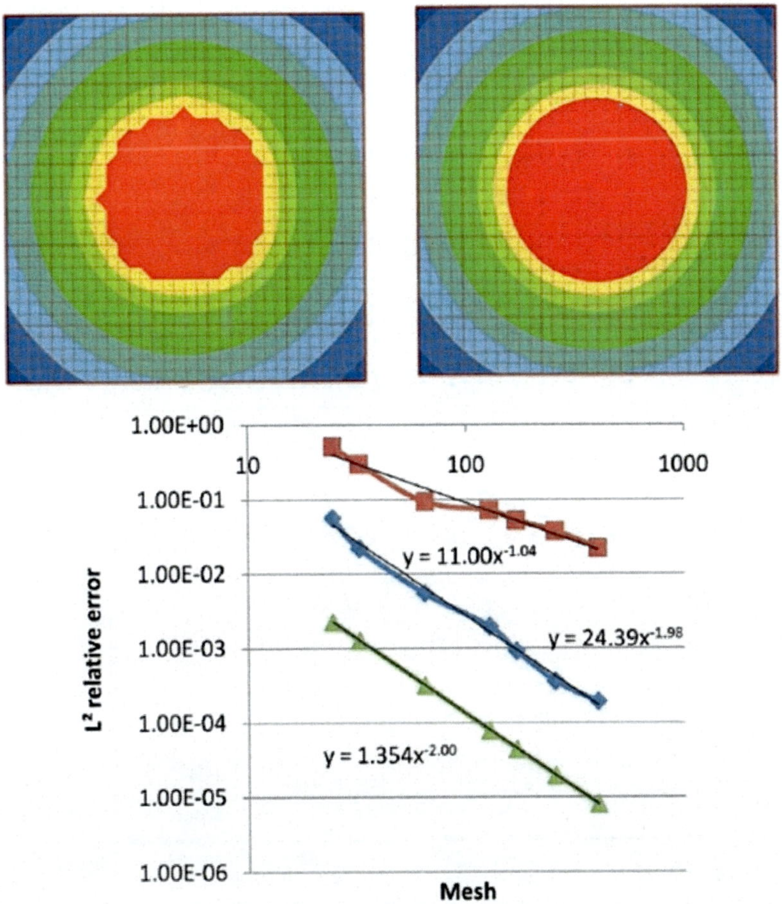


**Fig. 5.7** Simulation of Couette flow with penalty methods—Streamlines for VPM (top left) and SMP (top, right) are presented on a 32 × 32 mesh—Bottom picture provides a comparison of space convergence $L_2$ errors for VPM (red line), SMP (blue line) and curvilinear body-fitted (green line) methods

Computations are presented at a Reynolds number based on the free-stream velocity and cylinder diameter $D$ of 0.1. The simulation is performed on a 2$D$ grid. The computational domain equals $[0, 10] \times [-4.05, 4.05]$ which is discretized using a $572 \times 296$ grid. Uniform free stream condition with velocity $U_\infty = 1.0$ m/s are applied at the inlet boundary. Figure 5.8 shows the lift and drag coefficient for $Re = 100$.

The drag and lift coefficients as well as the Strouhal number $S_t$ are compared with experimental and other numerical results in Table 5.2 with success.

- Turbulent flow in a channel at $Re_{u_\tau} = 395$ [57]

The flow field simulated is a fully developed two-dimensional turbulent flow in a channel. The flow is statistically homogeneous both in the streamwise and span-

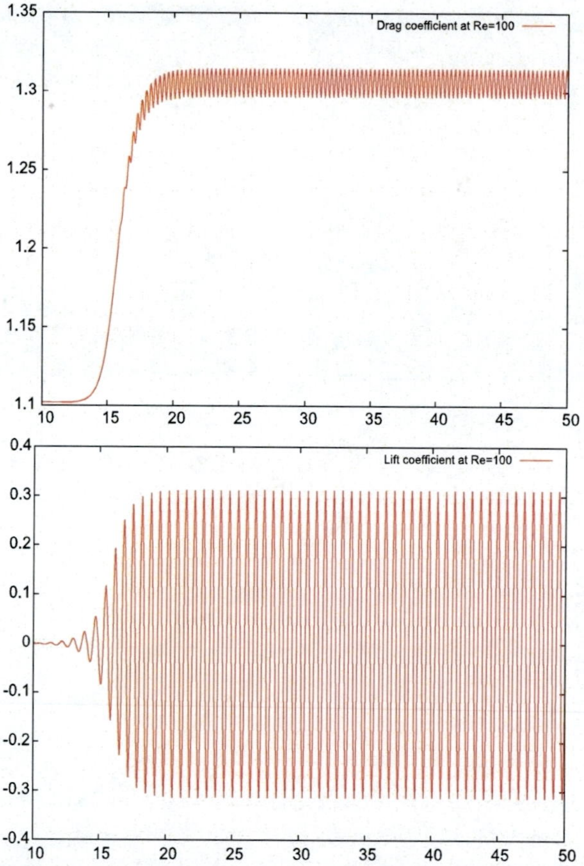


**Fig. 5.8** Time History of **a** Lift coefficient and **b** Drag coefficient at Re = 100

**Table 5.2** Comparison with other simulations and experiments

| | | $C_D$ | $C_L$ | $S_t$ |
|---|---|---|---|---|
| $Re = 100$ | Experiment [201] | 1.24-1.26 | – | 0.164–0.165 |
| | Kim et al. [202] | 1.33 | $\pm 0.32$ | 0.165 |
| | Le et al. [203] | $1.37 \pm 0.009$ | $\pm 0.323$ | 0.16 |
| | present | 1.306 | $\pm 0.314$ | 0.1617 |

wise directions and the statistics depend only upon the distance from the wall. The simulations of channel flows were performed at first time using the eddy detached simulation (DES) on fine grids, then the near-wall treatments were tested on a coarse grid. The simulation results were compared to the direct numerical simulation (DNS) data of Moser et al. [204] obtained for $Re_{u_\tau} = 395$ based on the friction velocity $u_\tau$ and the channel half-width.

Results for the channel simulation using (DES) model are presented in Fig. 5.9 and are compared to the Moser et al. [204] simulation. These results are obtained with $100 \times 130$ grid points. Also, the eddy viscosity of the Detached Eddy Simulation

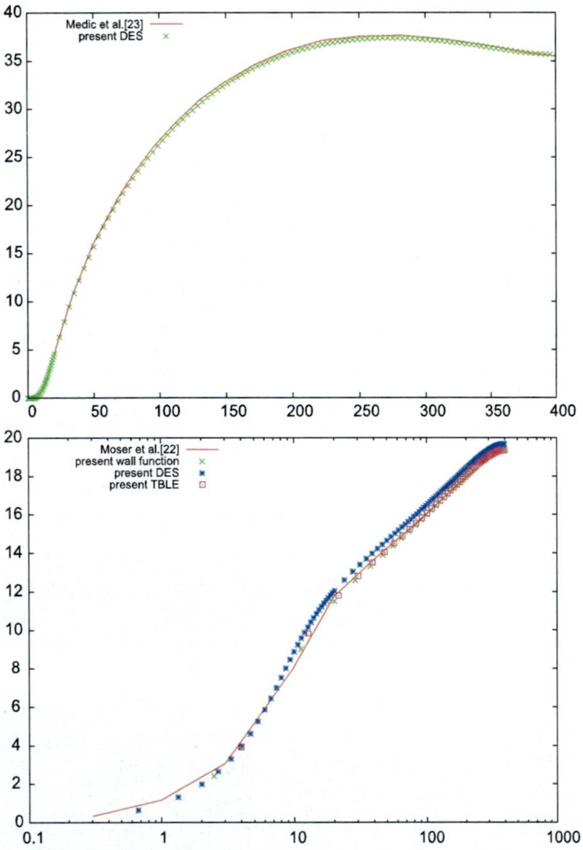


**Fig. 5.9** Simulation of the turbulent channel flow—top: Comparison between the eddy viscosity profile at $Re_\tau = 395$—bottom: Comparison between the mean velocity profile in wall units at $Re_\tau = 395$

(DES) model is compared with Medic et al. [205] RANS/LES simulation. In this case, no immersed boundary (penalized zone) is considered. When the walls are immersed solid boundaries in the fluid simulation domain, a coarse mesh $100 \times 50$ was used. On one hand, wall functions were imposed by VPM in fluid cells cut by the interface. On the same Eulerian mesh, the enhanced Turbulent Boundary Layer Equation (TBLE) approach was also used. In this case, a $1D$ skin mesh, with 27 points distributed in the viscous sub-layer, was generated in the fluid part, in the normal direction to the fluid/solid interface. The TBLE solution in the vicinity of the wall is coupled to the Navier-Stokes solution far from the solid wall by VPM. All approaches allow to recover the DNS results in a nice way. The important point here is that penalty methods, coupled to wall law treatments or skin mesh and TBLE model coupled to Navier-Stokes solving, allow to deal with turbulent flows and immersed interfaces.

**ITPM method for particulate flows**

- Settling of a cylindrical shape particle [23]

The settling of a cylindrical shape particle is first considered in two dimensions in order to characterize the effect of numerical parameters such as the solver, the

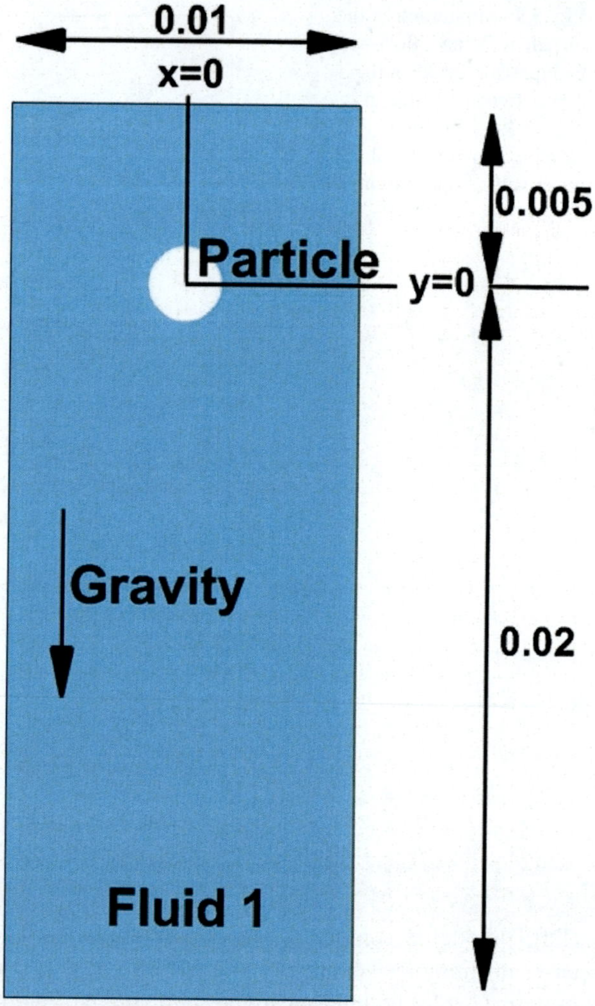


**Fig. 5.10** Definition sketch of the settling of a cylindrical shape particle

augmented Lagrangian parameter or the magnitude of the viscous penalty viscosity on the accuracy of the ITPM. The boundaries of the numerical domain are walls with a no slip condition. A convergence study is also provided. The definition sketch of the problem is illustrated in Fig. 5.10. At low Reynolds number, the analytical developments of Faxen [206] can be used to estimate the resistance force $F_r$ of the fluid on the cylindrical shape particle and its final settling velocity $U_\infty$. This reference velocity will be used to discriminate the most efficient numerical parameters of the ITPM and also to estimate the convergence order of the method. The resistance force $F_r$ reads

$$F_r = \frac{4\pi \mu_f U_\infty}{ln(1/a_r) - 0.9157 + 1.7244a_r^2 - 1.7302a_r^4} \tag{5.36}$$

with $a_r$ the aspect ratio between the particle diameter and the width of the fluid container. Index $f$ denotes the fluid properties. The fluid resistance to the particle is compensated by buoyancy forces when the particle has reached its settling velocity $U_\infty$. In this way,

$$F_r = \left(\rho_f - \rho_p)\right) \pi R_p^2 \, g \tag{5.37}$$

where $g$ is the magnitude of gravity, $R_p$ the particle radius and index $p$ refers to the particle properties. Finally, the particle settling velocity can be obtained as

$$U_\infty = \frac{\left(\rho_f - \rho_p)\right) R_p^2 \, g \left[ln(1/a_r) - 0.9157 + 1.7244a_r^2 - 1.7302a_r^4\right]}{4\mu_f} \tag{5.38}$$

The characteristics of the problem are the following concerning the fluid and particle properties

$$\rho_p = 200 \text{ kg m}^{-3} \tag{5.39}$$

$$50 \leq \mu_p \leq 50{,}000 \text{ Pa s} \tag{5.40}$$

$$\rho_f = 100 \text{ kg m}^{-3} \tag{5.41}$$

$$\mu_f = 5 \text{ Pa s} \tag{5.42}$$

whereas $R_p = 0.001$ m, $a_r = 0.2$ and $g = 9.81$ m s$^{-2}$. The magnitude of the analytical particle settling velocity is $U_\infty = 0.37275 \times 10^{-4}$ m s$^{-1}$. The corresponding particle Reynolds number is $Re_p = \dfrac{2\rho_p R_p U_\infty}{\mu_f} = 1.5 \times 10^{-6}$, which is in agreement with the Stokes approximation used by Faxen.

Five grids are used in the numerical simulations with a constant space step and the viscous law $NVL3$. They are referred to as $G_1$, $G_2$, $G_3$, $G_4$ and $G_5$ and are associated respectively to $50 \times 125$, $100 \times 250$, $200 \times 500$, $400 \times 1000$ and $800 \times 2000$ grid cells. Three time steps are also considered, namely $\Delta t_1 = 10s$, $\Delta t_2 = 1s$ and $\Delta t_3 = 0.1s$ and 25 virtual particles per direction for shape projection are used.

The effect of the solver used to obtain the solution of the momentum conservation equations approximated by the adaptive augmented Lagrangian method is now considered with grid $G_4$, time step $\Delta t_2$ and penalty viscosity $\mu_p = 5000$ Pa s. The PARDISO direct solver [207] and the iterative BiCGSTAB II solver [208] are first compared, with various values of $\mathcal{K}$ used in equation (5.22). It is well known that the 3AL approach [56] is part of penalty methods and that the higher $\mathcal{K}$ is, the higher the conditioning of the discretization matrix is. The direct solver provides reference values as it is not sensitive to the conditioning of the matrix, contrary to the iterative solver. The first parametric study is investigated on the effect of the viscosity ratio across the interface and also on $\mathcal{K}$ on the efficiency of the iterative solver, as presented in Fig. 5.11. In this case, the number of convergence iterations at each time step is imposed to be 100, to maintain a computational time comparable to the one induced by the direct solver. It is observed that for large values of viscosity ratios and $\mathcal{K}$, the

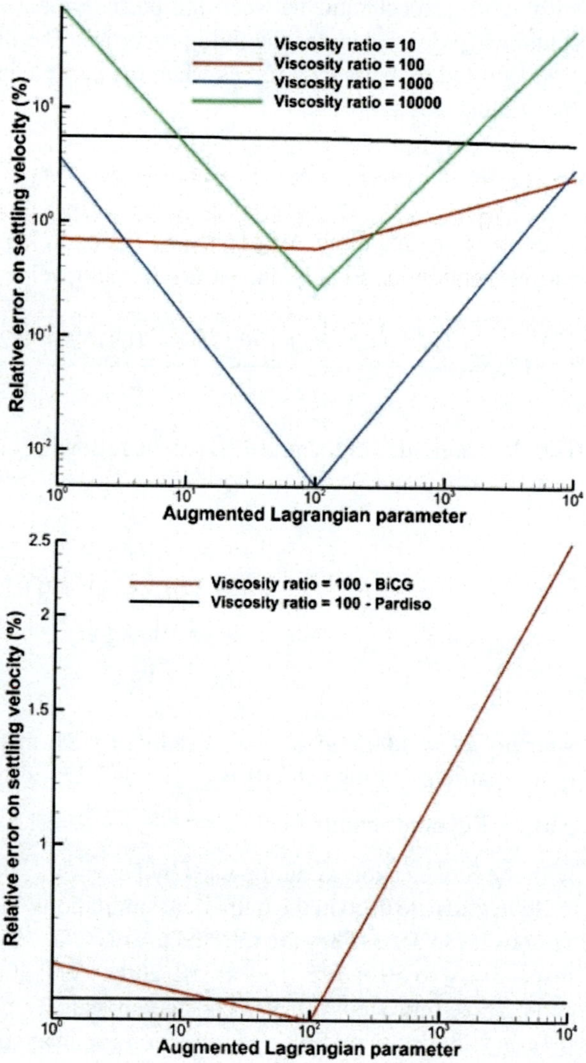


**Fig. 5.11** Comparison of relative errors on settling velocity: left, results obtained with the iterative solver for various viscosity ratios and augmented Lagrangian parameter $\mathcal{K}$ and right, comparison of PARDISO and BiCGSTAB II for a viscosity ratio of 100

iterative solver is not able to provide a correct settling velocity in 100 iterations, as the conditioning is too important in these cases. The best solution is obtained for a viscosity ratio of 100 and a augmented Lagrangian parameter $\mathcal{K} = 100$. For small values of $\mathcal{K}$, the relative error on $U_\infty$ is larger as the divergence free constraint is not maintained in this case. For $\mathcal{K} = 100$ and a viscosity ratio of 100, the error brought by the two solvers are comparable. These values will be used in the following sections to simulate in particular three dimensional flows.

The second interesting and crucial parameter which is studied is the viscosity ratio at the interface. A PARDISO solver is used on the $G_4$ grid with $\Delta t_2$ and $\mathcal{K} = 10^8$ in this case, in order to reach an almost computer error divergence accuracy at

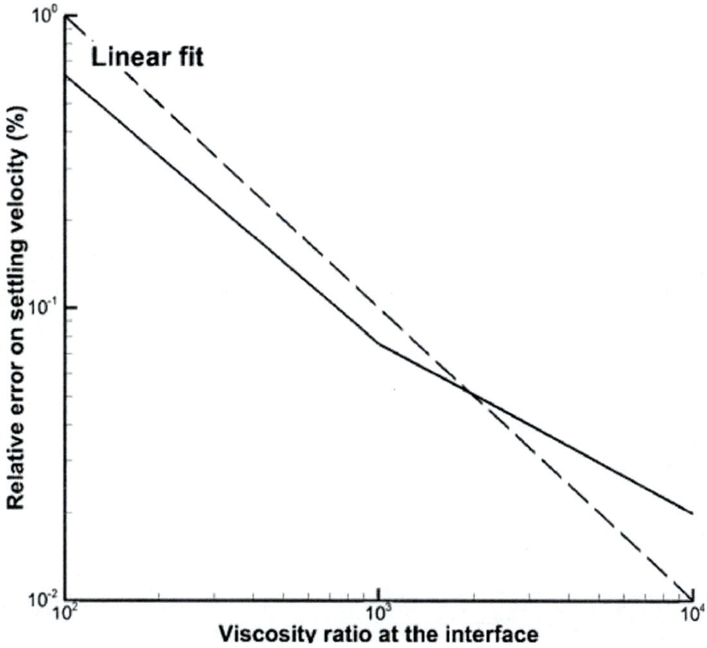


**Fig. 5.12** Sensitivity of the relative error on $U_\infty$ obtained with the PARDISO solver for various viscosity ratios

each calculation time. The relative errors are plotted in Fig. 5.12. As expected, it is observed that these errors decrease linearly for increasing viscosity ratios at the interface. With the selected numerical parameters, the error is reduced from almost 0.6 to almost 0.02 percents by multiplying by 100 the viscosity ratio.

To finish with investigations of numerical behavior of ITPM, the performances of the global method (Navier-Stokes solving, particle tracking, ...) are evaluated on a parallel computer. The following numerical parameters are chosen, the same as those that will be used in the validations and applications presented later in the present article: grid $G_5$, time step $\Delta t_2$, iterative solver, viscosity ratio of 100 and $\mathcal{K} = 10^2$. In Fig. 5.13 are presented the results obtained for simulations driven on 8 to 64 processors during one hour of CPU time. The theoretical speed-up curve is provided in long dashed line, corresponding to a multiplication of the reference number of time iterations on 8 processors by the number of used processors divided by 8. It is observed that the numerically obtained speed-up is 1 until 32 processors, and remains equal to 0.94 for 64. This last value corresponds to parallel blocs of 25000 Eulerian grid cells in this case, which appears to be the limit of bloc size to maintain a good parallel acceleration of computational time. Hence, the ITPM is nicely implemented on parallel architectures.

A space convergence study is carried out for estimating if a high order (second order can be expected) convergence rate of the ITPM is reached and if it is consistent with the second order of the centered schemes used to discretize the fluxes in the

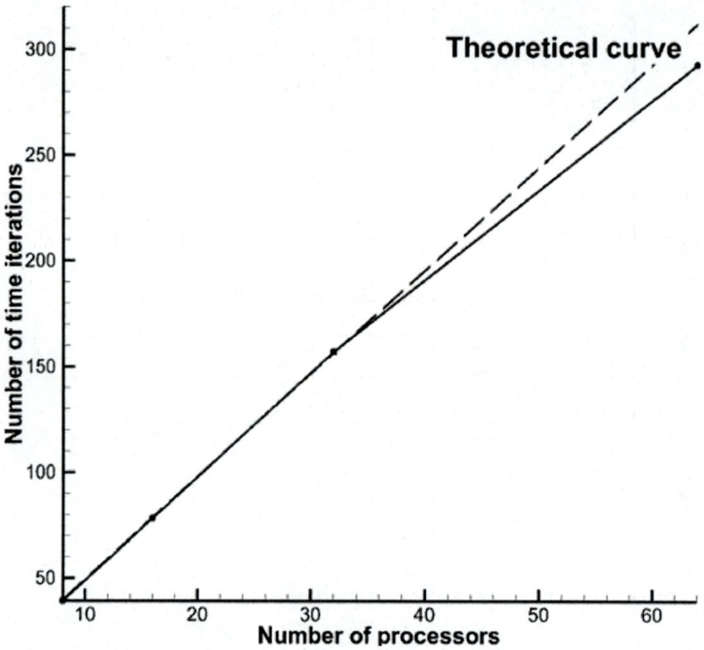


**Fig. 5.13** Evolution of the time iterations calculated in one hour of CPU time according to the number of processors

finite volume method. The following numerical parameters are considered for the simulations: a PARDISO solver is utilized with $\Delta t_2$ and $\mathcal{K} = 10^8$. The effect of the viscosity ratio is estimated in the range 100 to 10,000 whereas the grids $G_1$ to $G_5$ are also considered. For a given viscosity ratio, the space convergence order $\alpha_s$ is calculated by means of the Richardson extrapolation method on four solutions as follows [209]:

$$\alpha_s = \frac{ln\left(\left|\frac{|U_\infty|_{G_2} - |U_\infty|_{G_4}}{|U_\infty|_{G_3} - |U_\infty|_{G_5}}\right|\right)}{ln\,(2)} \tag{5.43}$$

where the constant 2 in relation (5.43) corresponds to the ratio between two consecutive numbers of grid points in each direction of space. The settling velocities of the particle for the various viscosity ratios and Eulerian grids are given in Table 5.3. It is observed that the finer the mesh is, the nearer is the numerical settling velocity to the analytical velocity $U_\infty = -3.72753 \times 10^{-5}$ m s$^{-1}$ from the Faxen theory. In addition, it can be concluded that increasing the viscosity ratio allows to fulfill with better accuracy the deformation free constraint in the solid particle. The resulting settling velocities are also nearer to the reference velocity for increasing viscosity ratios on a given Eulerian grid. Concerning the space convergence order, $\alpha_s$ slightly depends on the value of the viscosity ratio, ranging from 1.8 to 2.4 in the investigated

**Table 5.3** Settling velocity (in m s$^{-1}$) according to grid and viscosity ratio—the reference velocity is $-3.72753 \times 10^{-5}$ m s$^{-1}$

| Eulerian grid | Viscosity ratio = 100 | Viscosity ratio = 1000 | Viscosity ratio = 10,000 |
|---|---|---|---|
| $G_1$ | $-3.8189 \times 10^{-5}$ | $-3.7895 \times 10^{-5}$ | $-3.7860 \times 10^{-5}$ |
| $G_2$ | $-3.7699 \times 10^{-5}$ | $-3.7448 \times 10^{-5}$ | $-3.7420 \times 10^{-5}$ |
| $G_3$ | $-3.7552 \times 10^{-5}$ | $-3.7330 \times 10^{-5}$ | $-3.7306 \times 10^{-5}$ |
| $G_4$ | $-3.7510 \times 10^{-5}$ | $-3.7303 \times 10^{-5}$ | $-3.7282 \times 10^{-5}$ |
| $G_5$ | $-3.7497 \times 10^{-5}$ | $-3.7298 \times 10^{-5}$ | $-3.7278 \times 10^{-5}$ |
| Convergence order $\alpha_s$ | 1.8 | 2.2 | 2.3 |

**Table 5.4** Settling velocity according to time step - the reference velocity is $-3.72753 \times 10^{-5}\ m.s^{-1}$

| $\Delta t_1$ | $\Delta t_2$ | $\Delta t_3$ | Convergence order $\alpha_t$ |
|---|---|---|---|
| $-3.7615 \times 10^{-5}$ | $-3.7298 \times 10^{-5}$ | $-3.7297 \times 10^{-5}$ | 2.1 |

numerical conditions. It can be concluded that a second order convergence in space is obtained whatever the viscosity ratio, as expected initially when the ITPM was designed.

To finish with, it can be important for our unsteady applications to simulate particle flows with a high order in time numerical method. The time convergence order is so calculated on $G_5$ with the PARDISO solver, $\mathcal{K} = 10^8$ and a viscosity ratio of 1000. The finest mesh is used in order to reduce to the maximum the space truncation error due to the space discretizations. Three time steps are considered, namely $\Delta t_1$, $\Delta t_2$ and $\Delta t_3$. The time convergence order $\alpha_t$ is estimated with a Richardson extrapolation on three solutions which reads

$$\alpha_t = \frac{ln\left(\left|\frac{|U_\infty|_{\Delta t_1} - |U_\infty|_{\Delta t_2}}{|U_\infty|_{\Delta t_2} - |U_\infty|_{\Delta t_3}}\right|\right)}{ln\,(10)} \tag{5.44}$$

The results are presented in Table 5.4. A second order convergence in time is obtained as for the space convergence order.

- Settling of a spherical particle [23]

The numerical simulation of the settling of a spherical particle for various Reynolds numbers is considered under the experimental conditions of [210]. Compared to the Faxen test case where $Re << 1$, the interest of the present configuration is to investigate a particle flow in which inertial effects exist and also to compare numerical results to measurements. This problem has already been simulated by other authors with a first order penalty method [211], a Boltzmann approach [210, 211], a fractional step fully resolved simulation technique [212] or an immersed boundary method

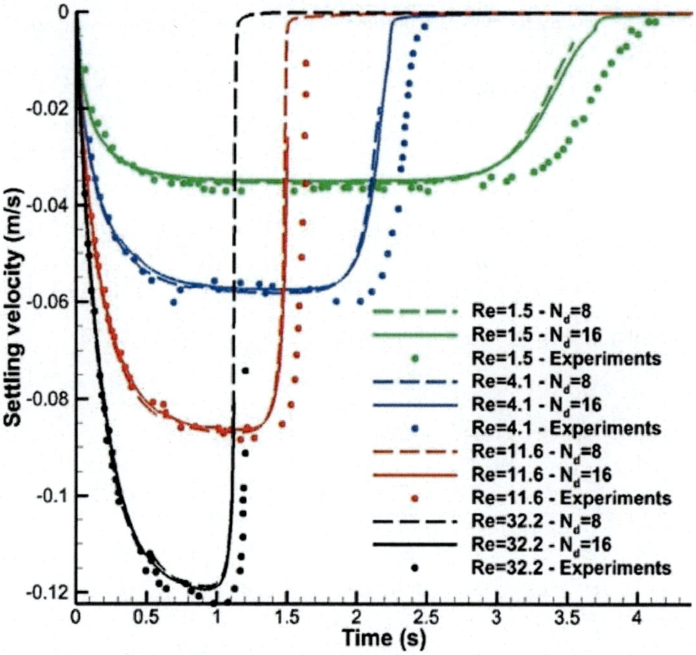


**Fig. 5.14** Settling velocity over time for the test case of [210]—Long dashed line is associated to $N_d = 8$, straight line to $N_d = 16$ and symbols to experiments [210]. Green, blue, red and black colors are related respectively to $Re = 1.5$, $Re = 4.1$, $Re = 11.6$ and $Re = 32.2$

[213]. The numerical parameters used in this section are a penalty viscosity 100 times larger than the fluid viscosity, 25 virtual particles per direction, viscous law $NVL3$, $\mathcal{K} = 100$ and an iterative solver.

The definition sketch of the particle sedimentation have been presented into details in [210]. A single spherical particle is released without initial velocity in a parallelepiped box and it reaches a limit sedimentation velocity after a characteristic time. The Reynolds 1.5, 4.1, 11.6 and 32.2 are considered here. Results in terms of settling velocity over time are reported in Fig. 5.14 according to the number of Eulerian grid cells for a particle diameter $N_d = D_p/\Delta x$. The simulated velocities are compared to the measurements. It can be observed that the unsteady behavior of the particle velocity as well as the limit settling velocity are nicely recovered by the simulations for all investigated Reynolds numbers. After a certain time, the numerical velocity tends to zero when the particle reaches the bottom of the tank whereas this deceleration is measured later in the experiments. These differences have previously been obtained by other authors [211, 213] whereas these discrepancies are not reported in the works of [210, 212]. It seems that the initial position of the particle is not exactly the same in our simulations than in the reference experiments. However, the acceleration of the particle as well as the settling velocity are correct and weakly dependent on the number of Eulerian grid cell used to simulate the problem, which is an important feature of the ITPM approach as three-dimensional simulations of finite size particles are limited by the number of flow grid cells that can be investigated in real particle flow problems.

# Chapter 6
# Compressible (Low-Mach) Two-Phase Flows

The present model has been first presented in [214]. It is devoted to the prediction of macroscopic fields such as pressure, temperature and velocity, when different equations of state govern the fluids in interaction. Each fluid is governed by its own equation of state and is characterized by temperature and pressure dependent physical properties, such as density, viscosity, thermal conductivity, specific heat, dilatation and compression coefficients. The surface tension coefficient of the fluids is supposed to be known. The main difficulty to overcome in a multi-phase one-fluid approach is the dependance of the physical properties on the pressure and temperature, but also on the discontinuous phase function C. The fluid nature induces huge density variations ($10^4$ kg m$^{-3}$) while temperature variations for instance results in very weak changes in density ( $10^{-5}$ kg m$^{-3}$). These two scales of density fluctuation constrain the global model in a noticeably different way. Thus, novel models and numerical methods unifying these effects are at stake. The only unknowns of the incompressible model are the velocity and the temperature, since the pressure is a numerical parameter allowing the system to meet the incompressibility constrain $\nabla \cdot \mathbf{u} = 0$. The other quantities are the physical characteristics among which the viscosity $\mu$, the thermal conductivity $\lambda$, the specific heat $c_p$ and the density $\rho$ are very common. They are independent of the Navier-Stokes equations resolution, and can be calculated according to the fluid nature, state equations, or color function C in multi-phase flows. Extending a one-fluid model to compressible flows requires to build a new mass conservation equation to account of incompressibility in the liquid and dilatation in the gas or the plasma.



S. Vincent et al., *Small Scale Modeling and Simulation of Incompressible Turbulent Multi-Phase Flow*, CISM International Centre for Mechanical Sciences 607,
https://doi.org/10.1007/978-3-031-09265-7_6

## 6.1 Mass Conservation

Generally speaking, the flow is compressible ($\nabla \cdot \mathbf{u} \neq 0$) and a state equation must be applied to the fluids. For an interfacial flow, the mass conservation equation follows:

$$\frac{d}{dt}\iiint_D \rho \, dv = \iiint_D \left[\frac{\partial \rho}{\partial t} + \nabla \cdot (\rho \, \mathbf{u})\right] dv + \iint_\Sigma [[\rho \; (\mathbf{u} - \mathbf{w})]] \cdot \mathbf{n}_i \; ds \tag{6.1}$$

and

$$[[\rho \, (\mathbf{u} - \mathbf{w})]] \cdot \mathbf{n}_i = \left(\rho_{D_1} \left(\mathbf{u}_{D_1} - \mathbf{w}\right)\right) \cdot \mathbf{n}_i - \left(\rho_{D_2} \left(\mathbf{u}_{D_2} - \mathbf{w}\right)\right) \cdot \mathbf{n}_i \tag{6.2}$$

where $\mathbf{w}$ is the interface velocity, $\mathbf{n}_i$ the normal to the interface $\Sigma$ which separates the sub-domains $D_1$ et $D_2$. This can be written in local formulation:

$$\begin{cases} \dfrac{\partial \rho}{\partial t} + \nabla \cdot (\rho \, \mathbf{u}) = 0 \quad \text{in } D = D_1 \cup D_2/\Sigma \\ [[\rho \, (\mathbf{u} - \mathbf{w})]] \cdot \mathbf{n}_i = 0 \quad \text{on } \Sigma \end{cases} \tag{6.3}$$

If the relative interface velocity is $\mathbf{u}_r = (\mathbf{u} - \mathbf{w})$, the jump condition associated to the mass conservation provides:

$$[[\rho V_r]] = 0 \tag{6.4}$$

The mass flow rate $\rho \; V_r = m$ through a discontinuity surface $\Sigma$ is continuous. If the interface is supposed to be a contact surface, with no mass flux through the discontinuity surface $\Sigma$, i.e. no phase change occurs, the jump equation can be simplified:

$$[[\mathbf{u}]] \cdot \mathbf{n}_i = 0 \quad \text{on } \Sigma \tag{6.5}$$

which establishes the continuity of the normal component of the velocity. In our case, the tangent component is also supposed to be zero, which is verified experimentally on a macroscopic point of view. The mass conservation equation in $D$ can be reformulated as an equation for the pressure by developing the derivative of the pressure according to time in function of temperature and density:

$$\frac{dp}{dt} = \left.\frac{\partial p}{\partial T}\right|_\rho \frac{dT}{dt} + \left.\frac{\partial p}{\partial \rho}\right|_T \frac{d\rho}{dt}$$

$$\frac{dp}{dt} = \frac{\beta}{\chi_T}\frac{dT}{dt} + \left.\frac{\partial p}{\partial \rho}\right|_T \frac{d\rho}{dt}$$

Using the compressible mass conservation $\frac{d\rho}{dt} + \rho\nabla \cdot \mathbf{u} = 0$, we obtain:

$$\frac{dp}{dt} = \frac{\beta}{\chi_T}\frac{dT}{dt} + \left.\frac{\partial p}{\partial \rho}\right|_T (-\rho\nabla \cdot \mathbf{u})$$

and finally

$$\frac{\partial p}{\partial t} = \frac{\beta}{\chi_T}\frac{dT}{dt} - \frac{1}{\chi_T}\nabla \cdot \mathbf{u} - \mathbf{u} \cdot \nabla p \tag{6.6}$$

and after a semi-discretization in time:

$$p^{n+1} = p^n - \Delta t\left(\frac{1}{\chi_T}\nabla \cdot \mathbf{u}^{n+1} + \mathbf{u}^{n+1} \cdot \nabla p^n - \frac{\beta}{\chi_T}\frac{dT}{dt}^{n+1}\right) \tag{6.7}$$

where $\beta$ is the coefficient of thermal expansion and $\chi_T$ is the isothermal compressibility coefficient. The relative importance of the different terms in Eq. (6.6) is function of the physical properties and constrains. It is achieved through its non dimensional expression, involving the following non-dimensional variables:

$$p^* = \frac{p}{\rho\, V_0^2}; \quad \mathbf{u}^* = \frac{\mathbf{u}}{V_0}; \quad t^* = \frac{t\, V_0}{H}; \quad T^* = \frac{T}{\delta T} \tag{6.8}$$

where $V_0$ is a reference velocity, $\delta T$ a reference temperature, H a characteristic length

$$\frac{\partial p^*}{\partial t^*} = -\frac{1}{M^2}\nabla \cdot \mathbf{u}^* - \mathbf{u}^* \cdot \nabla p^* + \frac{\beta\,\delta T}{M^2}\frac{dT^*}{dt^*} \tag{6.9}$$

noticing that the sound velocity is $c = Cte\sqrt{1/\rho\,\chi_T}$. The Gay-Lussac number $G_a = \beta\delta T$ is introduced, this parameter quantifies the density differences of the working fluid under the temperature field. The non-dimensional expression becomes:

$$\frac{\partial p^*}{\partial t^*} = -\frac{1}{M^2}\nabla \cdot \mathbf{u}^* - \mathbf{u}^* \cdot \nabla p^* + \frac{Ga}{M^2}\frac{dT^*}{dt^*} \tag{6.10}$$

where M is the Mach number. In the dimension analysis (6.8), the time scale of the pressure fluctuations is supposed to be $H/V_0$. In order to capture the pressure waves, a time scale $\tau$ defined as $\frac{H}{c}$, where $c$ is the sound velocity, seems more appropriate. The dimensionless set of parameters is written again as:

$$p^* = \frac{p}{\rho\, V_0^2}; \quad \mathbf{u}^* = \frac{\mathbf{u}}{V_0}; \quad t^* = \frac{t\, c}{H}; \quad T^* = \frac{T}{\delta T} \tag{6.11}$$

and the non-dimensional equation for pressure becomes:

$$\frac{\partial p^*}{\partial t^*} = -\frac{1}{M}\nabla \cdot \mathbf{u}^* - M\,\mathbf{u}^* \cdot \nabla p^* + \frac{Ga}{M^2}\frac{dT^*}{dt^*} \tag{6.12}$$

Noticing that $\delta T/T \approx \delta p/p \approx M^2$ for weakly compressible flows, the last term of Eq. 6.12 is of first order $O(1)$. This can be shown by considering the following Eq. 6.13:

$$\frac{dT}{dt} = a\,\nabla^2 T \tag{6.13}$$

$$\frac{\partial p^*}{\partial t^*} = -\frac{1}{M}\nabla \cdot \mathbf{u}^* - M\,\mathbf{u}^* \cdot \nabla p^* + \frac{1}{M}\frac{1}{Pe} Ga \nabla^2 T^* \tag{6.14}$$

where a is the thermal diffusivity and $Pe = \dfrac{H V_0}{a}$ is the Peclet number. For low Mach number highly inertial flows, the relation (6.15) is generally satisfied:

$$M \to 0, \quad Ga << 1, \quad Pe >> 1 \tag{6.15}$$

Thus, the two last terms in Eq. (6.12) can be neglected in zero Mach number limit, including highly compressible flows, involving fluids with a high compressibility coefficient. The resulting equation for pressure is:

$$\frac{\partial p^*}{\partial t^*} = -\frac{1}{M}\nabla \cdot \mathbf{u}^* \tag{6.16}$$

which is written in real variables, after a semi-discretization in time:

$$p^{n+1} = p^n - \frac{\Delta t}{\chi_T}\nabla \cdot \mathbf{u} \tag{6.17}$$

where n is a time index such as $t = n\Delta t$ and $\Delta t$ is the time step. The density is updated as a function of the temperature and pressure:

$$\rho = f(p, T) \tag{6.18}$$

The multiphase compressible model is based on the formulation of the mass conservation equation which will be partially or completely integrated in the one-fluid momentum equation. This equation replaces $\nabla \cdot \mathbf{u} = 0$ in the one-fluid model even for incompressible flows, which will be related to high values of $\chi_T$.

## 6.2 Momentum Conservation

The goal is to interpret equation (6.6) in order to establish the relation between mass and momentum conservation equation in the Navier-Stokes equations. The Stokes equation (6.19), is deduced from the continuum approach assuming that the volume viscosity coefficient is equal to zero:

$$3\,\lambda + 2\,\mu = 0 \tag{6.19}$$

The compression viscosity coefficient $\lambda$ is unknown and the validity of this relation is limited to mono-atomic gases. For other fluids, especially for incompressible liquids, it is not demonstrated that this coefficient has a physical sense [187]. For a compressible and an incompressible flow as well, the pressure gradient in the Navier-Stokes equation can be associated to the term $\nabla \cdot (\lambda\, \nabla \cdot \mathbf{u} Id) = \nabla(\lambda\, \nabla \cdot \mathbf{u})$ as:

$$-\,\nabla\,(p - \lambda\,\nabla \cdot \mathbf{u}) \tag{6.20}$$

The incompressibility, usually formalized by the divergence free condition $\nabla \cdot \mathbf{u} = 0$, can be here associated to an infinite compression viscosity coefficient. The association of this coefficient to a viscosity effect is rather artificial. Indeed, even if its dimension is similar to a viscosity, it is closely related to the fluid compressibility coefficient and can be written as:

$$\lambda = \frac{\tau}{\chi_T} \tag{6.21}$$

where $\tau$ is the time scale of the physical phenomenon, which extends from a wave propagation with a small time scale to large fluid motions. For instance, at a fixed time scale and in normal pressure and temperature conditions, the compression resistance of water is $10^5$ higher than for air. In a classical approach, air would be considered as a compressible fluid, for which $\lambda = -2/3\,\mu$, whereas the water would be strictly incompressible with a non physical $\lambda$ coefficient and would require a second step to ensure the divergence-free constraint. Moreover, the assumption $\lambda = -2/3\,\mu$ for air would result in a highly compressible non-physical behavior forcing the fluid to collapse under its own weight. With Eqs. (6.7), (6.20) and (6.21), the momentum equations in semi-discrete form are written as:

$$\rho\left(\frac{(\mathbf{u}^{n+1} - \mathbf{u}^n)}{\Delta t} + \mathbf{u}^n \cdot \nabla \mathbf{u}^{n+1}\right)$$
$$= -\nabla\left(p^n - \frac{\tau}{\chi_T}\nabla \cdot \mathbf{u}^{n+1} - \tau\,\mathbf{u}^{n+1} \cdot \nabla p^n + \frac{\tau\,\beta}{\chi_T}\frac{dT}{dt}^{n+1}\right)$$
$$+\rho\,\mathbf{g} + \nabla \cdot \left(\mu\left(\nabla \mathbf{u}^{n+1} + \nabla^t \mathbf{u}^{n+1}\right)\right) + F_{TS}$$

This formulation is self-consistent since the local constraint accounting for the degree of fluid compressibility is directly included in the equation. The parameter $\lambda = \tau / \chi_T$ behaves as a compression resistance which depends on the time scale of the phenomenon. Indeed, a medium behaves differently to low and high frequency stresses which makes the time scale a crucial notion in this formulation. The total mass conservation issue differs drastically from the incompressible approaches and is a matter of volume mass transfers in $D$ or surface mass transfers through the $\Sigma$ boundary.

For a low Mach number assumption, this equation is:

$$\rho \left( \frac{\partial \mathbf{u}}{\partial t} + \mathbf{u} \cdot \nabla \mathbf{u} \right) = -\nabla \left( p - \frac{\tau}{\chi_T} \nabla \cdot \mathbf{u} \right) + \rho\, \mathbf{g} \\ + \nabla \cdot \left( \mu \left( \nabla \mathbf{u} + \nabla^t \mathbf{u} \right) \right) + F_{TS} \tag{6.22}$$

Equation (6.23), coupled with (6.17), provide a modeling framework for two-phase compressible isothermal flows. This version of the compressible one-fluid model is simpler to implement than the complete formulation and is useful for many problems. Several examples are provided in the validation section. The simplified one-fluid model (6.23–6.17) is a generalization of the formulation obtained with the augmented Lagrangian method [52, 53] for incompressible flows which calculates the pressure as follows, after a semi-discretization in time:

$$p^{n+1} = p^n - r\, \nabla \cdot \mathbf{u}^{n+1} \tag{6.23}$$

where $r$ is a numerical parameter which enables to locally ensure $\nabla \cdot \mathbf{u} = \varepsilon$ with the chosen accuracy $\varepsilon \to 0$. In this case, the pressure $p$ can be seen as a compressibility storage variable. The extension of this method to compressible flows results in the expression:

$$p^{n+1} = p^n - \frac{\Delta t}{\chi_T}\, \nabla \cdot \mathbf{u}^{n+1} \tag{6.24}$$

where pressure is the thermodynamic pressure or the dynamic pressure if the total pressure only depends on time. The physical parameters involved in the equation system are the isothermal compressibility $\chi_T$, expansion coefficients $\beta$, imposed or calculated by a state equation, and the density $\rho(p, T)$ calculated by a state equation.

## 6.3 Energy Conservation

The energy equation is written under its non-conservative form with compressibility terms:

$$\rho c_p \left( \frac{\partial T}{\partial t} + \mathbf{u} \cdot \nabla T \right) = \nabla \cdot (\lambda \nabla T) + \beta T \frac{dp}{dt} + q + \phi \tag{6.25}$$

where q is the volume production and $\phi$ the dissipation function.

Another form of the energy equation is obtained by developing the derivative of the pressure according to time in function of temperature T and density $\rho$:

$$\frac{dp}{dt} = \left.\frac{\partial p}{\partial T}\right|_\rho \frac{dT}{dt} + \left.\frac{\partial p}{\partial \rho}\right|_T \frac{d\rho}{dt} \tag{6.26}$$

with $\frac{d\rho}{dt} = -\rho \nabla \cdot \mathbf{u}$ according to the compressible mass conservation. For the term $\frac{\partial p}{\partial T}$, the thermodynamic expression including the specific heat is used:

$$\beta T \left.\left( \frac{\partial p}{\partial T} \right)\right|_\rho = -\frac{T}{\rho} \left( \frac{\partial \rho}{\partial T} \right)_P \left( \frac{\partial p}{\partial T} \right)_\rho = \rho (c_p - c_v) \tag{6.27}$$

With expressions (6.26) and (6.27), the term $\beta T \frac{dp}{dt}$ of (6.25) is written as:

$$\begin{aligned} \beta T \frac{dp}{dt} &= \rho (c_p - c_v) \frac{dT}{dt} - \frac{\beta T}{\chi_T} \nabla \cdot \mathbf{u} \\ &= \rho (c_p - c_v) \left( \frac{\partial T}{\partial t} + \mathbf{u} \cdot \nabla T \right) - \frac{\beta T}{\chi_T} \nabla \cdot \mathbf{u} \end{aligned} \tag{6.28}$$

Finally, the new formulation of the conservation energy equation is obtained by introducing (6.28) in (6.25):

$$\rho c_v \left( \frac{\partial T}{\partial t} + \mathbf{u} \cdot \nabla T \right) = \nabla \cdot (\lambda \nabla T) - \frac{\beta T}{\chi_T} \nabla \cdot \mathbf{u} + q + \phi \tag{6.29}$$

A phase change source term $F_p$ can be added in (6.29) to take into account the related interface jump conditions involving latent heat. For more details, the reader can consult the work of [215]. This specific physical aspect will not be considered in the present article.

The form (6.29) of the conservation energy equation is more interesting than (6.25) as the derivative of the pressure according to time in the second member. A $\nabla \cdot \mathbf{u}$ term is now in the equation. Furthermore, now the equation depends on the constant volume specific heat and no the constant pressure one.

## 6.4 Comparison with Classical "Low Mach Number" Model

The compressible Navier-Stokes equations in their conservative and non-dimensional form are given in [216]. The term $\frac{1}{M^2}$ present in the momentum equation complicates the resolution for low Mach number as this term tends to 0 for weak flow velocities. This difficulty is avoided by performing asymptotic expansions of the unknowns of Navier-Stokes equations [217].

$$
\begin{aligned}
\rho(x,t,\varepsilon) &= \rho_0(x,t) + \varepsilon\rho_1(x,t) + \varepsilon^2\rho_2(x,t) + O(\varepsilon^3)\\
T(x,t,\varepsilon) &= T_0(x,t) + \varepsilon T_1(x,t) + \varepsilon^2 T_2(x,t) + O(\varepsilon^3)\\
\mathbf{u}(x,t,\varepsilon) &= \mathbf{u}_0(x,t) + \varepsilon\mathbf{u}_1(x,t) + \varepsilon^2\mathbf{u}_2(x,t) + O(\varepsilon^3)\\
p(x,t,\varepsilon) &= p_0(x,t) + \varepsilon p_1(x,t) + \varepsilon^2 p_2(x,t) + O(\varepsilon^3)
\end{aligned}
$$

with $\varepsilon = \gamma M^2$.

These asymptotic expansions are substituted into the compressible Navier Stokes equations and the power of M are equated. The terms of order $M^{-2}$ in the momentum equation imply $\nabla p_0 = 0$. The quantity $p_0$ is thus constant in space.

At the zero order, the "low Mach number" Navier-Stokes equations are obtained:

$$
\left\{
\begin{aligned}
&\frac{\partial \rho_0}{\partial t} + \nabla\cdot(\rho_0\mathbf{u}_0) && = 0\\
&\frac{\partial \rho_0\mathbf{u}_0}{\partial t} + \nabla\cdot(\rho_0\mathbf{u}_0\otimes\mathbf{u}_0) && = -\nabla p_1\\
&+\frac{1}{Re}\nabla\cdot\left[-\frac{2}{3}\mu(\nabla\mathbf{u}_0)\cdot\overrightarrow{\overrightarrow{I}} + \mu(\nabla\mathbf{u}_0 + \nabla^t\mathbf{u}_0)\right] &&\\
&\frac{\partial p_0}{\partial t} + \gamma\nabla\cdot(p_0\mathbf{u}_0) && = \frac{\gamma}{RePr}\nabla\cdot(\mu\nabla T_0)
\end{aligned}
\right.
\tag{6.30}
$$

Preconditioners and stable resolution methods are necessary to solve (6.30). Their development and implementation is rather complicated. It lays on a low Mach preconditioning of iterate solvers [216]. These models are used for low compression rate and weak speed. The model presented in this article with $Ma < 1$ is not a low Mach compressible model, no asymptotic expansions are leaded. Our model is more general because it can support a high compression rate and a strong density variation even with very low speeds. A simplified model can be obtained by assuming that Ma $\rightarrow 0$ but without developing pressure according to the Mach number M as in the low Mach number approach. Moreover, the one-fluid compressible model can be used in its general form to treat various compressible configurations with $Ma > 1$. Specific advection schemes devoted to the hyperbolic character of the non linear term in the momentum equations must be used in this case [218].

## 6.5 Synthesis of Models

In the general case, the equation system of the Compressible One-Fluid (COF) model is written as:

$$\begin{cases} \rho\left(\frac{\partial \mathbf{u}}{\partial t}+\mathbf{u}\cdot\nabla\mathbf{u}\right)=-\nabla\left(p-\frac{\tau}{\chi_T}\nabla\cdot\mathbf{u}-\tau\,\mathbf{u}\cdot\nabla p+\frac{\tau\,\beta}{\chi_T}\frac{dT}{dt}\right) \\ +\rho\,\mathbf{g}+\nabla\cdot\left(\mu\left(\nabla\mathbf{u}+\nabla^t\mathbf{u}\right)\right)+F_{TS} \\ \frac{\partial p}{\partial t}=-\frac{1}{\chi_T}\nabla\cdot\mathbf{u}-\mathbf{u}\cdot\nabla p+\frac{\beta}{\chi_T}\frac{dT}{dt} \\ \rho C_v(\frac{\partial T}{\partial t}+(\mathbf{u}\cdot\nabla)T)=\nabla\cdot(\lambda\nabla T)-\frac{\beta T}{\chi_T}\nabla\cdot\mathbf{u}+q+\phi \\ \frac{dC}{dt}=\frac{\partial C}{\partial t}+\mathbf{u}\cdot\nabla C=0 \\ \rho=f(p,T,C),\lambda=f(T,C),\mu=f(T,C),C_p=f(T,C) \end{cases} \tag{6.31}$$

where $\tau=\Delta t$ in a discrete formulation. In equation (6.31), the compressible effects are implicitly acting on the interface evolutions through the velocity field which does not verify $\nabla\cdot\mathbf{u}$.

When the Mach number tends towards zero, the previous equation system, called Simplified Compressible One-Fluid (SCOF) model, becomes:

$$\begin{cases} \rho\left(\frac{\partial \mathbf{u}}{\partial t}+\mathbf{u}\cdot\nabla\mathbf{u}\right)=-\nabla\left(p-\frac{\tau}{\chi_T}\nabla\cdot\mathbf{u}\right)+\nabla\cdot\left(\mu\left(\nabla\mathbf{u}+\nabla^t\mathbf{u}\right)\right) \\ \rho\,\mathbf{g}+F_{TS} \\ \frac{\partial p}{\partial t}=-\frac{1}{\chi_T}\nabla\cdot\mathbf{u} \\ \rho C_v(\frac{\partial T}{\partial t}+(\mathbf{u}\cdot\nabla)T)=\nabla\cdot(\lambda\nabla T)-\frac{\beta T}{\chi_T}\nabla\cdot\mathbf{u}+q+\phi \\ \frac{dC}{dt}=\frac{\partial C}{\partial t}+\mathbf{u}\cdot\nabla C=0 \\ \rho=f(p,T,C),\lambda=f(T,C),\mu=f(T,C),C_p=f(T,C) \end{cases} \tag{6.32}$$

This equation system will be considered in the rest of the article.

The model is characterized by time and space dependent physical properties $\rho, \mu, \beta, \chi_T$ which are functions of the phase function C. More than two fluids can be compressible and dilatable. They have their own state equation The local constraint of compression resistance is implicitly taken into account as a function of the phenomenon time scale. The initial conservation equations are not subjected to any assumption. Any time scale can be resolved, extending from wave propagation in a continuum to common flows on large time scales. It aims at unifying the resolution of the equations in compressible configurations with Mach number $Ma < 1$ or incompressible flow configurations, in the presence of several phases or not. It enables resolving the propagation and more generally the constraints and the motions in an elastic medium through the definition of the viscosity coefficients $\mu$ et $\lambda$ from Young's modulus and the Poisson's coefficient. The extension to fluid-structure interaction is straightforward.

## 6.6 Validation of Isothermal Compressible One-Fluid Model

The validations presented in this section have previously been published in [214].

**Compression of an air bubble**

- Presentation of the case

An air bubble, with a radius of $R = 30$ mm, is compressed by water in a closed cavity of $L = 0.1$ m sides, as presented in Fig. 6.1. The water flow is very slow, with $V_0 = 0.0025$ m s$^{-1}$. The gravity and surface tension forces are not taken into account. The Reynolds number is $Re = 250$.

- Analytical solution

The initial bubble volume is known, $V_{ol} = \dfrac{m_0}{\rho_0}$, as is the injected water flow. The theoretical equation of the density during time is:

$$\rho(t) = \frac{\rho_0}{1 - \dfrac{q\,t}{\pi\, R^2}} \tag{6.33}$$

with $q$ the volumetric flow rate of injected water. Here, $q = 0.001$ m$^3$ s$^{-1}$.

- Numerical simulations

The bubble volume is very quickly reduced until the collapse is obtained and the mesh effect becomes more significant. The collapse takes place at $t = 2.82$ s, i.e. $\rho(t) \rightarrow +\infty$, in accordance with the theory. In order to test the grid convergence, the simulation is repeated with different meshes: $32 \times 32$, $64 \times 64$, $128 \times 128$, $256 \times 256$, $512 \times 512$. Numerical density for each mesh is given in Table 6.1.

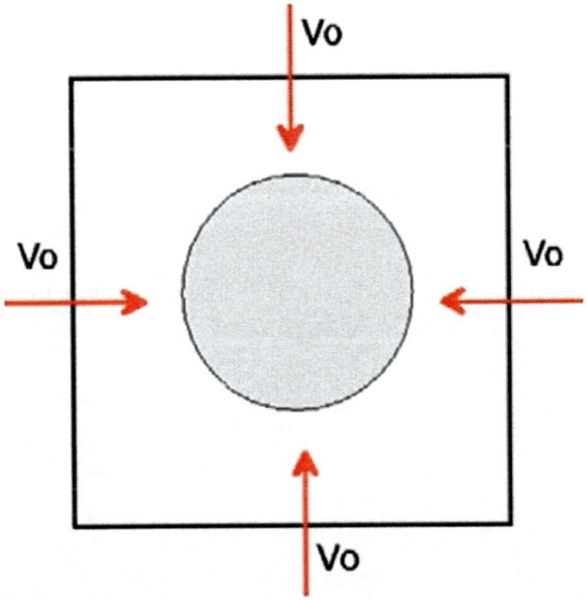


**Fig. 6.1** Compression of an air bubble by water

**Table 6.1** Density at t = 1.5 s

| Mesh | $\rho(1.5s)$ |
|---|---|
| 32 * 32 | 2.29658 |
| 64 * 64 | 2.45194 |
| 128 * 128 | 2.46224 |
| 256 * 256 | 2.49142 |
| 512 * 512 | 2.49481 |

The theoretical density is 2.50665 kg/m$^3$

The comparison between simulations on different meshes and the analytical solution is given in Fig. 6.2 for the density and in Fig. 6.3 for the pressure. The numerical results converge towards the analytical solution when the number of grid cells is increased.

The order $p$ of the method, calculated by the Richardson method [219], is $p = \dfrac{\ln(\frac{\rho_4-\rho_2}{\rho_3-\rho_1})}{\ln(k)} = 2.07$, where $k = \frac{1}{2}$. The asymptotic numerical density is $\rho = \rho_4 - \dfrac{\rho_4 - \rho_3}{(1-(\frac{N_3}{N_4})^{-p})} = 2.50055$ kg/m$^3$, i.e. a relative variation of 0.24% with the theoretical density. In spite of a very weak speed, we obtain a good precision. The index 1, 2, 3 and 4 are related to grids $64 \times 64$, $128 \times 128$, $256 \times 256$ and $512 \times 512$.

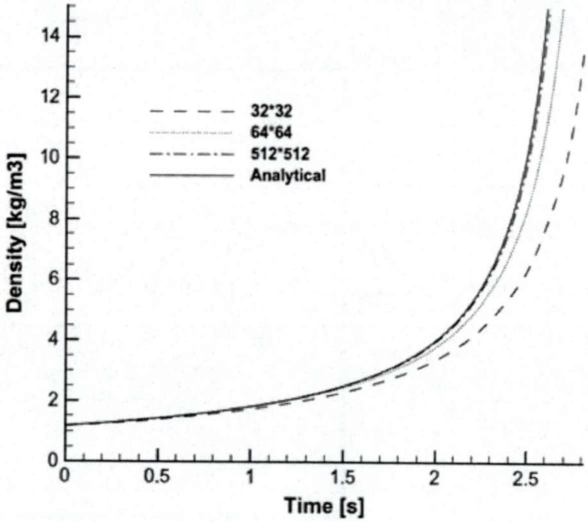


**Fig. 6.2** Evolution of the air density for different meshes

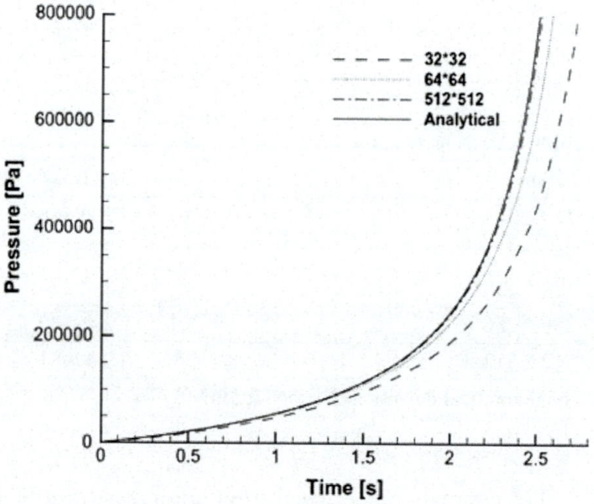


**Fig. 6.3** Evolution of the pressure in the air bubble

## Liquid injection in a closed cavity

- Presentation of the case

A gas volume initially placed in a closed cavity is compressed by a liquid water piston which is quasi-incompressible. The mass of air is constant and its volume regularly varies over time as water is injected at a constant speed. The cavity is a square of 0.1 m sides (see Fig. 6.4). The Reynolds number is $Re = 10{,}000$. The simulations are carried out for several time steps and meshes.

**Fig. 6.4** Water injection in a closed cavity initially full of air

- Analytical solution

The constant air mass ,which is equal to $m_0$, is compressed by a liquid piston, which moves at a constant speed $V_0 = 0.1$ m s$^{-1}$. The density varies then as:

$$\rho(t) = \frac{m_0}{\Omega_0 - V_0\,\Delta t\,S} = \frac{\rho_0\,H_0\,S}{H_0\,S - V_0\,\Delta t\,S} = \frac{\rho_0\,H_0}{H_0 - V_0\,\Delta t} = \frac{\rho_0}{1 - \frac{V_0\,t}{H_0}} \tag{6.34}$$

The density after $0.5s$ must be equal to $\rho = 3.5304$ kg/m$^3$. The pressure must be equal to $p = \rho\, r\, T_0 = 3.5304 \cdot 287 \cdot 300 = 303967$ Pa.

- Numerical simulations

The pressure increases over time (Fig. 6.5) but it is uniform throughout the cavity, in both water and air. Without gravity, the pressure is strictly uniform. A correct behavior of the numerical solution is noticed. The liquid remains incompressible with a constant density of 1000 kg m$^3$ (Fig. 6.6) while the density in air is linear.

Different meshes are tested. In Table 6.2, the density is given for these meshes as is the relative error with the analytical solution (in %). A regular convergence towards the theoretical value is observed. On the finest grid, the difference between the analytical and simulated density is around 0.1%. In this case, the free surface is moving and can be subjected to deformations. However, since a coarse $64 \times 64$ grid, the error on density in air is less than 1% in this case.

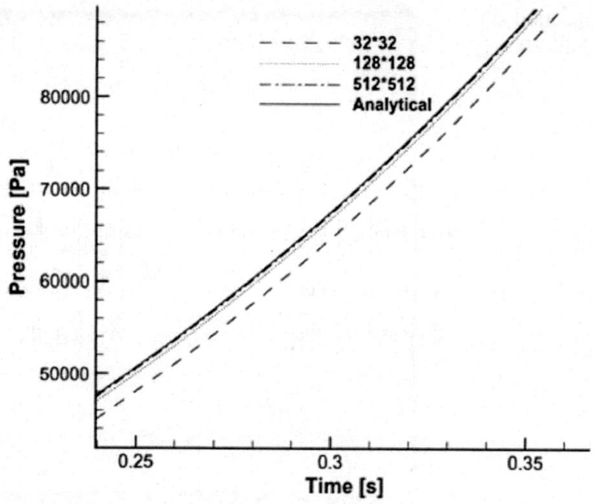


**Fig. 6.5** Evolution of the pressure in air for different meshes

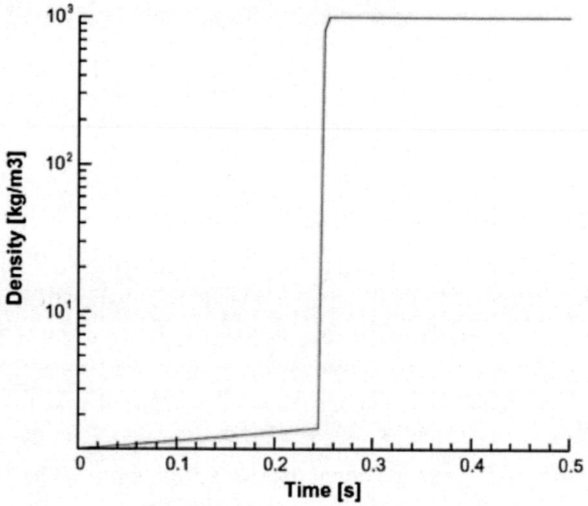


**Fig. 6.6** Evolution of the density at the center of the cavity

**Table 6.2** Density at $t = 0.3s$

| Mesh | $\Delta t$ | $\rho(0.3s)$ | $E_r$ |
|---|---|---|---|
| 32 * 32 | 0.008 | 1.92831 | 1.6861 |
| 64 * 64 | 0.004 | 1.94444 | 0.8637 |
| 128 * 128 | 0.002 | 1.95269 | 0.4431 |
| 256 * 256 | 0.001 | 1.95696 | 0.2254 |
| 512 * 512 | 0.0005 | 1.95916 | 0.1132 |

The theoretical density is 1.96138 kg/m$^3$

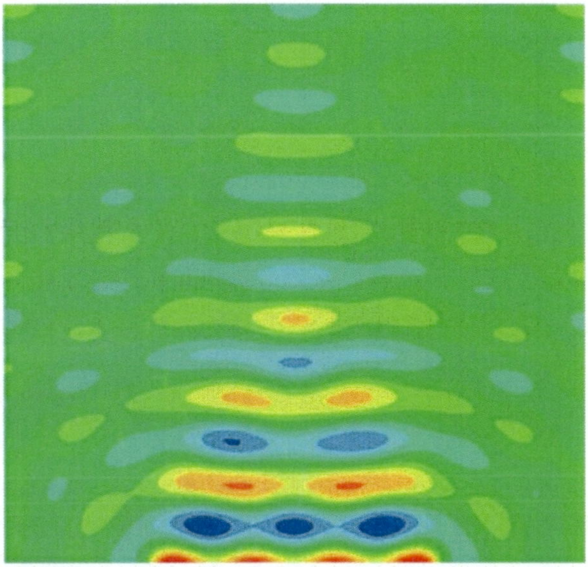

**Fig. 6.7** Pressure in the cavity

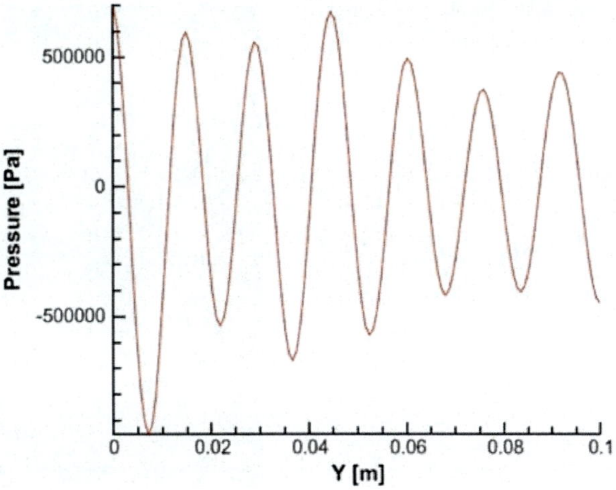


**Fig. 6.8** Axial cut (x = 0.05 m) of the pressure field

## Bubble subjected to an ultrasound field

- Presentation of the case

A transducer emits a signal of frequency $f = 100$ kHz with an amplitude of $e = 10^{-6}$ m on the lower wall of the cavity of 0.1 m length, i.e. a sinusoidal celerity with an amplitude $e$ and a frequency $f$ is imposed on the lower wall of the cavity. The case corresponds to the action of power ultra-sounds on an air bubble of radius R = 0.03 m in a water cavity. The mesh is $128 \times 128$ and the time step is $10^{-7}s$.

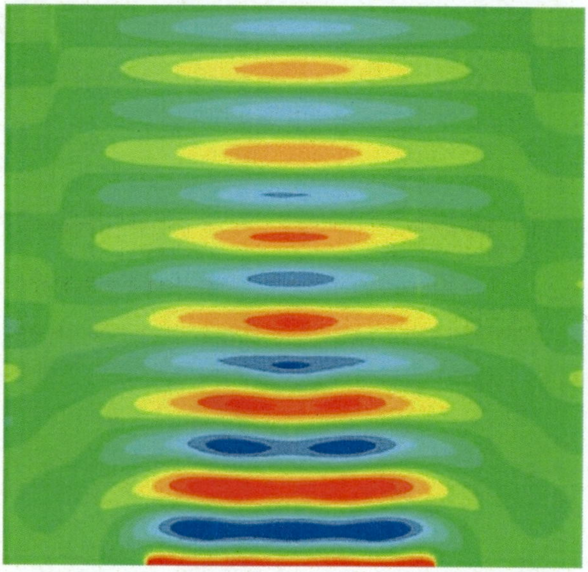

**Fig. 6.9** Velocity in the cavity

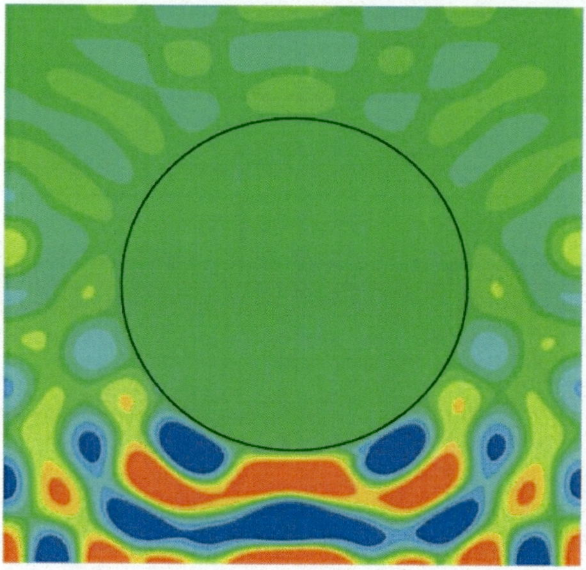

**Fig. 6.10** Bubble subjected to an ultrasound field

- Numerical simulations

The propagation of an ultrasonic wave in water is first studied. The ultrasonic wave propagation in a liquid creates phases of compression and depression (Figs. 6.7 and 6.8). Numerically, the wave length is 0.015298 m on the $128 \times 128$ grid. The speed of sound in water is equal to 1500 m $s^{-1}$. The wave length is proportional to the sound celerity: $\lambda = \dfrac{c}{f} = 0.015$ m. Thus there is a variation of 1.99% between the numerical and analytical wave length. The velocity varies between $\pm 0.6$ m s$^{-1}$ (Fig. 6.9).

Now an air bubble is placed in the cavity. The bubble is represented by a black line. The amplitude and frequency characteristics remain equal to $10^{-6}$ m and 100 kHz. The new pressure variations reach $p \pm 15 bar$ (Fig. 6.10). For such pressures, we are aware that cavitation effects take place. However, these effects [220] are not considered in this test case, but they were already studied with the same numerical tool [221]. The present test case demonstrates that the compressible two phase model also makes it possible to take into account the acoustic waves interaction with fluid interfaces.

# Chapter 7
# Large Eddy Simulation of Resolved Scale Interfacial Flows

There are two basic families of LES subgrid modelling, functional and structural models [38]. The functional models are originally designed for single-phase LES where only convective SGS terms are accounted for. Although there have been numerous studies in developing functional models for single-phase LES (usually based on eddy viscosity approach such as the Smagorinsky model [39]), their applicability to two-phase LES is still under question. This makes developing functional models for two-phase LES an active area of research. The structural models have purely mathematical nature and are established with no prior knowledge of the physics of the interactions between the resolved and subgrid scales [38]. The main strategy in this approach is to reconstruct an approximation of SGS terms from the information provided by resolved quantities. There are different types of structural models among which Approximate Deconvolution Model (ADM) [222, 223] and Bardina model [40] are well-known.

In the context of LES of turbulent interfacial flows, several works have utilized the functional LES models coupled with an interface capturing methods (mainly VOF or Level Set and in a few examples Front Tracking) with application in bubbly flows [224], fuel atomization [225, 226], slag layer dynamics in continuous casting process [227, 228] and other industrial flows [229]. But these studies are typically based on a turbulence cascade process hypothesis which neglects the contribution of the subgrid surface tension effect and interface curvature [230]. In other words, they extend the classical single-phase LES to two-phase flows by only modeling the subgrid convective term and employ the interface capturing method based on the filtered (resolved) velocity field. Therefore, the subgrid effects of interface curvature and consequently the unresolved surface tension forces are neglected. But this approach may be questionable in the context of surface tension-driven dynamics of interface for example in later stages of phase inversion problem [15] or when the density and viscosity difference of two immiscible fluids is small (e.g. water/oil systems) [230].



S. Vincent et al., *Small Scale Modeling and Simulation of Incompressible Turbulent Multi-Phase Flow*, CISM International Centre for Mechanical Sciences 607,
https://doi.org/10.1007/978-3-031-09265-7_7

The modeling and simulation of multi-phase flows in the vicinity of resolved scale interfaces, i.e. under the Gibbs assumption the transition between two phases or fluids is sharp and the molecular interfacial thickness is assumed to be zero, is a complex task as soon as interfacial rupture or coalescence occur, together with the action of capillary forces and interactions with a surrounding turbulent flow motion. Pretending to be able to simulate all the time and space scales of this type of flow (DNS) is almost impossible, even for moderate Reynolds number, due to the large range of scales brought by the interface, that can be smaller than the Kolmogorov scale corresponding to the often used smallest scale in a turbulent single phase flows. Based on these considerations, authors have recently tried to use Large Eddy Simulation models [38, 231] of turbulence to simulate multiphase flows [9, 14, 15, 19, 20, 41, 42, 232–240] . The purpose of this work is to derive new LES models for multiphase flows, based on Approximate Deconvolution Models (ADM) [222] and statistical analysis of the DNS of various multi-phase flows such as the phase inversion in a closed box [15, 241, 242]. All the studies in the literature have been investigated with continuous media and the incompressible Navier-Stokes equations.

The starting point of this chapter is to reformulate the filtering of the motion equations so as to explicitly account for two filtering scales, i.e. the implicit filter imposed by the mesh and numerical methods and an explicit larger scale filter controlled by the model, as exposed for single phase flows by [243]. Two types of motion equations will be considered, the classical incompressible Navier-Stokes equations formulated as a 1-fluid model (1FNS) [77] and also the new Discrete Mechanics (DM) framework of motion equations proposed by [244–246]. At the end, the subgrid scale (SGS) unclosed contributions induced by the LES and the filtering will be formulated and compared between 1FNS and DM.

A second part of this chapter, once the unresolved filtered contribution will have been identified for both 1FNS and DM models, the purpose and second original contribution of this chapter is to try to use ADM approaches to approximate unclosed terms coming from the LES framework formulated before. A last part of the work is devoted to a priori filtering of DNS of multiphase flows in order to evaluate the magnitude of the SGS-terms.

## 7.1 Filtering 1-Fluid Navier-Stokes Equations—Continuous Media Framework

### Standard single implicit filter LES

Based on all the works [38, 231] that have been achieved concerning single phase LES, the mass, momentum and energy equations are filtered with a single implicit filter, basically active at the mesh size. All the space scales larger than this scale $\bar{\Delta}$ are assumed resolved whereas all scales smaller than $\bar{\Delta}$, i.e. referred to as subgrid scale (SGS) terms, are assumed unresolved and their contributions have to be modeled.

### Filtering of mass and momentum conservation equations

The formulation of the Navier-Stokes equations valid for the Direct Numerical Simulation (DNS) of multi-phase flows with sharp interfaces can be formulated as follows [35, 36, 247]

$$\nabla \cdot \mathbf{u} = 0, \tag{7.1a}$$

$$\frac{\partial}{\partial t}(\rho \mathbf{u}) + \nabla \cdot (\rho \mathbf{u} \otimes \mathbf{u}) - \rho \mathbf{g} + \nabla p - \nabla \cdot (2\mu \mathbf{S}_D) - \sigma \kappa \nabla \chi = 0, \tag{7.1b}$$

$$\frac{\partial \chi}{\partial t} + \mathbf{u} \cdot \nabla \chi = 0, \tag{7.1c}$$

where $\mathbf{g}$ is the gravity, $\mathbf{u}$ and $p$ are the flow velocity and pressure fields, $\mathbf{S}_D$ the stress tensor whereas $\rho$ and $\mu$ are the density and the dynamic (shear) viscosity respectively. According to the one fluid formulation of [35], $\chi$ is defined as the indicator function or color function for each phase ($\chi = 1$ in the dispersed phase and $\chi = 0$ in the surrounding or carrier fluid). Various approaches exist [248] for solving $\chi$, namely among the most popular methods, the volume of fluid (VOF) technique [73, 76, 77, 249], the level set approach [8–10, 112], the front tracking method [65, 81, 82] or also diffuse interface approaches and Euler-Euler models using fluid fractions [233, 235, 250, 251]. With all these methods, at the end, you are able to define $\chi$ so that the 1FNS model can be solved. According to [44], $\sigma$ is the surface tension and $\kappa = \nabla \cdot \left( \frac{\nabla \chi}{\|\nabla \chi\|} \right)$ the local curvature of the interface. By definition, for a Newtonian fluid, $\mathbf{S}_D = \frac{1}{2}\left(\nabla \mathbf{u} + (\nabla \mathbf{u})^T\right)$ is the second-order deviatoric stress tensor.

The DNS formulation of the Navier-Stokes equations (7.1a–7.1c) for multiphase flows with interfaces are rarely possible because of limitations in computing resources and computing time. An alternative approach is to try to resolve the larger flow motion and interfacial scales while the smaller ones will be modeled. This is the framework of LES [38, 231]. We now consider the following filtered Navier-Stokes equations for incompressible turbulent two-phase flows and resolved scale interfaces defined on an open bounded domain $V$ of $R^d$ ($d = 3$ being the space dimension) and over a time interval $I = [0, T]$ of $R$

$$\nabla \cdot \bar{\mathbf{u}} = 0, \tag{7.2a}$$

$$\frac{\partial}{\partial t}(\overline{\rho \mathbf{u}}) + \nabla \cdot (\overline{\rho \mathbf{u} \otimes \mathbf{u}}) - \bar{\rho} \mathbf{g} + \nabla \bar{p} - \nabla \cdot (2\overline{\mu \mathbf{S}_D}) - \sigma \overline{\kappa \nabla \chi} = 0, \tag{7.2b}$$

$$\frac{\partial \overline{\chi}}{\partial t} + \overline{\mathbf{u} \cdot \nabla \chi} = 0, \tag{7.2c}$$

where $\bar{\cdot}$ refers to a filtered variable in the LES sence [38]. On a practical point view, the resolvable variables are $\overline{\mathbf{u}}$, $\overline{P}$, $\overline{\rho}$, $\overline{\mu}$ and $\overline{\chi}$. By introducing two-phase mixing laws in Representative Elementary Volumes (REV), $\overline{\rho}$ and $\overline{\mu}$ are obtained directly according to $\overline{\chi}$ as follows

$$\overline{\rho} = \overline{\chi}\rho_1 + (1 - \overline{\chi})\rho_0 \tag{7.3}$$

$$\overline{\mu} = \overline{\chi}\mu_1 + (1 - \overline{\chi})\mu_0 \tag{7.4}$$

with $\rho_0$, $\rho_1$, $\mu_0$ and $\mu_1$ the properties of each phase (here we only consider a two-phase flow, this can be easily generalized) assumed constant if the flow is isothermal.

The set of equations (7.2) can be reformulated according to $\bar{\mathbf{u}}$, $\overline{P}$, $\overline{\rho}$, $\overline{\mu}$ and $\overline{\chi}$ as

$$\nabla \cdot \bar{\mathbf{u}} = 0, \tag{7.5a}$$

$$\frac{\partial \bar{\rho}\bar{\mathbf{u}}}{\partial t} + \boldsymbol{\tau}_{\text{time}} + \nabla \cdot (\bar{\rho}\bar{\mathbf{u}} \otimes \bar{\mathbf{u}} + \boldsymbol{\tau}_{\text{conv}}) - \bar{\rho}\mathbf{g}$$
$$+ \nabla \bar{p} - \nabla \cdot (2\bar{\mu}\overline{\mathbf{S}_D} + \boldsymbol{\tau}_{\text{diff}}) - (\sigma \bar{\kappa} \nabla \overline{\chi} + \boldsymbol{\tau}_{\text{surf}}) = 0, \tag{7.5b}$$

$$\frac{\partial \overline{\chi}}{\partial t} + \bar{\mathbf{u}} \cdot \nabla \overline{\chi} + \tau_{\text{interf}} = 0, \tag{7.5c}$$

where $\boldsymbol{\tau}_{\text{time}}$, $\boldsymbol{\tau}_{\text{conv}}$ $\boldsymbol{\tau}_{\text{diff}}$, $\boldsymbol{\tau}_{\text{surf}}$ and $\tau_{\text{interf}}$ are the unresolved (subgrid) contributions coming from inertia (time evolution and convection), diffusion (due to viscosity), surface tension (due to capillary) and interface tracking effects, respectively defined by

$$\boldsymbol{\tau}_{\text{time}} = \frac{\partial \overline{\rho \mathbf{u}}}{\partial t} - \frac{\partial \bar{\rho}\bar{\mathbf{u}}}{\partial t}, \tag{7.6a}$$

$$\nabla \cdot \boldsymbol{\tau}_{\text{conv}} = \nabla \cdot \left(\overline{\rho \mathbf{u} \otimes \mathbf{u}} - \bar{\rho}\bar{\mathbf{u}} \otimes \bar{\mathbf{u}}\right), \tag{7.6b}$$

$$\nabla \cdot \boldsymbol{\tau}_{\text{diff}} = \nabla \cdot \left(\overline{2\mu \mathbf{S}_D} - 2\bar{\mu}\overline{\mathbf{S}_D}\right), \tag{7.6c}$$

$$\boldsymbol{\tau}_{\text{surf}} = \sigma \left(\overline{\kappa \nabla \chi} - \bar{\kappa} \nabla \overline{\chi}\right), \tag{7.6d}$$

$$\tau_{\text{interf}} = \overline{u \cdot \nabla \chi} - \bar{\mathbf{u}} \cdot \nabla \overline{\chi}. \tag{7.6e}$$

As previously explained by [14, 15, 41, 233, 235], the filtering of the two-phase formulation of the Navier-Stokes equations reveals new turbulent and two-phase coupling terms that have to be modeled in the framework of LES. This is the objective of this work, by considering statistical analysis of DNS simulations and trying to propose representative stochastic models for $\boldsymbol{\tau}_{\text{interf}}$, $\boldsymbol{\tau}_{\text{time}}$, $\boldsymbol{\tau}_{\text{conv}}$, $\boldsymbol{\tau}_{\text{diff}}$ and $\boldsymbol{\tau}_{\text{surf}}$.

**Filtering of Kinetic Energy Equation**

The conservation equation for the DNS of the kinetic energy is formulated by taking the scalar product of the velocity $\mathbf{u}$ with the momentum equation (7.1b) as follows

$$\mathbf{u} \cdot \left( \frac{\partial}{\partial t}(\rho \mathbf{u}) + \nabla \cdot (\rho \mathbf{u} \otimes \mathbf{u}) - \rho \mathbf{g} + \nabla p - \nabla \cdot (2\mu \mathbf{S}_D) - \sigma \kappa \nabla \chi \right) = 0 \tag{7.7}$$

or also

$$\frac{\partial}{\partial t}\left(\frac{1}{2}\rho\mathbf{u}^2\right) + \nabla\cdot(\rho\mathbf{u}^2\mathbf{u}) - \rho\mathbf{g}\cdot\mathbf{u} + \nabla p\cdot\mathbf{u} - \nabla\cdot(2\mu\mathbf{S}_D)\cdot\mathbf{u} - \sigma\kappa\nabla\chi\cdot\mathbf{u} = 0. \tag{7.8}$$

Introducing the filtering of equation (7.8) leads to the classical form of the LES equation for the kinetic energy conservation [231]

$$\begin{aligned}\frac{\partial}{\partial t}\overline{(\frac{1}{2}\rho\mathbf{u}^2)} + \nabla\cdot\overline{(\frac{1}{2}\rho\mathbf{u}^2\mathbf{u})} &= \overline{\rho\mathbf{g}\cdot\mathbf{u}} - \nabla\cdot\overline{p\mathbf{u}}\\ &+\overline{p\nabla\cdot\mathbf{u}} + \overline{\mathbf{u}\cdot\nabla\frac{1}{2}\rho\mathbf{u}^2} + \nabla\cdot\overline{(2\mu\mathbf{S}_D\cdot\mathbf{u})} - \overline{2\mu\mathbf{S}_D:\nabla\mathbf{u}} + \overline{\sigma\kappa\nabla\chi\cdot\mathbf{u}}\end{aligned} \tag{7.9}$$

The different terms of the kinetic energy equation (7.9) can be associated to transport by the filtered velocity field (left hand side) and for the right hand side terms, transport of Archimède effects, power of the filtered external pressure forces in the filtered motion, filtered pressure-dilatation energy transfer (which is strictly null if the filtered velocity field $\bar{\mathbf{u}}$ is solenoidal), transport of the filtered gradient of the kinetic energy, energy exchange with the turbulent kinetic energy (shear production), power of the filtered external viscous stresses in the filtered motion, viscous heat production associated with mechanical dissipation by the filtered flow and filtered capillary kinetic energy production. All these contributions corresponds to important variables for the interpretation of the turbulence and LES modeling [38]. However, the form (7.9) is not convenient in practice as it involves the acting of the filter on the product of DNS variables that are not known. Different forms are considered here:

- by considering that the resolved variables are $\bar{\mathbf{u}}$ and $\frac{1}{2}\bar{\rho}\,\bar{\mathbf{u}}^2$ and introducing them in the kinetic equation (7.9):

$$\begin{aligned}&\frac{\partial}{\partial t}\frac{1}{2}\bar{\rho}\,\bar{\mathbf{u}}^2 + \nabla\cdot(\frac{1}{2}\bar{\rho}\,\bar{\mathbf{u}}^2\bar{\mathbf{u}}) - \bar{\rho}\mathbf{g}\cdot\bar{\mathbf{u}} + \nabla\cdot\bar{p}\bar{\mathbf{u}} - \bar{p}\nabla\cdot\bar{\mathbf{u}}\\ &-\bar{\mathbf{u}}\cdot\nabla\frac{1}{2}\overline{\bar{\rho}\mathbf{u}^2} - \nabla\cdot(2\bar{\mu}\overline{\mathbf{S}_D}\cdot\bar{\mathbf{u}}) + 2\bar{\mu}\overline{\mathbf{S}_D}:\nabla\bar{\mathbf{u}} - \sigma\bar{\kappa}\nabla\bar{\chi}\cdot\bar{\mathbf{u}}\\ &= -\frac{\partial}{\partial t}\overline{(\frac{1}{2}\rho\mathbf{u}^2)} - \frac{\partial}{\partial t}\frac{1}{2}\bar{\rho}\,\bar{\mathbf{u}}^2 - \left[\nabla\cdot\overline{(\frac{1}{2}\rho\mathbf{u}^2\mathbf{u})} - \nabla\cdot(\frac{1}{2}\bar{\rho}\,\bar{\mathbf{u}}^2\bar{\mathbf{u}})\right]\\ &+\overline{\rho\mathbf{g}\cdot\mathbf{u}} - \bar{\rho}\mathbf{g}\cdot\bar{\mathbf{u}} - \nabla\cdot\bar{p}\bar{\mathbf{u}} + \nabla\cdot\overline{p\mathbf{u}} + \overline{p\nabla\cdot\mathbf{u}} - \bar{p}\nabla\cdot\bar{\mathbf{u}}\\ &+\overline{\mathbf{u}\cdot\nabla\frac{1}{2}\rho\mathbf{u}^2} - \bar{\mathbf{u}}\cdot\nabla\frac{1}{2}\overline{\bar{\rho}\mathbf{u}^2} + \nabla\cdot\overline{(2\mu\mathbf{S}_D\cdot\mathbf{u})} - \nabla\cdot(2\bar{\mu}\overline{\mathbf{S}_D}\cdot\bar{\mathbf{u}})\\ &-\overline{2\mu\mathbf{S}_D:\nabla\mathbf{u}} + 2\bar{\mu}\overline{\mathbf{S}_D}:\nabla\bar{\mathbf{u}} + \overline{\sigma\kappa\nabla\chi\cdot\mathbf{u}} - \sigma\bar{\kappa}\nabla\bar{\chi}\cdot\bar{\mathbf{u}}\end{aligned} \tag{7.10}$$

  where the right hand side terms are subgrid energy terms to be estimated or modeled.
- by introducing in equation (7.8) the decomposition of the instantaneous velocity and pressure, i.e. $\mathbf{u} = \bar{\mathbf{u}} + \mathbf{u}_{SGS}$ and $p = \bar{p} + p_{SGS}$ with $\mathbf{u}_{SGS}$ and $p_{SGS}$ the subfilter (or subgrid in practice) contributions:

$$
\begin{aligned}
&\frac{\partial}{\partial t}\frac{1}{2}\overline{\rho}\,\overline{\mathbf{u}}^2 + \nabla\cdot(\frac{1}{2}\overline{\rho}\,\overline{\mathbf{u}}^2\overline{\mathbf{u}}) \\
&- \overline{\rho}\mathbf{g}\cdot\overline{\mathbf{u}} + \nabla\cdot\overline{p\mathbf{u}} - \overline{p}\nabla\cdot\overline{\mathbf{u}} \\
&- \overline{\mathbf{u}}\cdot\nabla\frac{1}{2}\overline{\rho\mathbf{u}^2} - \nabla\cdot(2\overline{\mu}\overline{\mathbf{S}_D}\cdot\overline{\mathbf{u}}) + 2\overline{\mu}\overline{\mathbf{S}_D}:\nabla\overline{\mathbf{u}} - \sigma\overline{\kappa}\nabla\overline{\chi}\cdot\overline{\mathbf{u}} \\
&= -\mathbf{u}_{SGS}\cdot\left(\frac{\partial}{\partial t}(\rho\mathbf{u}_{SGS}) + \nabla\cdot(\rho\mathbf{u}_{SGS}\otimes\mathbf{u}_{SGS})\right) \\
&- \mathbf{u}_{SGS}\cdot\left(\rho\mathbf{g} + \nabla p_{SGS} - \nabla\cdot(2\mu\mathbf{S}'_D) - \sigma\kappa\nabla\chi'\right) \\
&- \overline{\mathbf{u}}\cdot\left(\frac{\partial}{\partial t}(\rho\mathbf{u}_{SGS}) + \nabla\cdot(\rho\mathbf{u}_{SGS}\otimes\mathbf{u}_{SGS} + \rho\overline{\mathbf{u}}\otimes\mathbf{u}_{SGS}\right) \\
&+ \overline{\mathbf{u}}\cdot\left(\rho\mathbf{u}_{SGS}\otimes\overline{\mathbf{u}}) + \nabla p_{SGS} - \nabla\cdot(2\mu\mathbf{S}'_D) - \sigma\kappa\nabla\chi'\right) \\
&- \mathbf{u}_{SGS}\cdot\left(\frac{\partial}{\partial t}(\rho\overline{\mathbf{u}}) + \nabla\cdot(\rho\overline{\mathbf{u}}\otimes\overline{\mathbf{u}} + \rho\overline{\mathbf{u}}\otimes\mathbf{u}_{SGS}\right) \\
&+ \mathbf{u}_{SGS}\cdot\left(\rho\mathbf{u}_{SGS}\otimes\overline{\mathbf{u}}) + \nabla\overline{p} - \nabla\cdot(2\mu\overline{\mathbf{S}_D}) - \sigma\kappa\nabla\overline{\chi}\right)
\end{aligned} \tag{7.11}
$$

The form of this kinetic energy conservation equation is interesting as it provides resolved scaled terms acting on $\mathbf{u}$ and terms to be modeled formulated in terms of coupling between filtered resolved and subgrid contributions as well as subgrid contributions only. The SGS terms will have to be studied in future works (right hand part of equation (7.11).

- by considering a double filtering of equation (7.8) to provide an Approximate Deconvolution framework for LES [222], i.e. the implicit filter $\overline{\mathbf{u}}$ and the explicit larger filter $\tilde{\mathbf{u}}$ such that $\mathbf{u} = \tilde{\mathbf{u}} + \overline{\mathbf{u}} + \mathbf{u}_{SGS}$. These equations, as well as the momentum double filtered equations, are not written here. They will be considered in future works.

After the filtering of the instantaneous Navier-Stokes equations, we have explained that the instantaneous (or DNS) velocity $\mathbf{u}$ is in fact $\mathbf{u} = \overline{\mathbf{u}} + \mathbf{u}_{SGS}$, i.e. $\overline{\mathbf{u}}$ bringing the large scale motions or Large eddy structures and $\mathbf{u}_{SGS}$ describing the fluctuating and small scales (subgrid). Based on this definition, a conservative energy equation can be formulated for subgrid scale velocity fluctuation $\mathbf{R} = \mathbf{u}_{SGS}\otimes\mathbf{u}_{SGS}$ (or equivalently $R_{ij} = u'_i u'_j$) as follows

$$
\begin{aligned}
\frac{\partial R_{ij}}{\partial t} + \frac{\partial}{\partial x_k}\left(\overline{u_k}R_{ij}\right) = &-\left(R_{jk}\frac{\partial\overline{u_i}}{\partial x_k} + R_{ik}\frac{\partial\overline{u_j}}{\partial x_k}\right) \\
&-\frac{\partial}{\partial x_k}(\overline{u'_i u'_j u'_k}) - \left(\frac{\partial}{\partial x_i}(\overline{p_{SGS}u'_j}) + \frac{\partial}{\partial x_j}(\overline{p_{SGS}u'_i})\right) \\
&+2\overline{p_{SGS}S'_{Di,j}} + \sigma\left(\overline{\kappa[\nabla\chi]'_i u'_j} + \overline{\kappa[\nabla\chi]'_j u'_i}\right) \\
&+2\mu\left(\overline{u'_j\frac{\partial}{\partial x_k}S'_{Di,k} + u'_i\frac{\partial}{\partial x_k}S'_{Dj,k}}\right)
\end{aligned} \tag{7.12}
$$

In this equation, the left-hand side term is the advection of the Reynolds tensions whereas the right-hand side terms correspond respectively to

- production/destruction by transfer between $\mathbf{u}_{SGS}$ and $\overline{\mathbf{u}}$;
- turbulent diffusion or transport by $\mathbf{u}_{SGS}$;
- space diffusion induced by pressure fluctuations;
- production/destruction by $p_{SGS}$;
- power of the fluctuations of the capillary forces;
- viscous diffusion and viscous dissipation.

This equation is also interesting to be studied as it provides interesting information on the time evolution of fluctuating (or subgrid) velocity contributions. The Reynolds tensions are often used for modeling in RANS and LES turbulence approaches as being representative in some sense of dissipation and energy cascade at small scale.

**Single implicit filtering with Favre's averaging for mass and momentum conservation equations**

The velocity is now filtered according to a Favre's averaging, as usually considered in compressible flows or more generally in motions with density varying fluids [38]. This choice could be viewed as cleaver in our problems as in two-phase flows with interfaces, the density is sharply varying across interface. The Favre's averaging of velocity $\mathbf{u}$ reads $\widetilde{\mathbf{u}}^{FA}$. Here, other variables such as $P$ or $\chi$ are still filtered according with the $\overline{G}$ implicit filter as they are not conservative variables unlike $\rho\mathbf{u}$.

The advection equation for the indicator function $\chi$ is first filtered with $\overline{G}$ and the Favre's averaged velocity $\widetilde{\mathbf{u}}^{FA}$ is considered for it. We have

$$\frac{\partial \rho}{\partial t} + \nabla \cdot (\rho \mathbf{u}) = \frac{\partial \rho}{\partial t} + \rho \nabla \cdot \mathbf{u} + \mathbf{u} \cdot \nabla \rho = 0 \tag{7.13}$$

Which is written in an equivalent way for multiphase flows as

$$\rho \nabla \cdot \mathbf{u} = (\rho_0 - \rho_1) \left( \frac{\partial \chi}{\partial t} + \mathbf{u} \cdot \nabla \chi \right) = 0 \tag{7.14}$$

$$\text{with } \rho = \rho_1 \chi + (1 - \chi)\rho_0 \tag{7.15}$$

where $\rho_0$ is the density in the phase located by $\chi = 0$ and similarly $\rho_1$ the density associated to $\chi = 1$. AS $\rho_0$ and $\rho_1$ are not zero, the mass conservation is satisfied for a two-phase flow (this can be extended to $n - 1$ advection equation on indicator functions for a n-phase flow) if the resulting equations are satisfied:

$$\rho \nabla \cdot \mathbf{u} = 0 \tag{7.16}$$

$$\frac{\partial \chi}{\partial t} + \mathbf{u} \cdot \nabla \chi = 0 \tag{7.17}$$

The filtering of the mass conservation equation with Favre's averaging leads to the following considerations

$$\overline{\rho\nabla\cdot\mathbf{u}} = (\rho_0-\rho_1)\left(\frac{\partial\overline{\chi}}{\partial t}+\overline{\mathbf{u}\cdot\nabla\chi}\right)=0 \tag{7.18}$$

$$\overline{\rho\nabla\cdot\mathbf{u}} = \overline{\nabla\cdot(\rho\mathbf{u})-\mathbf{u}\cdot\nabla\rho} \tag{7.19}$$

$$= \nabla\cdot\left(\overline{\rho}\widetilde{\mathbf{u}}^{FA}\right)-\overline{\mathbf{u}\cdot\nabla\rho} \tag{7.20}$$

$$= \overline{\rho}\nabla\cdot\widetilde{\mathbf{u}}^{FA}+(\rho_1-\rho_0)\,\widetilde{\mathbf{u}}^{FA}\cdot\nabla\overline{\chi}-\overline{(\rho_1-\rho_0)\,\mathbf{u}\cdot\nabla\chi}=0 \tag{7.21}$$

and

$$\frac{\partial\overline{\chi}}{\partial t}+\widetilde{\mathbf{u}}^{FA}\cdot\nabla\chi+\widetilde{\tau_{\text{interf}}}^{FA}=0 \tag{7.22}$$

It can be observed that for a constant density flow, i.e. if $\rho=\rho_0=\rho_1$, the $\overline{\rho}=\widetilde{\rho}^{FA}=\rho$. According to equation (7.21), the divergence of the Favre's average velocity is obtained

$$\nabla\cdot\widetilde{\mathbf{u}}^{FA}=\frac{(\rho_1-\rho_0)}{\overline{\rho}}\widetilde{\tau_{\text{interf}}}^{FA} \tag{7.23}$$

with $\widetilde{\tau_{\text{interf}}}^{FA}=\overline{u\cdot\nabla\chi}-\widetilde{\mathbf{u}}^{FA}\cdot\nabla\overline{\chi}$. As with the bar implicit filtered velocity, the divergence of the Favre's filtered incompressible two-phase flow is no more free.

The Favre's LES filtering of the momentum conservation equations is now undertaken. Similarly as what has been obtained for the bar implicit filter, it reads

$$\frac{\partial\bar{\rho}\widetilde{\mathbf{u}}^{FA}}{\partial t}+\nabla\cdot(\bar{\rho}\widetilde{\mathbf{u}}^{FA}\otimes\widetilde{\mathbf{u}}^{FA}+\widetilde{\boldsymbol{\tau}_{\text{conv}}}^{FA})-\bar{\rho}\mathbf{g}+\nabla\bar{p} \tag{7.24a}$$

$$-\nabla\cdot(2\bar{\mu}\widetilde{\mathbf{S}_D}^{FA}+\widetilde{\boldsymbol{\tau}_{\text{diff}}}^{FA})-\left(\sigma\bar{\kappa}\nabla\overline{\chi}+\widetilde{\boldsymbol{\tau}_{\text{surf}}}^{FA}\right)=0 \tag{7.24b}$$

with

$$\widetilde{\boldsymbol{\tau}_{\text{time}}}^{FA}=\frac{\partial\overline{\rho\mathbf{u}}}{\partial t}-\frac{\partial\bar{\rho}\widetilde{\mathbf{u}}^{FA}}{\partial t}=\frac{\partial\overline{\rho\mathbf{u}}}{\partial t}-\frac{\partial\bar{\rho}\frac{\overline{\rho\mathbf{u}}}{\bar{\rho}}}{\partial t}=0, \tag{7.25a}$$

$$\nabla\cdot\widetilde{\boldsymbol{\tau}_{\text{conv}}}^{FA}=\nabla\cdot\left(\overline{\rho\mathbf{u}\otimes\mathbf{u}}-\bar{\rho}\widetilde{\mathbf{u}}^{FA}\otimes\widetilde{\mathbf{u}}^{FA}\right), \tag{7.25b}$$

$$\nabla\cdot\widetilde{\boldsymbol{\tau}_{\text{diff}}}^{FA}=\nabla\cdot\left(\overline{2\mu\mathbf{S}_D}-2\bar{\mu}\widetilde{\mathbf{S}_D}^{FA}\right), \tag{7.25c}$$

$$\widetilde{\boldsymbol{\tau}_{\text{surf}}}^{FA}=\sigma\left(\overline{\kappa\nabla\chi}-\bar{\kappa}\nabla\overline{\chi}\right), \tag{7.25d}$$

$$\widetilde{\tau_{\text{interf}}}^{FA}=\overline{u\cdot\nabla\chi}-\widetilde{\mathbf{u}}^{FA}\cdot\nabla\overline{\chi}. \tag{7.25e}$$

The main interest of the Favre's averaging, compared to the bar filtering, is that it involves one SGS term less, namely the time SGS term $\widetilde{\boldsymbol{\tau}_{\text{time}}}^{FA}$. It has to be noticed also that the interfacial SGS term $\widetilde{\tau_{\text{surf}}}^{FA}$ is the same with the bar and Favre's filtering $\tau_{\text{surf}}$. If a non-conservative formulation is chosen for the Favre's averaging (this is generally the case in practice as incompressible two-phase flows are considered here and the variable are the pressure and velocity), the filtered time discretization term becomes

$$\frac{\partial \bar{\rho}\widetilde{\mathbf{u}}^{FA}}{\partial t} = \bar{\rho}\frac{\partial \widetilde{\mathbf{u}}^{FA}}{\partial t} + \widetilde{\mathbf{u}}^{FA}\frac{\partial \bar{\rho}}{\partial t} = \bar{\rho}\frac{\partial \widetilde{\mathbf{u}}^{FA}}{\partial t} + (\rho_1 - \rho_0)\,\widetilde{\mathbf{u}}^{FA}\frac{\partial \bar{\chi}}{\partial t}$$
$$= \bar{\rho}\frac{\partial \widetilde{\mathbf{u}}^{FA}}{\partial t} + (\rho_0 - \rho_1)\left[\widetilde{\mathbf{u}}^{FA} \cdot \nabla \chi + \widetilde{\tau_{\text{interf}}}^{FA}\right] \tag{7.26}$$

which involves the interfacial subgrid term in the momentum equation.

The Favre's averaging and $\widetilde{G}^{FA}$ filter will not be considered in the present work as they are more appropriate for conservation equations formulated in conservative form and applied to conservative variables. In our case, the framework is incompressible or low Mach number flows and the variables are the pressure and the velocity. These are non conservative variables and our formulation is also non-conservative. If we were working with models coming from compressible flows, the Favre's averaging will be more appropriate, even for two-phase flows with separated phases, as in [235, 252].

## 7.2 Filtering Discrete Mechanics Equations

### Reminder of general equations

The discrete mechanics (DM) is a new modeling framework introduced in preliminary works by Caltagirone [244–246] in order to unify the mechanics of solids and incompressible or compressible flows. The motion law in discrete mechanics is formulated as a Hodge-Helmholtz decomposition of acceleration vector:

$$\gamma = -\nabla\phi + \nabla \times \psi \tag{7.27}$$

Each component of acceleration can be split into a solenoidal or divergence free component $\nabla\psi$ and an irrotational one $\nabla \times \phi$. They both act without interaction on the edges of each DM element because they are derived from orthogonal fields. The equation of discrete motion fo DM is derived from the conservation equation of acceleration by expressing the deviations of the potentials $\phi$ and $\psi$ as a function of the velocity $\mathbf{u}$. These deviators are obtained on the basis of the physical analysis of the storage-destocking processes of compression and shear energies; the first is written as the divergence of velocity and the second as a dual curl velocity. The physical modeling of these terms is developed in [245]. The vectorial equation of motion and its updates are formulated as the following

$$\gamma = \frac{\partial \mathbf{u}}{\partial t} + \nabla \left( \frac{\|\mathbf{u}\|^2}{2} \right) - \nabla \times \left( \frac{\|\mathbf{u}\|^2}{2} \mathbf{n} \right)$$
$$= -\nabla \left( \phi^0 - dt c_l^2 \nabla \cdot \mathbf{u} \right) + \nabla \left( \psi^0 - dt c_t^2 \nabla \times \mathbf{u} \right) \tag{7.28}$$
$$\alpha_l \phi^0 - dt c_l^2 \nabla \cdot \mathbf{u} \longmapsto \phi^0 \tag{7.29}$$
$$\alpha_t \psi^0 - dt c_t^2 \nabla \times \mathbf{u} \longmapsto \psi^0 \tag{7.30}$$
$$\mathbf{u}^0 - dt \gamma \longmapsto \mathbf{u}^0 \tag{7.31}$$

where $\phi^0$ and $\psi^0$ are the potentials of acceleration defined at a reference initial instant $t_0$. These potentials are advanced in time at moment $t$ to pass from one mechanical equilibrium to another over a period of time $dt$. This time upgrade is represented by symbol $\longmapsto$. The quantities $c_l$ and $c_t$ are the longitudinal and transverse celebrities (written here for a Newtonian fluid). In the motion equation, $\mathbf{n}$ is the normal to the local DM element.

For a fluid, the factor $\alpha_l = 1$ whereas $\alpha_t \to 0$. In solids, $\alpha_t = 1$. Upgraded quantities $(\phi^0, \psi^0)$ should be advected by the velocity field. This advection, which independent of the equation (7.28), can be realized in an Eulerian framework by the resolution of a transport equation or in a Lagrangian frame. Similarly, if density is not constant, for two-phase flows for example, $\rho$ must be advected by an appropriate technique. The form of the inertial terms in (7.28) differs from that of the continuum mechanics, and also corresponds to a Hodge-Helmholtz decomposition. These terms may be partially implicited for better time discretization and stability.

**LES approach based on single filtering scale**

Let us now consider an unresolved flow field modeled by DM approach. It is assumed that the discrete mapping of the physical space brought by DM is associated to characteristic size of DM elements larger than the Kolmogorov scale $\eta$. As for the 1FNS, the equations for motion of DM, i.e. equation (7.28), has to be filtered in the sense of LES. Here, only the implicit filter $\bar{G}$ imposed by the mesh and associated numerical methods is considered.

$$\overline{\gamma} = \frac{\partial \overline{\mathbf{u}}}{\partial t} + \nabla \left( \frac{\overline{\|\mathbf{u}\|^2}}{2} \right) - \nabla \times \left( \overline{\frac{\|\mathbf{u}\|^2}{2} \mathbf{n}} \right)$$
$$= -\nabla \left( \overline{\phi^0} - dt c_l^2 \nabla \cdot \overline{\mathbf{u}} \right) + \nabla \left( \overline{\psi^0} - dt c_t^2 \nabla \times \overline{\mathbf{u}} \right) \tag{7.32}$$

By applying the implicit filter $\bar{G}$ described in Sect. 7.3 to equation (7.28) and by remembering that the filtering operator is not commutative with respect to products, we have:

$$\overline{\|\mathbf{u}\|^2} = \|\overline{\mathbf{u}}\|^2 + \overline{\|\mathbf{u}\|^2} - \|\overline{\mathbf{u}}\|^2 = \|\overline{\mathbf{u}}\|^2 + d_{m,1}(\mathbf{u}, \overline{\mathbf{u}}) \tag{7.33}$$
$$\overline{\frac{\|\mathbf{u}\|^2}{2} \mathbf{n}} = \|\overline{\mathbf{u}}\|^2 \overline{\mathbf{n}} + \overline{\frac{\|\mathbf{u}\|^2}{2} \mathbf{n}} - \|\overline{\mathbf{u}}\|^2 \overline{\mathbf{n}} = \|\overline{\mathbf{u}}\|^2 \overline{\mathbf{n}} + \mathbf{d}_{m,2}(\mathbf{u}, \overline{\mathbf{u}}) \tag{7.34}$$

The final implicit filtered DM model IFDM reads

$$
\begin{aligned}
\overline{\gamma} &= \frac{\partial \overline{\mathbf{u}}}{\partial t} + \nabla \left( \frac{\|\overline{\mathbf{u}}\|^2}{2} \right) - \nabla \times \left( \frac{\|\overline{\mathbf{u}}\|^2}{2} \overline{\mathbf{n}} \right) + d_{m,1}(\mathbf{u}, \overline{\mathbf{u}}) - \mathbf{d}_{m,2}(\mathbf{u}, \overline{\mathbf{u}}) \\
&= -\nabla \left( \overline{\phi^0} - dt c_l^2 \nabla \cdot \overline{\mathbf{u}} \right) + \nabla \left( \overline{\psi^0} - dt c_t^2 \nabla \times \overline{\mathbf{u}} \right)
\end{aligned} \tag{7.35}
$$

It has to be noted that in (7.35), even in the multiphase flow context, only the non-linear inertial terms are inducing SGS terms $d_{m,1}$ and $\mathbf{d}_{m,2}$ that have to be modeled. Indeed, in the DM approach, the density and viscosity, or $c_l$ and $c_t$, can be assumed constant by part on edges or elements in a kind of 2-fluid discontinuous approach. No averages nor mixing are considered in the DM approach. We recall here that the resolved variable is $\overline{\mathbf{u}}$. As a consequence, the modeling of $d_{m,1}$ and $\mathbf{d}_{m,2}$ will be based on the use of structural LES to approximate $\mathbf{u}$ and so to calculate $\overline{\|\mathbf{u}\|^2}$ and $\overline{\|\mathbf{u}\|^2 \mathbf{n}}$. This will be developed with ADM methods in the next section.

The implicit filter IFDM model can be considered in a different way, by introducing the decomposition of the variable described classically in Sect. 7.3 and introducing this decomposition in (7.28). The following is obtained

$$
\begin{aligned}
\overline{\gamma} + \gamma_{SGS} &= \frac{\partial \left[\overline{\mathbf{u}} + \mathbf{u}_{SGS}\right]}{\partial t} + \nabla \left( \frac{\| \left[\overline{\mathbf{u}} + \mathbf{u}_{SGS}\right] \|^2}{2} \right) - \nabla \times \left( \frac{\| \left[\overline{\mathbf{u}} + \mathbf{u}_{SGS}\right] \|^2}{2} \mathbf{n} \right) \\
&= -\nabla \left( \overline{\phi^0} - dt c_l^2 \nabla \cdot \left[\overline{\mathbf{u}} + \mathbf{u}_{SGS}\right] \right) + \nabla \left( \overline{\psi^0} - dt c_t^2 \nabla \times \left[\overline{\mathbf{u}} + \mathbf{u}_{SGS}\right] \right)
\end{aligned} \tag{7.36}
$$

The filtered IFDM equation, acting on $\bar{\gamma}$ or $\overline{\mathbf{u}}$, reads

$$
\begin{aligned}
\overline{\gamma} &= \frac{\partial \overline{\mathbf{u}}}{\partial t} + \nabla \left( \frac{\|\overline{\mathbf{u}}\|^2}{2} \right) - \nabla \times \left( \frac{\|\overline{\mathbf{u}}\|^2}{2} \mathbf{n} \right) - \gamma_{SGS} \\
&= -\nabla \left( \overline{\phi^0} - dt c_l^2 \nabla \cdot \overline{\mathbf{u}} \right) + \nabla \left( \overline{\psi^0} - dt c_t^2 \nabla \times \overline{\mathbf{u}} \right) - \gamma_{SGS}
\end{aligned} \tag{7.37}
$$

$$
\begin{aligned}
\gamma_{SGS} &= - \left[ \frac{\partial \mathbf{u}_{SGS}}{\partial t} + \nabla \left( \frac{\|\mathbf{u}_{SGS}\|^2}{2} \right) - \nabla \times \left( \frac{\|\mathbf{u}_{SGS}\|^2}{2} \mathbf{n} \right) \right] \\
&= - \left[ \nabla \left( dt c_l^2 \nabla \cdot \mathbf{u}_{SGS} \right) - \nabla \left( dt c_t^2 \nabla \times \mathbf{u}_{SGS} \right) \right]
\end{aligned} \tag{7.38}
$$

According to equation (7.35), we can observe that $\gamma_{SGS} = -d_{m,1}(\mathbf{u}, \overline{\mathbf{u}}) + \mathbf{d}_{m,2}(\mathbf{u}, \overline{\mathbf{u}})$. To conclude with IFDM, the SGS terms to be modeled are $\gamma_{SGS}$ or equivalently $d_{m,1}$ and $\mathbf{d}_{m,2}$.

## 7.3 Structural LES and Approximate Deconvolution Models (ADM)

### General definitions and notations

This ADM model was first proposed by Adams and Stolz [222, 253]. This structural LES approach is based on the approximate evaluation of the inverse of a filter G through a deconvolution operator, called $Q_N$. This operator can be used to find an approximation of the DNS quantities before applying filter G. The approximated DNS quantities $\Phi$ estimated by the ADM are denoted $\Phi^\star$. They read:

$$\Phi^\star = Q_N * \breve{\Phi} = Q_N * G * \Phi \approx \Phi \tag{7.39}$$

with $*$ the convolution operator and $\breve{\cdot}$ the implicit $\overline{\cdot}$ or explicit $\widehat{\cdot}$ filtered variables. The approximate inverse filter $Q_N$ is so obtained

$$Q_\infty = \sum_{l=0}^{\infty} (\underline{\underline{Id}} - G)^l \approx G^{-1} \tag{7.40}$$

$$Q_N = \sum_{l=0}^{N} (\underline{\underline{Id}} - G)^l \approx G_N^{-1} \tag{7.41}$$

Here, $G$ can either be the implicit or the explicit filter. By definition, $N$ represents the order of the ADM model. In their publication, Adams and Stolz [253] obtained a good agreement by choosing $N = 3$ and showed that an order larger than 5 did not improve the results. In this section, $\underline{\underline{Id}}$ (and any kind of other double underline notation) is related to the identity tensor and not to a quantity filtered twice. Thus, for single-phase flows, $N$ is usually taken equal to 4 or 5. For multiphase flows, recent a priori analysis have demonstrated that using ADM with orders up to 7 can improve the accuracy of the modelled SGS terms [233, 235]. It has to be reminded that all SGS terms are the consequence of the non commutative property of filter $G$ with respect to products of 2 variables $\Phi_1$ and $\Phi_2$. In a general formulation, each previously formulated SGS term, noted $F_S GS$, can take the form

$$F_{SGS}(\Phi_1, \Phi_2) = \breve{\Phi_1 \Phi_2} - \breve{\Phi}_1 \breve{\Phi}_2 \tag{7.42}$$

With ADM modeling approach, we assume that the DNS variable $\Phi$ can be approximated or estimated by means of the deconvoluted variable $\Phi^\star$. This way, the mathematical nature of the SGS terms can be respected (no functional modeling is required) and all SGS terms can be rewritten as follows:

$$F_{SGS}(\Phi_1, \Phi_2) = \breve{\Phi_1^\star \Phi_2^\star} - \breve{\Phi}_1 \breve{\Phi}_2 \tag{7.43}$$

with $\breve{\Phi}_1$ and $\breve{\Phi}_2$ the filtered resolved variables and $\Phi_1^\star$ and $\Phi_2^\star$ the deconvoluted variables approximated with ADM.

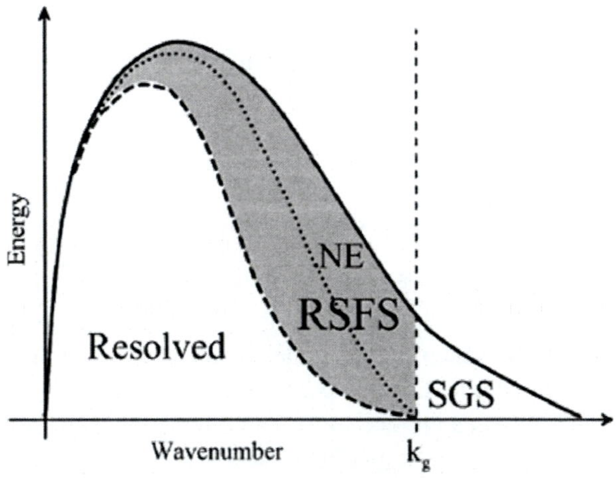


**Fig. 7.1** Schematic view of the velocity energy spectrum with linear axis, RSFS refers to the resolvable subfilter-scales, NE to the numerical errors done during the deconvolution process (located above the dotted line) and SGS to the subgrid scales, the grid filter corresponds to the dashed line at the wavenumber kg and the explicit filter to the curved dashed line [254]

In practice, one can note that replacing the DNS variable $\Phi$ by the deconvoluted variable $\Phi^\star$ is not sufficient to model the effect of the subgrid scales on the resolved ones. Chow and Street [254] proposed a very interesting illustration to understand the effect of deconvolution models. As shown in Fig. 7.1, the total energy spectrum can be split into three parts: the resolvable structures, the Resolvable Sub Filter-Scale (RSFS) that are submitted to an explicit filter and the Unresolvable SubFilter-Scale (USFS) induced by the implicit filter, also called Sub Grid Scale (SGS). The RSFS can be theoretically reconstructed by the inverse filter operation. This procedure is limited by the Numerical Errors (NE) which increase close to the grid cutoff [255]. The SGS terms are in fact not reconstructed after applying deconvolution models, even though the considered filter is the implicit filter (imposed by the mesh). Indeed, to deconvolute the implicit filtered quantity $\overline{\Phi}$, an explicit filter $G = \widehat{\Phi}$ is used. So, by using directly the resolved variable $\overline{\Phi}$ to estimate $\Phi^\star$ according to $Q_N$, that depends on the explicit filter, large explicit filtering scales and small implicit filtering scales are mixed in a wrong manner. Practically, only $\overline{\Phi}$ is known and resolved. Thus, an additional relaxation term in the momentum balance equation is required to reproduce the energy dissipation of the subgrid scales [236, 256, 257]. This term has the following structure $\Lambda_R \left(\underline{\underline{Id}} - Q_N * G\right) * \overline{\Phi}$. Different methods have been proposed to implement this term and especially to evaluate the value of the parameter $\Lambda_R$ [257]. One of these methods consist in applying a Smagorinsky formulation with a Germano procedure [258]. This choice requires to adapt the subgrid term expression (7.43) so that the predicted model coefficient has a non zero value:

$$F_{SGS}(\Phi_1, \Phi_2) = \breve{\Phi_1^\star \Phi_2^\star} - \breve{\Phi_1^\star}\, \breve{\Phi_2^\star} \tag{7.44}$$

The expression of the dynamic Smagorinsky coefficient $C_S$ is changed compared to functional LES dynamic models and is written as:

$$C_S = \sqrt{\frac{\left(\overline{\overline{L}}_{ij} - \overline{\overline{H}}_{ij}\right)\overline{\overline{M}}_{ij}}{2\overline{\overline{M}}_{ij}^2}} \tag{7.45}$$

To obtain equation (7.45), Germano et al. [258] proposed to vary dynamically the coefficient $C_S$ in order to cancel the eddy viscosity near walls. In what follows, the dynamic version of the Smagorinsky model is referred as the Smagorinsky model with the dynamic procedure. For this purpose, they consider a second filter, called test filter and noted $\check{\Phi}$ when applied to $\Phi$. This filter is chosen twice the size of the implicit mesh filter, i.e. $\frac{\check{\Delta}}{\overline{\Delta}} = 2$. With this regards, if $C_S$ is used in a vectorial equation whose variable is the velocity $\mathbf{u}$, the different tensors in (7.45) are defined as

$$\overline{\overline{L}}_{ij} = \check{\overline{\mathbf{u}}}_i \, \check{\overline{\mathbf{u}}}_j - \check{\overline{\mathbf{u}}_i \, \overline{\mathbf{u}}_j} \tag{7.46}$$

$$\overline{\overline{H}}_{ij} = \check{\overline{\overline{\mathbf{u}}}}_i \, \check{\overline{\overline{\mathbf{u}}}}_j - \check{\overline{\overline{\mathbf{u}}}_i \, \overline{\overline{\mathbf{u}}}_j} \tag{7.47}$$

$$\overline{\overline{M}}_{ij} = \check{\Delta}^2 \check{\overline{\underline{\underline{S}}}}_{ij} \, \check{\overline{\underline{\underline{S}}}}_{ij} - \overline{\Delta}^2 \check{\overline{\underline{\underline{S}}}_{ij} \, \overline{\underline{\underline{S}}}_{ij}} \tag{7.48}$$

with $\underline{\underline{S}} = \frac{1}{2}\left[\nabla\mathbf{u} + (\nabla\mathbf{u})^t\right]$ the deformation tensor.

Following [239], a van Cittert approximation can be used for ADM estimate $\Phi^\star$ of the DNS variable $\Phi$. It consists in using $N$ iteration of Picard algorithm for the fixed point problem as follows:

$$\text{Given } \check{\Phi} \text{ solve } \Phi^\star = \Phi^\star + \left[\check{\Phi} - G * \Phi^\star\right] \text{ for finding } \Phi^\star \tag{7.49}$$

The van Cittert approximate deconvolution operator $Q_N$ is defined as

$$\text{Set } \Phi_0 = \check{\Phi} \tag{7.50}$$

$$\text{for } n = 1, 2, \ldots, N-1 \, \Phi_{n+1} = \Phi_n + \left[\check{\Phi} - G * \Phi_n\right] \tag{7.51}$$

$$\text{Define } Q_N * \check{\Phi} = \Phi_N \tag{7.52}$$

By definition, $\Phi^\star = Q_N * \check{\Phi} = \Phi_N$.

The ADM methodology and related relaxation terms have been implemented by different research groups mainly for single phase flows [254, 259–262], but also with some rare attempts concerning multiphase flows with resolved scale interfaces [233, 235, 236], all starting with a formalism of the filtered Navier-Stokes equations in a single implicit filter framework. An extension of the ADM, in which 2 filtering scales are considered, will be proposed in the next sections for multiphase flows and DM model, as an extension of pioneer works of [243]. The interest of using two filters (explicit and implicit) since the filtered equation formulation is that the coupling between of the resolvable, RSFS and SGS terms can be explicitly formulated and so used for designing models based on deconvolution approaches.

**Filtering operations**

Three levels of filtering have been introduced in the previous section devoted to general definitions of ADM. The first filter, that will be always present when dealing with LES is the implicit filter. It is imposed by the mesh size and its filtering action is also dependent on the action of numerical methods. If a subgrid model is used for SGS terms, the implicit filter is also correlated to this SGS models. To fix the implicit filter width, the assumption is usually made that the effect of the grid cell plays a leading role [38]. Therefore, for a three dimensional mesh, Deardoff [263] proposed the following filter width:

$$\overline{\Delta} = \frac{1}{3} (\Delta x \Delta y \Delta z)^{1/3} \tag{7.53}$$

The choice of the cell size is then a crucial issue especially when dispersed phases and interfaces are present in the flow. Indeed, the filter width must be small enough but higher than the diameter $d_{db}$ of the dispersed bubbles in the case of Euler/Euler models [224, 264]. Milelli [264] showed that the best ratio between the filter width and the bubble diameter is $1.2 \leq \frac{\overline{\Delta}}{d_{db}} \leq 1.8$. Moreover, when large interfaces are undertaken, the mesh has to be enough refined to solve both the turbulent scales and the interface deformation [265]. Therefore, a compromise between all theses limitations has to be defined in complex flows. In general, the finer as possible is used for resolved scales interfaces whereas for Euler/Euler approaches, the separation scale between the flow scale and the dispersed phase scale imposes the mesh refinement, together with the previous constraint due to the implicit filter.

The second level of filtering, called explicit filtering $\check{\cdot}$ or test filtering $\check{\cdot}$ (here, it is assumed that both explicit filters are the same, that is classically assumed in most of the literature works, even if these two levels of explicit filtering could be different), selects the smallest resolved scales. The choice for this filter is important since it is explicitly applied in structural turbulence models such as ADM but also for dynamic procedures encountered in functional LES ad used in ADM for relaxation terms as previously detailed. This explicit filtering is based on a scale-similarity assumption. To satisfy this assumption and so to limit numerical errors, the test filter, corresponding to the $\breve{G}$ operator, has to be similar to the implicit filter, referred to as the $\overline{G}$ filter. Usually, for a second order numerical scheme, the explicit filter width $\breve{\Delta}$ is twice the implicit filter width [266]. Different types of explicit filter have been used in the literature. They are reported here in 1D for a space variable $a$:

- Box or top hat filter:

$$\breve{G}(a) = \frac{1}{\breve{\Delta}} H\left(\frac{\breve{\Delta}}{2} - |a|\right) \tag{7.54}$$

with $H$ the Heaviside function and $\mathbf{x}$ the coordinate vector of the considered point. Because of its simplicity, this filter is widely used for a priori LES studies [14, 15, 41]. For a posteriori simulations, i.e. true LES, two main discrete box filters are generally considered: the trapezoidal filter based on the trapezoidal rule [254, 259, 267, 268] which has the following expression in 1D at second order:

$$\breve{\Phi} = \frac{1}{4}\left(\Phi_{i+1} + 2\Phi + \Phi_{i-1}\right) \tag{7.55}$$

and the Simpson filter using the Simpson rule [269]:

$$\breve{\Phi} = \frac{1}{6}\left(\Phi_{i+1} + 4\Phi + \Phi_{i-1}\right) \tag{7.56}$$

where $\Phi_i$ refers to the value of the quantity $\Phi$ to be filtered in cell $i$. The main advantage of the trapezoidal filter is that, in spectral space, the transfer function goes to zero at the grid cutoff contrary to the Simpson filter and therefore eliminates the highest wave-number that could be sustained by the grid [270].
- Gaussian filter:

$$\breve{G}(a) = \sqrt{\frac{6}{\pi \breve{\Delta}^2}} \exp\left(\frac{-6a^2}{\breve{\Delta}^2}\right) \tag{7.57}$$

In the Fourier space, the Gaussian filter is still Gaussian. Thus, it does not have a sharp cutoff in the Fourier space.
- Sharp spectral filter:

$$\breve{G}(a) = \frac{\sin\left(\frac{\pi a}{\breve{\Delta}}\right)}{\pi a} \tag{7.58}$$

This filter corresponds to an Heaviside function in the Fourier space with a sharp cutoff. It has been used in [271, 272].
- Padé or Compact discrete filter: Lele proposed in [273] different compact discrete filters. Such filters provide a greater control over the shape of the transfer function. The order and parameter numbers can be easily managed. Many research groups have chosen these filters at different orders as test/explicit filter for deconvolution procedures [253, 261, 262, 274].

Several researchers have worked on the filter choice. One main difference between all these available filters is the cutoff in the Fourier space which is either sharp or smooth. Thus, De Stefano and Vasilyev [275] showed that if LES was based on smooth filter, then subgrid scale models should also model the effect of the filter on large scales and remove (or add) energy at the resolved scales. Otherwise, to minimize the effect of filter on large scales dynamics and energy transfer, the filter has to be chosen with a sharp cut-off. Lund [271] also showed that non-sharp filters in the Fourier space lead to an incorrect dissipation, proportional to the degree to which the filter departs from a sharp-cutoff. A comparison of different explicit filters associated to ADM has been proposed on a single-phase flow by San et al. [269] and with the Smagorinsky model with a dynamic procedure by Gullbrand and Chow [259].

## 7.4 LES of Multiphase Flows

The numerical simulation of turbulent multiphase flows with resolved scale interfaces is still a challenging research topic. From numerical viewpoint, a reliable modeling strategy for interfacial turbulent flows must include proper computational treatment first for the modeling of the two-phase motion (i.e. in general the solution of the Navier-Stokes equations), also the interface capturing between phases and finally the modeling or solving of turbulence. This latter aspect is the subject of the present section. The modeling of turbulent multiphase flows has improved in many aspects in the past decade thanks to the advances in the high performance scientific computing. As the Direct numerical simulation (DNS) is nearly impossible to achieve even for moderate Reynolds numbers, the great majority of numerical simulation of industrial applications has adopted Reynolds Averaged Navier-Stokes (RANS) models which can predict the general behavior of the flow as well as statistical representation of turbulence based on time- or ensemble-averaged quantities. Due to its intrinsic averaged and steady framework, this approach is not well suited for modeling highly unsteady two-phase flows dealing with small scale interface fluctuations such as small droplet formation in liquid atomization or interface rupture and coalescence in water/oil phase inversion problem. Therefore, large eddy simulation (LES) is typically considered as an intermediate solution for modeling such processes with unsteady or unstationary considerations. The principal idea behind LES is to reduce the computational cost by filtering small scale motions spatially, and only resolve the largest ones. Therefore, the problem of addressing the small scale fluctuations is simply shifted to modeling the unresolved subgrid scale terms in the spatially-filtered governing equations [230].

Since late 2000s, a variety of a priori studies on DNS results have been done investigating the contribution of all subgrid terms appearing in the filtered Navier-Stokes equation, into the turbulent kinetic energy budget [15, 276]. Labourasse and co-workers [41] presented the mathematical background and formulation of subgrids for LES of interfacial flow which is then followed by [15] to perform a DNS study on phase inversion problem. The latter concluded that the effect of subgrid terms of surface tension and interface cannot be neglected. For instance, the contribution of subgrid scale surface tension force could be up to 20% of the maximum filtered inertia during the latest stage. Following this work, [14] provided an a priori analysis of the subgrid terms by means of a parametric study to investigate for various surface tension coefficients, density and dynamic viscosity ratios. The observations of this parametric study on the density, viscosity and surface tension coefficient highlight the predominant role of two subgrid terms associated to viscous stress tensor and surface tension force. The work of [277] introduced a dual-scale modeling approach for liquid-gas turbulent interface. This approach aimed to maintain a fully resolved representation of the phase interface geometry on a separate high-resolution grid. Following this approach, they proposed subgrid models for the dynamics of immiscible interfaces when sub-filter surface tension forces are dominant. This is the case during the final stages of topology change events, where length scales approach zero, or for immiscible fluids of similar densities and viscosities, like water/oil systems [230, 278]. Shirani and coworkers [279] extended their previous model in correlating the pressure fluctuations and interface in RANS modelling of two immiscible flows [280] to model the unclosed subgrid terms of surface tension and VOF equation. However, they have not provided much details of derivation of the closure terms as well as implementation in the CFD solver. They also validated their LES against the DNS results by Vincent et al. [15] and demonstrated comparable results with DNS. But this comparison was only done for few initial seconds of simulation and no validation is reported for the latest stages of the problem when the subgrid surface tension effects become predominant. Recently, [235] investigated the influence of each subgrid term and evaluated the performance of different existing single-phase LES models for the phase inversion problem. They highlighted two important conclusions: first, the convective SGS term is not always predominant in two-phase flows with fragmentation and rupture of interface and secondly, only structural LES approaches are able to model all subgrid terms including those corresponding to the two-phase nature of the flow. Among structural models, ADM has shown clearly a better performance. According to this a priori study, we have considered the ADM as the most feasible approach for a posteriori large eddy simulation of interfacial turbulence in the present study.

The ADM has been proposed by [222] and since early 2000s, this method has been further developed for large eddy simulation of incompressible and compressible flows with different applications including turbulent wall-bounded flow [223], shock capturing [281], turbulent jet [282], atmospheric boundary layer [254], channel flow [283] ... Later on, [260] proposed a coupled ADM with a functional dynamic model, and tested for cavity flow. Also, [284] has accentuated the applicability of ADM in the subgrid closure procedure of LES studies. More recently, San [269, 285] has improved the potentials of ADM by studying the effect of different spatial filters that are of the great importance in this mathematical approach. However, less attention is paid to the application of ADM in the context of multiphase flow modelling. Apart from the recent investigation by Schneiderbauer and Saeedipour [286] on the application of ADM in turbulent gas-solid flows, there are only few studies in the literature which utilized this structural approach to model SGS terms of interfacial flows. ADM was employed by [287] to reconstruct the subgrid surface tension from the solution of filtered equation. This work is further developed to propose the ADM-$\tau$ subgrid surface tension model in [288]. The validation of this method was done against the DNS result of water/oil phase inversion problem [235]. But the main focus

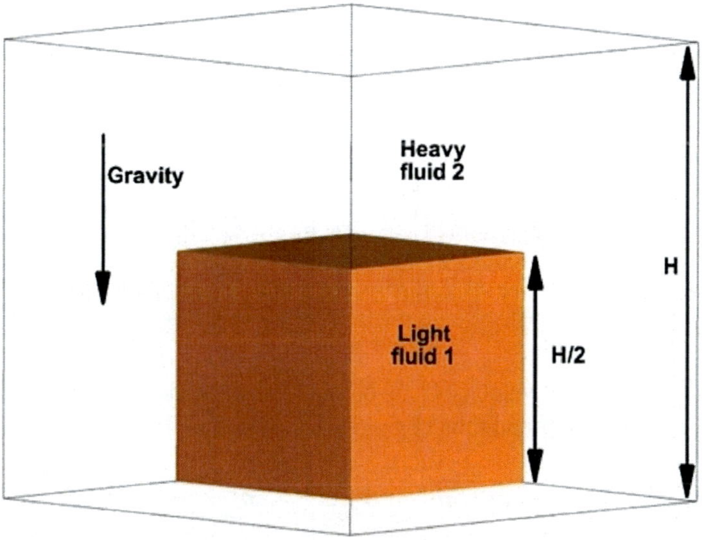


**Fig. 7.2** Definition sketch for the phase inversion benchmark in a closed box [14, 15, 289]

of these studies was only on the surface tension subgrid term in the context of level set [70], and the other subgrid effects such as interface dynamics were neglected. In the present section, the concept of ADM is extended to obtain the ADM-VOF formulation for all SGS terms of turbulent two-phase flows.

### A Priori LES filtering

The results of this section are taken from [235]. Various configurations of phase inversion benchmark have previously been proposed and used for a priori LES filtering, comparison of interface tracking methods, characterization of multi-scale interfacial flow and comparison of DNS two-phase flow codes [14, 15, 287, 289–291]. A previous configurations was chosen here for its turbulent regime and large range of interfacial scales. The initial condition of the problem is described in Fig. 7.2. An initial cubic blob of light liquid, referred to as fluid 1, is placed in the bottom part of a cubic box filled with a heavier liquid, referred to as fluid 2. The size of the box is $(H, H, H)$, while the size of the blob of light fluid is $(H/2, H/2, H/2)$, with $H = 1m$. All outer walls are considered as free-slip impermeable walls. The gravity is oriented along the vertical axis and its magnitude is $-9.81$ m s$^{-2}$. The interest of the present configuration lies in simplicity of initial and final flow configurations, i.e. all light fluid is inside a cubic blob initially and all light fluid is in the top part of the cavity at long times. Another advantage is that phase inversion naturally provides multiple coalescence, break-up and fragmentation events in a kind of atomization process, while not depending on complex initial or forcing conditions such as those encountered in real turbulent two-phase flow configurations [292, 293]. The characteristics of fluid 1 and 2 are $\rho_1 = 900$ kg m$^{-3}$, $\rho_2 = 1000$ kg m$^{-3}$, $\mu_1 = 0.1$ Pa s and $\mu_2 = 0.001$ Pa s. The constant surface tension coefficient is chosen equal to 0.45 N m$^{-1}$.

The fluids and flow properties that control the phase inversion dynamics can be expressed in terms of Atwood, Reynolds and Weber numbers. These numbers are defined as

$$At = \frac{\rho_2 - \rho_1}{\rho_1 + \rho_2}, \tag{7.59}$$

$$Re_1 = \frac{\rho_1 H U_g}{\mu_1} \text{ and } Re_2 = \frac{\rho_2 H U_g}{\mu_2}, \tag{7.60}$$

$$We_1 = \frac{\rho_1 H U_g^2}{\sigma} \text{ and } We_2 = \frac{\rho_2 H U_g^2}{\sigma}, \tag{7.61}$$

where the gravitational velocity $U_g$ is obtained as the balance between the pressure forces on the light fluid and the net gravity force on this fluid [15, 291]. It reads

$$U_g = \frac{\rho_2 - \rho_1}{\rho_1}\sqrt{\frac{gH}{2}} = At^{1/2}\frac{\rho_1 + \rho_2}{\rho_1}\sqrt{\frac{gH}{2}} \tag{7.62}$$

Based on $U_g$ and a characteristic space scale of $H/2$, the characteristic time scale is

$$t_c = \frac{H}{2U_g} = At^{-1/2}\frac{\rho_1}{(\rho_1 + \rho_2)}\sqrt{\frac{H}{2g}}. \tag{7.63}$$

The characteristic parameters of phase inversion are $At = 0.053$, $U_g = 0.246$ m s$^{-1}$, $Re_1 = 2.214 \cdot 10^3$, $Re_2 = 2.460 \times 10^5$, $We_1 = 1.210 \times 10^2$, $We_2 = 1.345 \times 10^2$, and $t_c = 2.032s$. An example of simulation with these parameters is illustrated in Fig. 7.3 for a $512^3$ grid with a one-fluid model. Due to large Reynolds and Weber numbers, the flow is first dominated by inertial effects induced initially by gravity forces with negligible capillary forces. For larger times, inertia decreases and surface tension forces become predominant. It is observed that initially the light fluid blob goes to the top of the cavity while being subjected to large interface deformation with rupture of ligaments into droplets. In a second step, a kind of atomization process occurs with the phase separation being active inducing coalescence of many droplets with the top layer of light fluid.

It has to be noticed that a recent complete modeling of all SGS terms has been proposed by [294] on the basis of functional modeling. The idea is to complete the standard SGS eddy viscosity models by a capillary contributions. A Favre averaging and VOF interface tracking are used. The models are based on the mixing length assumption which includes an additional production mechanism of turbulent kinetic energy stemming from the interfacial work i.e. surface tension. The model performance is evaluated by means of the a-priori filtering of the fine grid simulation of phase inversion problem.

Thanks to DNS simulations, the two-phase LES subgrid terms can be directly calculated by applying filter $G$ to the various variables of the phase separation problem. Moreover, the various LES models applied to $\tau_c$, $\tau_d$, $\tau_{st}$ and $\tau_{ia}$ can also be calcu-

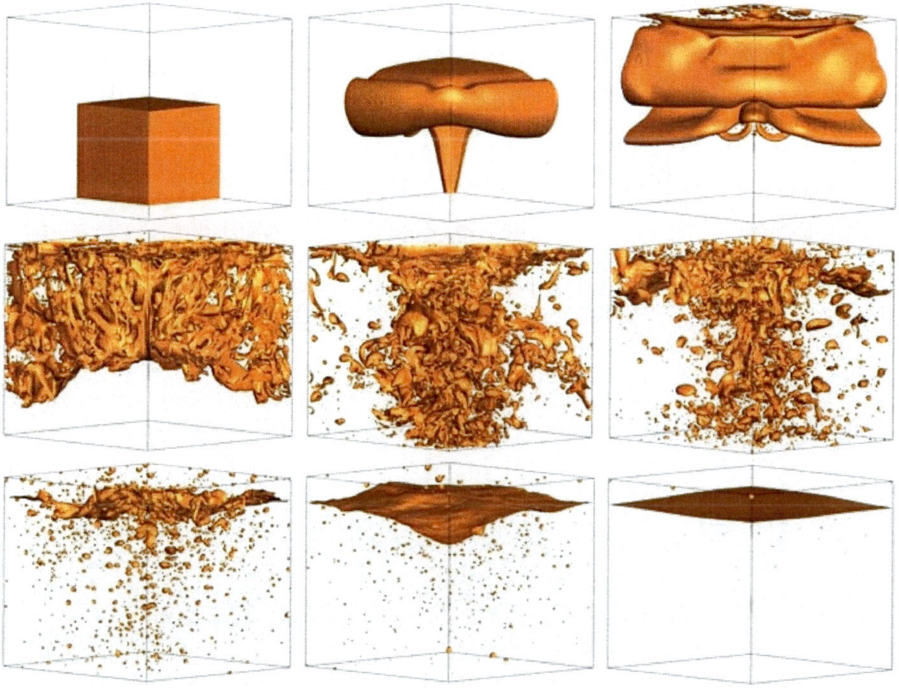

**Fig. 7.3** Simulation of a phase inversion in a closed box with a one-fluid model - The interface between the two liquids is plotted in orange - from left to right and top to bottom, the dimensionless time is 0, 0.246, 0.492, 0.738, 0.984, 1.329, 1.969, 3.396 and 15.305

lated with the DNS results. The main objective of the present section is to evaluate the representativeness of LES models presented in the previous section compared to filtered DNS subgrid terms. Results are presented in Fig. 7.4 in terms of correlation between a given model (Smagorinsky, Wale, Bardina, Mixed, ADM), for a given subgrid LES term, i.e. $\tau_c$, $\tau_d$, $\tau_{st}$ and $\tau_c$, and the filtered DNS of a corresponding subgrid term. The evaluation of LES models is provided for both OFM and MFA. On a general point of view, it is observed that compared to OFM, the models applied to MFA involve less dispersion of the correlation between LES models and filtered DNS. Indeed, the phase average $\alpha_k$ is present in all terms for MFA. It tends to eliminate the LES modeling far from the interfacial zone and it damps the gradients or filtering operations in the interfacial zone. On the contrary, in the OFM, all values of the models are a priori sharper and present everywhere, including zones far from the interface. In this case, a larger dispersion of correlation between LES models and filtered DNS subgrid terms is induced. If attention is now paid to the quality of models for a given subgrid term, several observations arise:

- only structural models are able to model all subgrid terms. Indeed, functional based models such as Smagorinsky, Wale or Mixed do rely on a physical interpretation of the behavior and interaction of under-resolved scales with large scales of

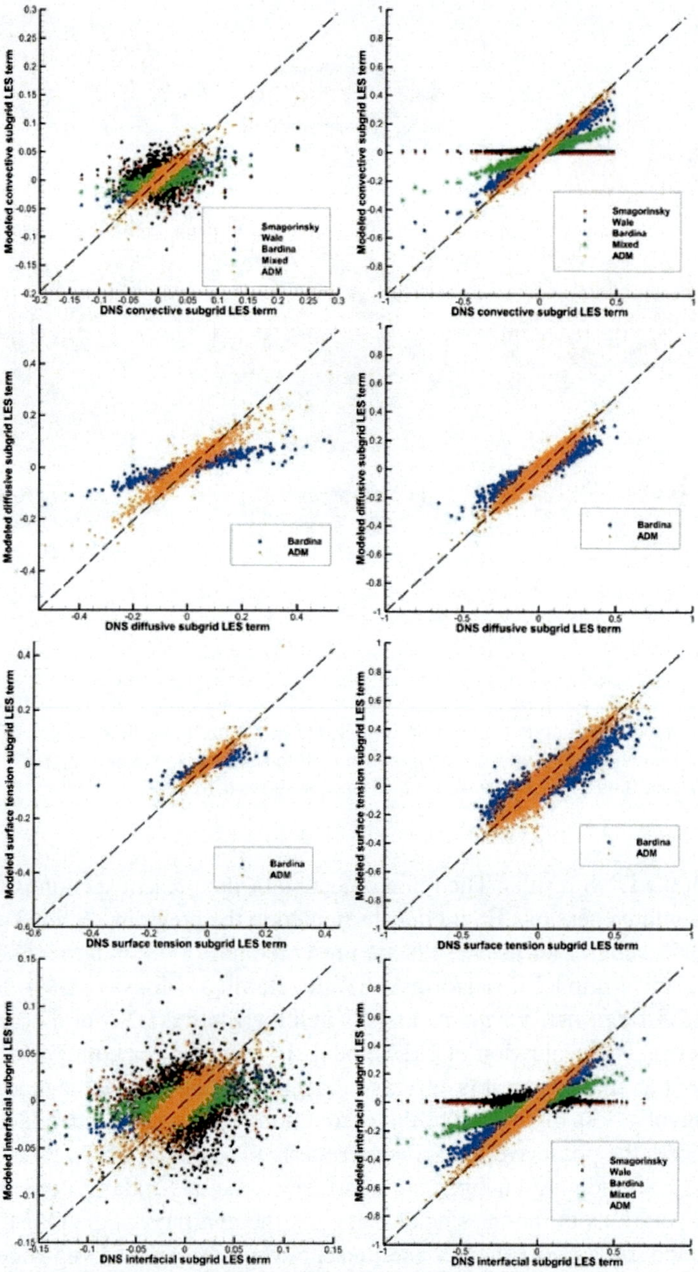


**Fig. 7.4** Spatial correlation of two-phase LES subgrid terms—The modeled terms are plotted according to the filtered DNS subgrid terms. A perfect modeling would be on the y = x line. Five LES models are considered: Smagorinsky [39], WALE [295], Bardina [40], Mixed model [296] and ADM [222]. Left column is for the one-fluid model whereas right column is for the multi-field approach

**Table 7.1** Average relative error (in %) of the modeled two-phase LES subgrid terms compared to the filtered DNS subgrid terms

| Subgrid term | Smagorinsky | Wale | Bardina | Mixed | ADM |
|---|---|---|---|---|---|
| *One-fluid model* | | | | | |
| $\tau_c$ | 93 | 108 | 40 | 64 | 17 |
| $\tau_d$ | – | – | 72 | – | 32 |
| $\tau_{st}$ | – | – | 54 | – | 24 |
| $\tau_{ia}$ | 105 | 88 | 62 | 78 | 58 |
| *Multi-field model* | | | | | |
| $\tau_c$ | 100 | 100 | 25 | 59 | 9 |
| $\tau_d$ | – | – | 43 | – | 20 |
| $\tau_{st}$ | – | – | 44 | – | 21 |
| $\tau_{ia}$ | 100 | 100 | 28 | 59 | 11 |

Both one-fluid and multi-field models are considered. Filter $\bar{\cdot}$ is applied to DNS, with a width of the low-pass filtering operator $G$ equal to 2 cells in each direction

turbulence (energy cascade of Kolmogorov, dissipative effect of small scale turbulent structures, ...). They are mainly based on interpretation of kinetic energy and inertial effects in turbulent flows. However, they do not consider the multi-scale character of interfaces when two-phase flows are dealt with. They so can only be applied to advection or inertial terms in conservation equations. This is the reason why functional LES approaches are not possible to be investigated for $\tau_d$ and $\tau_{st}$. If we pretend to perform real LES simulations accounting for all two-phase subgrid terms, using functional models seems to be inappropriate as they do not apply for all subgrid terms. This discussion is valid for both OFM and MFA.

- for the inertial and interfacial terms, all LES models can be compared. It is observed that for both one-fluid and multi-field approaches, the Smagorinsky and Wale models are not representative of filtered DNS terms. The Mixed model is better as it incorporates a contribution from the Bardina model that is working pretty well. The ADM approach is clearly the best for all terms and all modeling approaches (OFM and MFA). The functional based models involve a lot of dispersion and they do not respect the $y = x$ slope that should be obtained if a perfect correlation between LES models and filtered DNS will be reached. The interesting point here is that ADM based models (and also Bardina but with poorer results) are working for all subgrid terms and all two-phase flow modeling approaches (OFM and MFA). The average percentage error that is obtained for all models, all subgrid terms and all two-phase flow models is given in Table 7.1. It can be clearly inferred that Smagorinsky and Wale model are not suitable for two-phase subgrid terms, even concerning inertia. The representativeness of Mixed model is better than Smagorinsky, as it incorporates a part of Bardina modeling. The Bardina model is performing quite nicely for all subgrid terms. It behaves less well for $\tau_d$ and $\tau_{ia}$ when OFM is considered whereas its quality is worse for $\tau_d$ and $\tau_{st}$ with regard to MFA. It could be imagined to play on the Bardina constant $C_b$ so as to improve the performance of this model and to get closer to $y = x$ curve on the correlation plots. Finally, the ADM model is the best for all subgrid terms and all two-phase flow modeling approaches. The maximum error with MFA is at least 20% whereas this error is in the range 17–32% for OFM, with a peak of error of 58% for $\tau_{ia}$.

If true LES were simulated, the use of LES models for convective subgrid terms will be equivalent to add a turbulent viscosity in the viscous term of the momentum equations. It is interesting to evaluate the equivalent turbulent viscosity resulting from the use of a given LES model, compared to the equivalent viscosity that the filtering of the DNS provides when the two-phase convective LES subgrid term is considered. Classically [38, 41], we can write the turbulent viscosity $\mu_t$ for the OFM as follows:

$$\mu_t = \frac{\tau_c : \overline{\nabla \boldsymbol{u}}}{\overline{\boldsymbol{S}} : \overline{\nabla \boldsymbol{u}}} \tag{7.64}$$

And for the MFA:

$$\mu_t = \alpha_1 \frac{\tau_{c,1} : \overline{\nabla \boldsymbol{u}_1}}{\overline{\boldsymbol{S}}_1 : \overline{\nabla \boldsymbol{u}_1}} + \alpha_2 \frac{\tau_{c,2} : \overline{\nabla \boldsymbol{u}_2}}{\overline{\boldsymbol{S}}_2 : \overline{\nabla \boldsymbol{u}_2}} \tag{7.65}$$

Equivalent LES turbulent viscosities compared to equivalent filtered DNS one are plotted in Figs. 7.5 and 7.6. For OFM, the instantaneous turbulent viscosities are extracted in a vertical slice starting from point (0.6, 0.6, 0) whereas for MFA, the strating of the slice is (0.8, 0.8, 0). A time $t^* \approx 3$ is considered for the peak of enstrophy. The instantaneous simulations of OFM and MFA at a given time are not exhibiting the same interfacial and flow structures. The slices for turbulent viscosities have been chosen in order to cut different times the interface. If we focus on the equivalent DNS viscosity that we should obtained with LES models, we observe that negative values are obtained. This have previously been been reported by [41, 42]. This important feature of the effect of inertial subgrid term cannot be represented by functional LES approaches as they are intrinsically built on positive terms. This is illustrated in the top right and bottom plots of Figs. 7.5 and 7.6. This way, Smagorinsky and Wale models are again discriminated against structural ones. If we now compare Bardina, ADM and Mixed models, we notice that the first two are much better as expected. Concerning equivalent turbulent viscosities, we observe that ADM is globally nearer to DNS than Bardina, even if locally, Bardina can exhibit closer values to DNS reference.

To conclude on LES modeling of two-phase subgrid LES terms, it has been demonstrated that the structural approaches are the correct way of building models. The ADM formalism allows to recover the negative values of equivalent turbulent vis-

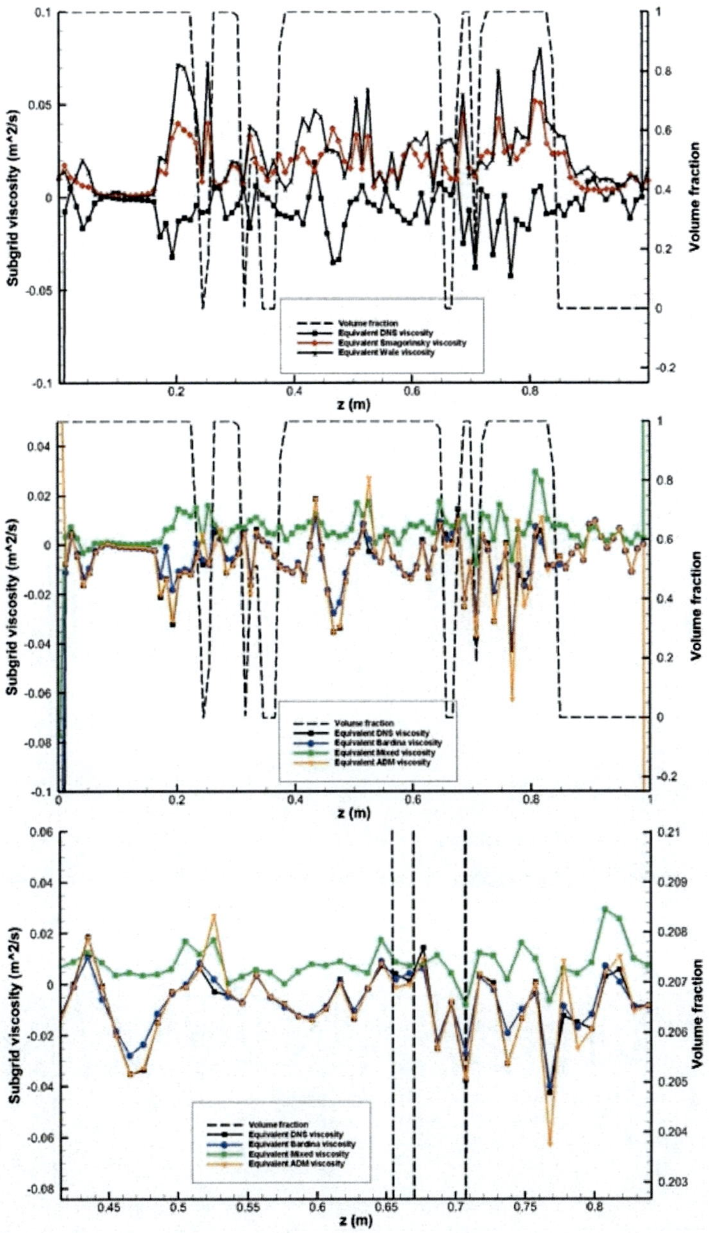


**Fig. 7.5** Comparison of equivalent turbulent viscosities obtained with DNS and LES OFM models, i.e. Smagorinsky [39], WALE [295], Bardina [40], Mixed model [296] and ADM [222]. The bottom line is a zoom on the middle plot. The results are plotted in a vertical line centered on $[x = 0.6, y = 0.6]$. The volume fraction is also plotted to locate interface position as $C = 0.5$

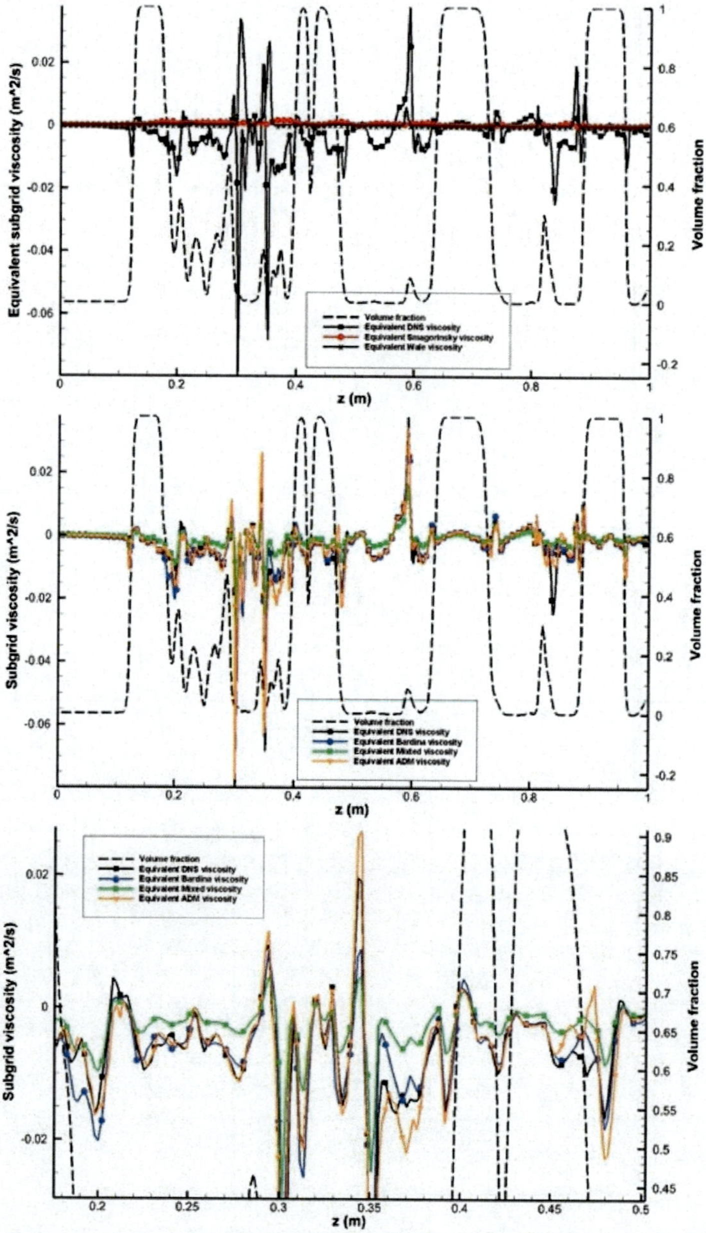


**Fig. 7.6** Comparison of equivalent turbulent viscosities obtained with DNS and LES MFA models, i.e. Smagorinsky [39], WALE [295], Bardina [40], Mixed model [296] and ADM [222]. The bottom line is a zoom on the middle plot. The results are plotted in a vertical line centered on $[x = 0.8, y = 0.8]$. The volume fraction is also plotted to locate interface position as $C = 0.5$

cosities in the good zones of the flow and they respect the behavior of filtered DNS subgrid terms with less than 30% discrepancy (50% for $\tau_{ia}$). We have to keep in mind that ADM modeling allows to reconstruct the fields between the explicit filter that we have and the implicit filter brought by the simulation mesh. In real LES, an additive relaxation term will have to be incorporated in the momentum equations [222] in order to obtain the effect of under-resolved scales smaller than the implicit filter size.

Based on direct numerical simulation of two-phase flow with resolved scale interfaces, i.e. a phase inversion problem, a priori filtering of two-phase subgrid LES terms has been investigated. The one-fluid model and multi-field approaches were considered and filtered. In addition, different LES models taken from the turbulent single-phase flow literature were utilized to propose functional and structural LES models for two-phase LES subgrid terms. The order of magnitude of all a priori filtered subgrid terms and LES models for these terms were compared to filtered DNS values for the first time on a same two-phase problem. The formulation of LES models for all subgrid terms was also proposed and discussed for the first time, in particular for a multi-field approach.

For the first time, the filtered LES formulation of one-fluid and multi-field approaches were compared in a same work and on a common phase inversion test problem with resolved scale interfaces and multi-scale interface topologies (kind of atomization process). An important result of the present study is to observe that the filtered LES subgrid terms obtained by both OFM and MFA can be (or not) modeled by the same models.

Concerning a priori filtering, the main conclusion is that the inertia term is not predominant in two-phase flows with fragmentation and rupture of interface that exhibit a kind of atomization process. This conclusion is different from that of the studies of [15, 20, 41–43]. As soon as small-scale interfacial structures are generated, correlation terms induced by the curvature in the subgrid surface tension terms or between the local interface presence and the velocity are predominant against inertial terms. The diffusive viscous term is always small in our phase inversion problem, whatever the two-phase flow model or the kind of filtering.

Five different LES models applied to two-phase subgrid LES terms have been compared against filtered DNS results. Two classes of models have been considered, the functional (Smagorinsky, Wale) and the structural (Bardina, ADM) approaches. A combined functional-structural model, called Mixed Bardina-Smagorinsky, have also been investigated. Conclusion are clear for both OFM and MFA: functional LES modeling do not correlate to filtered DNS results whereas structural approaches do. Bardina and ADM are clearly the good LES framework to consider for two-phase flows with resolved scale interfaces. ADM is clearly better than Bardina in our study. It has to be noticed that we do not have triggered the Bardina constant to try to improve the Bardina results. This could be useful to try to get closer to ADM. The improvement of Bardina, if possible, would be a nice point as ADM of $7^{th}$ is clearly more CPU time consuming. The Mixed approach, as expected, is better than functional models and worse than structural LES techniques. For the first time, we have been able to demonstrate that a common LES framework could be used for modeling all two-phase subgrid terms with resolved scale interfaces, i.e. the approximate deconvolution method. The main interest of structural approaches is their capability of representing negative equivalent turbulent viscosities whereas functional models are generating intrinsically positive viscosities. We have shown that filtered equivalent DNS viscosities are clearly negative in specific zones of the flow, not only near the interface. This negative feature is expected from LES models.

Future works will be oriented in two ways:

- performing a priori LES filtering and model comparisons on finer DNS meshes. A $2048^3$ grid is under consideration for phase inversion. The same analysis will be used on other two-phase flows with separated phases, such as the atomization of a liquid sheet in a HIT flow [8]. The idea is to control the turbulence forcing and to vary the properties of the fluid and interface properties to try to extract general conclusions on LES modeling frameworks for two-phase flows.
- investigating real ADM LES with both OFM and MFA. A coarse LES grid will be used in this case. The idea is to see if compared to DNS, the LES are able to capture the main features of the flow. The other important point is to compare implicit LES, i.e. LES without explicit modeling of the subgrid terms, and ADM models for all two-phase subgrid terms. In particular, the role of the introduction of a relaxation term [222] in the momentum equations will be discussed and evaluated compared to subgrid LES contributions. The starting test problem could be the phase inversion benchmark, for which different $Re$ and $We$ can be chosen in order to adjust inertial and capillary effects compared to viscous and gravity contributions. The interest of this case is that DNS can be generated and used for LES comparisons.

**A Posteriori LES filtering**

Employing the ADM technique for closing all the subgrid terms in the LES of interfacial flows within the framework of VOF, Front Tracking o LEvel Set interface tracking has been very little developed in the literature. Performing an a posteriori simulation with ADM for all the subgrid terms is now shortly commented. To the best of our knowledge, few works have been devoted to a posteriori, i.e. real LES, of multiphase flows. The main reason for that is that single LES turbulence models are not possible to be applied to all the SGS terms arising from the filtering of the motion equation for multiphase flows (generally 1FNS). Among the most advanced and interesting a posteriori LES of multiphase flows, we can cite:

- The SGS term related to capillary terms have been modeled by an ADM-$\tau$ approach [288] and has been investigated in terms of true LES simulations concerning phase inversion problem and droplet oscillation. The ADM-$\tau$ model consist in approximating the deconvoluted real normal and curvature based on ADM inverse filter approximations. A Level Set interface tracking method is considered in this work.
- The ADM approximation of all terms of the 1FNS have been used to perform LES of phase inversion by [233]. This is the first time that a posteriori LES simulation were conducted on multiphase flows with all SGS terms modeled. A discussion

on relaxation terms was also included. The same ADM-VOF model was applied to turbulent liquid jet interaction with an interface [297].

- The LES of jet atomization was also conducted by [238]. However, they did not use ADM approaches in their work. They used functional LES models such as the Smagorinsky approach. In addition, only the convective and surface tension contributions were considered.
- Toutant et al. [276] have proposed a real LES simulation based on a low order deconvolution model, applied to rising bubbles. Their study is based on a Bardina model, i.e. low order ($2^{nd}$ order) deconvolution model. In their work, they were able to demonstrate that negative equivalent eddy viscosities were required in the vicinity of the interface. This specific aspect of SGS terms is only possible to model with ADM, compared to functional model that generally enforce positive turbulent or subgrid viscosities.

# Chapter 8
# DNS of Particulate Flows

## 8.1 Fictitious Domain and Penalty Approaches

The basis of the DNS of particulate flows with resolved scale particles is in our case based on implicit fictitious domain methods developed in Sect. 5.2, and in particular the Darcy penalty method for fixed networks of particles and the ITPM for moving particles. Simple validations have also been presented for a single particle flow in two and three-dimensions in this section. Most of the results and developments of this chapter are taken from [23–31]. In particular, recent improvements of the ITPM have been proposed in [28, 29, 31] concerning the next choice of penalty parameters and the calculation of forces and heat transfers in the framework of fictitious domain and penalty methods.

Interactions between solid particles and a surrounding viscous fluid occurs in many environmental and industrial applications such as combustion devices [298], oil cracking [299], fluidized beds in coal stations and nuclear reactors [300], erosion of sediments on beaches [301] or hygiene and safety [302]. In all these applications, the interaction between the particles and the fluid flow is associated to unsteadiness or turbulence in a complex and not yet well understood manner concerning the drag laws of particles [303] or wall laws of the particulate flow as soon as the particle density is significant. On the one hand, leading simulations at the reactor scale for fluidized beds or combustion chambers require the implementation of macroscopic Eulerian-Eulerian models [304, 305] in which a separation scale is assumed between the particle and the flow scales. These models require drag laws and wall boundary conditions for the flow and the particles which are not yet well posed for dense particle flows. On the other hand, the Direct Numerical Simulation (DNS) of resolved particle flows, for which the grid size is smaller than the particle diameter, are generally restricted to academic configurations including a maximum of thousand of particles. The present section aims at presenting a numerical modeling strategy for the DNS of dense particle flows, validated on various physical test cases including particle-particle and particle-wall interactions or particle settling as well as numeri-



S. Vincent et al., *Small Scale Modeling and Simulation of Incompressible Turbulent Multi-Phase Flow*, CISM International Centre for Mechanical Sciences 607,
https://doi.org/10.1007/978-3-031-09265-7_8

cal convergence studies. This DNS particle flow solver will be devoted to simulating the interaction between unsteady or turbulent flows and dense whole of particles to understand the particle response to turbulent flows and the modification of turbulence due to the presence of the particles [306]. The DNS tool will allow to extract Lagrangian and Eulerian quantities such as the Lagrangian time scale of the particles, their drag laws or the auto-diffusion coefficients [22].

Simulating unsteady particle-laden flows where particle and fluid have the same characteristic scales, i.e. resolved scale particle flows, is the scope of the present work. The Direct Point Simulation (DPS) approaches based on a Lagrangian modeling of the particles does not remain valid for that purpose [307–309]. The field of numerical methods devoted to the simulation of particulate flows involving finite size particles was widely developed the last 20 years concerning the study of the flow over a small number of fixed or moving particle [202, 310–314] or a fixed arrangement of spheres [315]. All the previously mentioned works were developed on fixed grids. The unstructured Arbitrary Eulerian-Lagrangian (ALE) grid simulations were developed for particle flows in the 90s by Hu et al. [310] concerning two-dimensional flows involving two particles. The works of Maury [32] with the ALE method are also to be noticed concerning the flow of 1000 non spherical particles in a two-dimensional biperiodic domain. With the same type of approach, two-dimensional ALE simulations of particle motions involving 320 cylinders in a channel have been studied by Cho et al. [316]. However, few of the existing research works have been extended to the simulation of particle flows involving thousand of particles in three dimensions. The work of Hu et al. concerning three-dimensional ALE simulations of the segregation of 90 particles in a vertical channel are among the most advanced results with adapted unstructured grids [317]. A majority of the particulate flow simulations were carried out on structured grids to avoid the complexity of managing an evolving adapted mesh. Among them, the most relevant existing works concern the distributed Lagrangian multiplier (DLM) method of Glowinski and co-workers [318], the Physalis CFD code of Zhang and Prosperetti [319], the Immersed Boundary (IB) with direct forcing approach proposed by Uhlmann [320, 321], the lattice Boltzmann scheme [322, 323], the modified version of Uhlmann IB method published by Lucci et al. [324, 325] and the viscous penalty techniques of Ritz and Caltagirone [109] and Randrianarivelo et al. [21, 192].

To begin with, the DLM method [318] uses a variational formulation of the Navier-Stokes equations on a fixed grid and the Newton-Euler equations in the solid medium to model the particulate flow. On the interface between the particles and the fluid, Lagrange multipliers are integrated in the variational formulation to satisfy the rigid body motion inside the moving bodies. The force resulting from the interaction between particles is treated by an explicit modification of the particle positions after advection in order to avoid particle overlapping. A short range repulsive force is added in the right hand side of the Newton-Euler equations which is decreasing according to the distance between the center of mass of two particles [318]. The DLM method was utilized to simulate with success the fluidization of 1024 spheres in a square shape tank [326]. As for the Physalis approach [319], fixed grids for the flow solving are combined with analytical Stokes solutions in the vicinity of the particles in order to model the fluid-particle interaction. A cage surrounding the particle shape is used to match the flow solutions arising from a finite difference discretization with the spectral solutions of the Stokes flow. In their approach, Zhang and Prosperetti track in a Lagrangian manner the particles which are subjected to hydrodynamic and external forces. The interaction between the particles is not explicitly modeled. They have applied their numerical method to simulate the interaction between a Homogeneous Isotropic Turbulence (HIT) and particles in a periodic cubic box [319]. They have studied the effect of a two-way coupling procedure on the decaying of the turbulent kinetic energy and the mean particle displacement. Their works were restricted to low Taylor-microscale Reynolds number around 29 as their flow simulation grid was $256^3$. The largest particle simulations dealing with turbulent flow at a resolved particle scale have been proposed by Uhlmann concerning the sedimentation of 1000 particles in a periodic box on a $512^2 \times 1024$ grid [320] and the turbulent interactions inside a vertical particulate channel flow [321] on a $2048 \times 513 \times 1024$ grid. The numerical methods developed in these works are based on the addition of a direct forcing in the momentum equations to account of the particle presence. The particles are tracked in a Lagrangian manner and the forcing term is also estimated with a Lagrangian approach at the center of the particles. It is redistributed as an Eulerian force thanks to kernel or discrete Dirac functions, according to Peskin work [97]. The particle interaction is treated by means of the repulsive force introduced by Glowinski [318]. A modified version of the IB method of Uhlmann has been proposed by Lucci et al. [324, 325] to perform finite sized particle simulations. On a numerical point of view, their modifications concerned the time integration of the Navier-Stokes equations and the management of the two-way coupling force applied on the particles. Lucci and co-workers have investigated the modulation of a HIT flow by particles laying in the range 16–35 times the Kolmogorov scale at various Stokes numbers for Taylor-microscale Reynolds numbers of 75 and 110 respectively on $256^3$ and $512^3$ grids. Parametric studies on the density ratios between particles and fluid with a volume fraction in the range 0.01–0.1 have demonstrated for example that the turbulent kinetic energy was reduced by the particles compared to a reference single phase HIT flow. A different finite-size particle modeling has been proposed by Ten Cate et al. [322] which is related to lattice Boltzmann approaches. In their approach, the interaction between particles and fluid and the so the no-slip solid boundary condition is enforced through a body force added to the momentum equations. This force accounts for lubrication and repulsive effects during particle interactions. With the lattice Boltzmann model, Ten Cate and co-workers have studied the energy spectra and particle distribution functions for particles placed in a forced homogeneous turbulence for various Stokes numbers at a Taylor microscale $Re = 61$ on a $256^3$ grid. Other authors have used a classical bounce-back technique for simulating the sedimentation of turbulent particle-laden flows [323]. To finish with, a wide literature has been devoted to the numerical modeling of finite size particle flows by means of fictitious domains and penalty methods [109, 192, 211, 313]. The penalty techniques consist in playing on the magnitude of the dynamic viscosity to impose the local solid character of the medium. In our recent works, an Implicit Tensorial Penalty Method (ITPM), based on a decomposition of the viscous stress tensor, has been proposed to deal with particulate flows with an accuracy between 1 and 2 in space convergence [53, 54, 327]. The ITPM has been applied to the DNS of fluidized beds containing 2133 particles by Corre et al. [22].

In this chapter, the description of the second order in space penalty method for the DNS of particulate flows based on the ITPM is completed. The objective is to build an as implicit as possible approach in which the particle positions are not modified after tracking with the single fluid velocity obtained on structured Cartesian mesh, even in the case where particle collisions are present. The penalty method is chosen, instead of the DLM or direct forcing approaches, for its fully coupled character, robustness and easy implementation, as it is based on the existing physical terms in the momentum conservation equations. In addition, Lagrangian VOF approaches are implemented and the single fluid formulation of Kataoka [35] is extended to the framework of fluid-solid particle interactions. The whole algorithms are parallelized with MPI instructions for dealing with large grid simulations. To conclude, ITPM is applied to particulate flows with fixed and moving particle.

### *8.1.1 Physical Characteristics of the Equivalent Fluid*

The fictitious domain approach used in the present work considers the solid part of the particulate flow as a fluid with specific rheological properties whose evolutions are modeled by the Navier-Stokes equations. The present section explains how the characteristics of the solid equivalent fluid are defined to ensure a deformation free behavior in the related zones.

Classically, for any point **P** belonging to the solid zones of the calculation domain $\Omega$, i.e. the zones where $C = 1$, the solid behavior is characterized by a translation and a rotation. For any particle, constant translation and rotation velocities $\mathbf{u}_0$ and $\mathbf{w}$ exist such that

$$\forall \mathbf{P} \in \Omega^I, \exists \{\mathbf{w}, \mathbf{u}_0\}, \mathbf{u}(\mathbf{P}) = \mathbf{u}_0 + \mathbf{X_B^I P} \wedge \mathbf{w} \tag{8.1}$$

where $\mathbf{X_B^I P}$ is the barycenter of the considered particle $I$ and $\Omega^I$ the related solid subdomain. The solid constraint is intrinsically maintained if the deformation tensor is nullifed:

$$\forall \mathbf{P} \in \Omega^I, \nabla \mathbf{u} + \nabla^T \mathbf{u} = 0 \tag{8.2}$$

In other words, if we assume that (8.2) is satisfied, the following systems are obtained:

$$\begin{cases} \dfrac{\partial u}{\partial x} = 0 \\ \dfrac{\partial v}{\partial y} = 0 \\ \dfrac{\partial w}{\partial z} = 0 \end{cases} \tag{8.3}$$

$$\begin{cases} \dfrac{\partial u}{\partial y} = -\dfrac{\partial v}{\partial x} \\ \dfrac{\partial v}{\partial z} = -\dfrac{\partial w}{\partial y} \\ \dfrac{\partial w}{\partial x} = -\dfrac{\partial u}{\partial z} \end{cases} \tag{8.4}$$

Then, Eq. (8.3) give the direction dependencies of each velocity components as

$$\mathbf{u} = \mathbf{u}(y, z), \mathbf{v} = \mathbf{v}(x, z), \mathbf{w} = \mathbf{w}(x, y) \tag{8.5}$$

By using Eq. (8.4), three functions $\omega_1(x)$, $\omega_2(y)$ and $\omega_3(z)$ can be defined such that

$$\begin{cases} \dfrac{\partial u(y, z)}{\partial y} = -\dfrac{\partial v(x, z)}{\partial x} = \omega_3(z) \\ \dfrac{\partial v(x, z)}{\partial z} = -\dfrac{\partial w(x, y)}{\partial y} = \omega_1(x) \\ \dfrac{\partial w(x, y)}{\partial x} = -\dfrac{\partial u(y, z)}{\partial z} = \omega_2(y) \end{cases} \tag{8.6}$$

If it is assumed that the rotation velocity components have the necessary regularity, the Schwarz theorem gives a constant second order derivative of the velocities:

$$\begin{cases} \dfrac{\partial^2 u(y, z)}{\partial y \partial z} = \dfrac{\partial \omega_3(z)}{\partial z} = -\dfrac{\partial \omega_2(y)}{\partial y} = W_1 \\ \dfrac{\partial^2 v(x, z)}{\partial x \partial z} = \dfrac{\partial \omega_1(x)}{\partial x} = -\dfrac{\partial \omega_3(z)}{\partial y} = W_2 \\ \dfrac{\partial^2 w(x, y)}{\partial x \partial y} = \dfrac{\partial \omega_2(y)}{\partial y} = -\dfrac{\partial \omega_1(x)}{\partial x} = W_3 \end{cases} \tag{8.7}$$

It can be verified that the second order derivatives are equal to zero by solving the following system:

$$\begin{cases} W_2 = -W_3 \\ W_3 = -W_1 \\ W_1 = -W_2 \end{cases} \tag{8.8}$$

As a consequence, the functions $\omega_1$, $\omega_2$ and $\omega_3$ are constant and therefore continuous and derivable with continuous derivatives. Finally, by integrating equations (8.7), we obtain

$$\begin{cases} u = u_0 + \omega_3 y - \omega_2 z \\ v = v_0 + \omega_1 y - \omega_3 z \\ w = w_0 + \omega_2 y - \omega_1 z \end{cases} \tag{8.9}$$

The Eq. (8.1) is so verified if (8.2) is true. In this way, relations (8.1) and (8.2) are equivalent. Then, a solid constraint can be solved by computing $\nabla \boldsymbol{u} + \nabla^T \boldsymbol{u} = 0$. For the resolution of the momentum conservation Eq. (2.2) in the Navier-Stokes equations, this condition is asymptotically verified when $\mu \to +\infty$, as explained in Sect. 5.2.

According to the decomposition of the viscous stress tensor introduced in Sect. 5.2, an equivalent condition to (8.1) or (8.2) is to satisfy

$$\Lambda(\mathbf{u}) = 0 \tag{8.10}$$

$$\Gamma(\mathbf{u}) = 2\Theta(\mathbf{u}) \tag{8.11}$$

Indeed, it can be verified that $\nabla \boldsymbol{u} + \nabla^T \boldsymbol{u} = 2\Lambda(\mathbf{u}) + 2\Gamma(\mathbf{u}) - \Theta(\mathbf{u})$. For any Newtonian fluid or solid media, the relations $\kappa = 2\mu$, $\zeta = 2\mu$ and $\eta = \mu$ must be satisfied, as proposed in Sect. 5.2, together with $\mu \gg 1$ in the particles.

Numerically, Eqs. (8.2), (8.10) and (8.11) have to be implemented on the split viscous penalty grid described in Sect. 5.2. For internal solid nodes (those where $C = 1$) the solid viscosity is chosen to be 100–1000 times the viscosity of the carrier fluid. This choice will be discussed in the validation section concerning the accuracy of the simulations and the effect on the cost of the linear solvers. The previous remarks on the order of magnitude of the viscosity inside the solid parts stand for the continuous point of view.

On a discrete point of view, the flow grid cells cut by the fluid-solid interface $\Sigma$ must be distinguished compared to those entirely included in the particles. For these last nodes, the dynamic viscosity is chosen equal to $\mu_\infty \approx K \mu_{carrierfluid}$ with $100 \leq K \leq 1000$. For the fluid-solid cells, different methods can be used to define the homogenized viscosity. First, to avoid any direct interpolation of the viscosity coefficients associated to off diagonal viscous stress tensor components, an interpolation $C_\mu$ of the volume fraction is defined at the viscosity nodes for the split viscous penalty grid:

$$C_\mu = \frac{1}{4} \sum_N C_N \tag{8.12}$$

where $N$ denotes the indices of the pressure nodes located at the vertices of the cell to which $C_\mu$ belongs. Four different numerical viscous laws have been investigated according to $C$ for the diagonal viscous stress tensor terms $C_\mu$ and also a conditional indicator function $I_C$ satisfying $I_{C<0.5} = 1$ if $C < 0.5$ or $I_{C \geq 0.5} = 1$ if $C \geq 0.5$:

1. Numerical viscous law 1 (NVL1), the viscosity is defined in a discontinuous way according to

$$C_\mu : \kappa = 2\left[\mu_f I_{C<0.5} + \mu_s I_{C\geq 0.5}\right]$$
$$\zeta = 2\left[\mu_f I_{C_\mu<0.5} + \mu_s I_{C_\mu\geq 0.5}\right]$$
$$\eta = \mu_f I_{C_\mu<0.5} + \mu_s I_{C_\mu\geq 0.5}$$

2. Numerical viscous law 2 (NVL2), the viscosity is given by an arithmetic average:

$$\kappa = 2\left[(1-C)\mu_f + C\mu_s\right]$$
$$\zeta = 2\left[(1-C_\mu)\mu_f + C_\mu\mu_s\right]$$
$$\eta = (1-C_\mu)\mu_f + C_\mu\mu_s$$

3. Numerical viscous law 3 (NVL3), the viscosity is given by a harmonic average:

$$\kappa = 2\left[\frac{\mu_f\mu_s}{C\mu_f + (1-C)\mu_s}\right]$$
$$\zeta = 2\left[\frac{\mu_f\mu_s}{C_\mu\mu_f + (1-C_\mu)\mu_s}\right]$$
$$\eta = \frac{\mu_f\mu_s}{C_\mu\mu_f + (1-C_\mu)\mu_s}$$

4. Numerical viscous law 4 (NVL4), the viscosity is defined with a mixed average as proposed by Benkenida and Magnaudet [328]:

$$\kappa = 2\left[(1-C)\mu_f + C\mu_s\right]$$
$$\zeta = 2\left[\frac{\mu_f\mu_s}{C_\mu\mu_f + (1-C_\mu)\mu_s}\right]$$
$$\eta = \frac{\mu_f\mu_s}{C_\mu\mu_f + (1-C_\mu)\mu_s}$$

Concerning the density, an arithmetic average is used whatever its location on the discretization grid. The most clever choice of viscosity average for the ITPM will be discussed in the validation section.

### *8.1.2 Eulerian-Lagrangian VOF Method for Particle Tracking*

Once the particle center of mass is known in the computations, it has been explained that the ITPM requires to locate the interior and exterior of the particle thanks to a phase function $C$ in order to build the physical properties of the equivalent fluid such

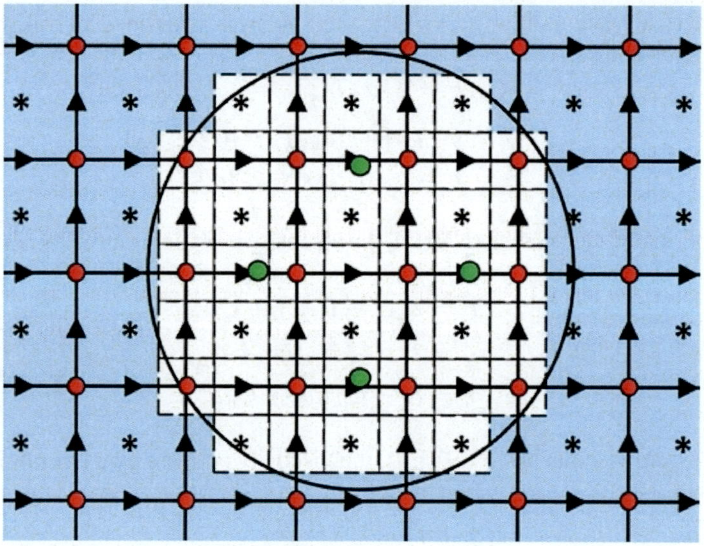

**Fig. 8.1** Position in two dimensions of the reference Lagrangian points (green points) used to interpolate the Lagrangian velocity at the barycenter of the particles

as the dynamic viscosity. This section explains how at each time step the finite size particles are advected and how the solid fraction $C$ is obtained on the Eulerian flow grid.

At each time step, the mass and momentum equations are solved by using the augmented Lagrangian approach explained before including the particle-particle collision forces $\boldsymbol{f}_{si}$ detailed in Sect. 8.1. Thanks to the ITPM, the one-fluid velocity field provides the solid velocity field inside the particles. Then, instead of using a classical Eulerian VOF approach [56, 73, 199] to obtain the new solid fractions in each Eulerian cells by solving Eq. (2.3), the Lagrangian velocities $\boldsymbol{V}^{I,L}$ of each particle, located in a Lagrangian way by their position $\boldsymbol{x}^{I,L}$, are computed at reference points (the four green points in Fig. 8.1, in the two-dimensional configuration) located inside each particle by using an interpolation method explained in Sect. 8.1. The new positions of the particles are then updated in a Lagrangian manner, as developed in Sect. 8.1 and finally the new Eulerian solid fractions are estimated by means of a projection of the exact shape of the particles on the flow grid. This last step is detailed in Sect. 8.1. The interest of using a Lagrangian scheme to advect the particles and to project their shape on the Eulerian mesh is to avoid any distortion of the particle shape over time, as would be induced by a classical Eulerian VOF scheme.

#### 8.1.2.1 Computation of Particle Velocity

The Lagrangian particle velocity located at the center of each solid sphere is computed by using 4 or 6 points respectively in two or three dimensions. The same procedure

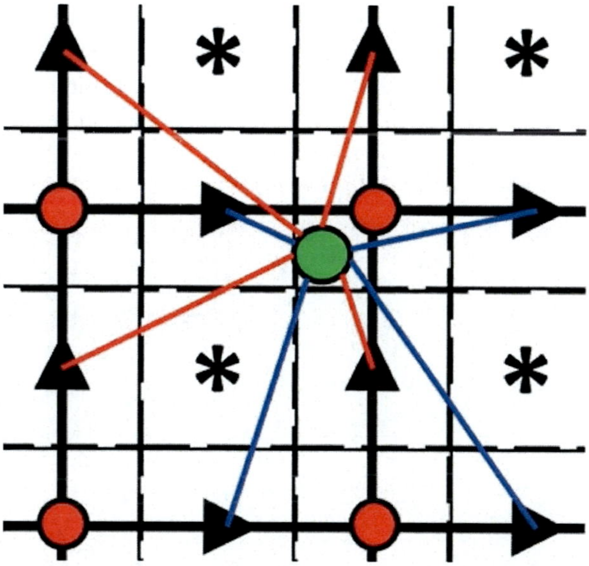

**Fig. 8.2** Eulerian velocity kernel (u-components in blue and v-components in red) used in two dimensions to estimate the components of the interpolation Lagrangian velocity points (green point)

allows to estimate the rotation velocity if required. In Fig. 8.1, the position of the velocity calculation point, colored in green, is represented for the two-dimensional case. These points are positioned on the axis parallel to the Cartesian coordinate system axis, at half the radius of the particles on each side of the center of the spheres. We have observed that it was more accurate to define the Lagrangian velocity of the particles by using fixed points located inside the frame of the particles compared to an Eulerian average of the velocity field inside each solid sphere. A lower accuracy has been also obtained with the use of a kernel based interpolation [329]. With the proposed four or six point approach, the idea is to estimate the Lagrangian velocity $\boldsymbol{V}^{I,L}$ at the barycenter of the spheres by interpolating it at the $\boldsymbol{x}_k^{I,L}$ locations according to the surrounding Eulerian solid velocity field. Index $I$ stands for the number of the particle and $k = 1..N$ for the number of the interpolation point associated to particle $I$, with $N = 4$ in 2D and $N = 6$ in 3D. The velocity is interpolated from the nearest Eulerian velocity node values as shown in Fig. 8.2 for the two-dimensional case. The relative positions $\tilde{X}$ of the interpolation points compared to the Eulerian velocity components $X_u$ and particle center $\boldsymbol{x}^{I,L}$ is first calculated as

$$\tilde{\boldsymbol{x}} = \begin{pmatrix} \tilde{x} \\ \tilde{y} \\ \tilde{z} \end{pmatrix} = \frac{\boldsymbol{x}_k^{I,L} - \boldsymbol{x}_u}{\boldsymbol{DX}^{I,L}} \tag{8.13}$$

where for a given interpolated Lagrangian velocity component, $\boldsymbol{DX}^{I,L}$ refers to the distance vector in each Cartesian coordinate between two opposite points of the box (parallelepiped in three dimensions) formed by the Eulerian velocity components used for the interpolation. In two dimensions, the surrounding Eulerian components used to interpolate $\boldsymbol{V}_k^{I,L}$ are represented in Fig. 8.2 with blue (u component) and red (v component) lines joining the Lagrangian interpolation position $\boldsymbol{x}_k^{I,L}$.

The basis $P_m$, $m = 1 \ldots N$ of the bilinear interpolation is then built according to the $n$ ($n = 4$ in 2D and $n = 8$ in 3D) Eulerian velocity components $V_n$ as follows

$$P_1 = V_1 \tag{8.14}$$
$$P_2 = V_2 - V_1 \tag{8.15}$$
$$P_3 = V_4 - V_1 \tag{8.16}$$
$$P_4 = V_1 + V_3 - V_2 - V_4 \tag{8.17}$$

in two dimensions and

$$P_1 = V_1 \tag{8.18}$$
$$P_2 = V_2 - V_1 \tag{8.19}$$
$$P_3 = V_4 - V_1 \tag{8.20}$$
$$P_4 = V_5 - V_1 \tag{8.21}$$
$$P_5 = V_3 + V_1 - V_2 - V_4 \tag{8.22}$$
$$P_6 = V_6 + V_1 - V_2 - V_5 \tag{8.23}$$
$$P_7 = V_8 + V_1 - V_4 - V_5 \tag{8.24}$$
$$P_8 = V_7 - V_6 - V_8 + V_5 \tag{8.25}$$

in three dimensions. The $l$th component of the Lagrangian interpolated velocity is obtained by

$$V_{k,l}^{I,L} = P_1 + P_2\tilde{x} + P_3\tilde{y} + P_4\tilde{x}\tilde{y} \text{ in 2D} \tag{8.26}$$
$$V_{k,l}^{I,L} = P_1 + P_2\tilde{x} + P_3\tilde{y} + P_4\tilde{z} + P_5\tilde{x}\tilde{y} + P_6\tilde{x}\tilde{z} + P_7\tilde{z}\tilde{y} + P_8\tilde{x}\tilde{y}\tilde{z} \text{ in 3D} \tag{8.27}$$

Finally, the Lagrangian velocity at the barycenter of the particle reads

$$\boldsymbol{V}^{I,L} = \frac{\sum_N \boldsymbol{V}_k^{I,L}}{N} \tag{8.28}$$

#### 8.1.2.2 Transport of the Particles

Once the Lagrangian particle velocity is known by (8.28), the particles are advected in a Lagrangian manner by an update of the position of their mass center $\boldsymbol{x}^{I,L}$ as follows

$$\boldsymbol{x}^{I,L,n+1} = \boldsymbol{x}^{I,L,n} + \Delta t \boldsymbol{V}^{I,L,n+1} \tag{8.29}$$

with the second order Lagrangian velocity extrapolation at time $(n+1)\Delta t$ being given by

$$V^{I,L,n+1} = \frac{V^{I,L,n} + V^{I,L,n+1/2}}{2} \tag{8.30}$$

The intermediate Lagrangian particle velocity is defined according to the interpolated position of the particle $I$ at time $(n+1/2)\Delta t$, which is given by $x^{I,L,n+1/2} = x^{I,L,n} + \Delta t V^{I,L,n}$, as follows

$$V^{I,L,n+1/2} = \frac{\sum_N V_k^{I,L,n+1/2}}{N} \tag{8.31}$$

The Lagrangian particle advection scheme (8.29–8.31) corresponds to a second order accurate in time Runge-Kutta scheme.

#### 8.1.2.3 Update of the Solid Fraction

The one-fluid model used to simulate the particle flow does not explicitly use the Lagrangian positions or velocities of the solid spheres. As a fictitious domain approach, it requires the accounting of the Eulerian solid fraction $C$ in each Eulerian grid cell. The phase function $C$ have to be computed according to the Lagrangian position of particles at time $(n+1)\Delta t$ instead of solving explicitly the Eulerian advection equation on $C$. This is achieved by generating a set of virtual test points on a regular network inside each pressure control volume, as described for the two-dimensional case in Fig. 8.3. On a statistical point of view, the local solid fraction in a given control volume is naturally defined by the number of virtual points belonging to the interior of the particle normalized by the total number of virtual points used in the control volume.

The effect of the number of virtual points on the accuracy of the surface of the particle projected on the Eulerian flow grid is proposed in Fig. 8.4. Various Eulerian

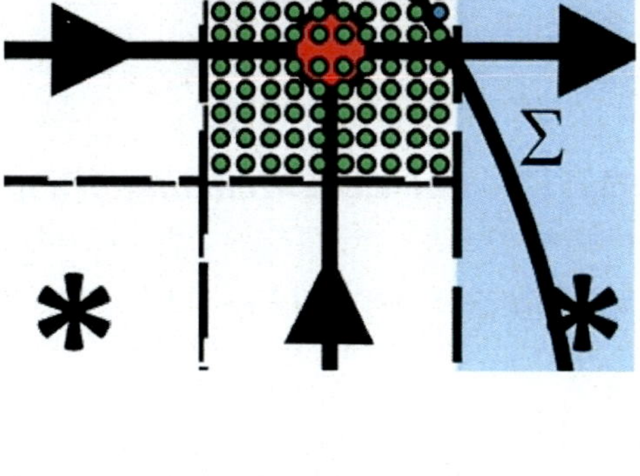


**Fig. 8.3** Two dimensional network of $10^{Ndim}$ virtual points used to estimate the local solid fraction $C^{n+1}$ thanks to a Lagrangian position $x^{I,L,n+1}$ of a particle, whose interface is noted $\Sigma$. In green are plotted the virtual points belonging to the particle and in blue the other ones. In the example, $C^{n+1} = 92/100 = 0.92$

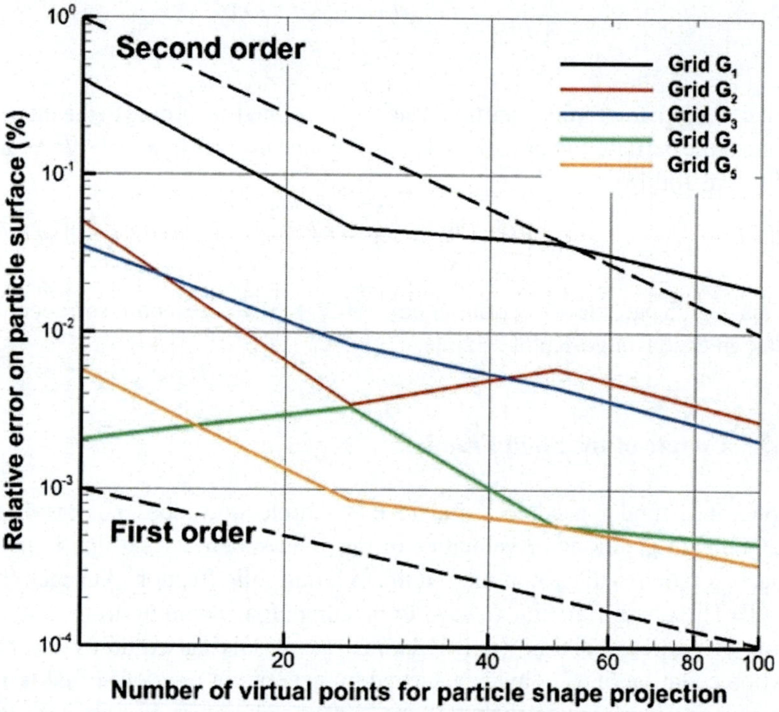


**Fig. 8.4** Error on the projected surface of a circular shape particle according to the number of virtual points in each space direction and to the Eulerian grid used

grids are considered, i.e. $G_1$, $G_2$,$G_3$,$G_4$ and $G_5$, containing respectively 5, 10, 20, 40 and 80 discretization points along a particle diameter. It is observed that even with only 5 Eulerian grid points for a particle diameter, the relative error on the particle shape is less than $0.1\%$. Increasing the number of virtual points reduces the error on the particle surface, with a convergence order between 1 and 2. The network of virtual points which will be considered in our simulations will be composed of $25^{Ndim}$ points, with $Ndim$ the number of dimensions of the problem. Indeed, it can be observed that in this case, the error on the solid fraction estimate is always less than $0.1\%$ whatever the computational flow grid. This constraint is important to deal with DNS of particulate flows as only a moderate number of Eulerian cells along a particle diameter can be used in three-dimensional simulations to simulate realistic problems.

### *8.1.3 Numerical Modeling of Particle Interaction*

Leading fully resolved particulate simulations requires to manage the contact and collisions between particles in order to prevent overlapping between particles or particles and walls. This spurious numerical phenomenon is in particular induced by the approximation of the particle trajectories and the projection of the spherical shape on the Eulerian grid. For volume ratios of particles lower than 0.1%, collisions are often assumed negligible [330, 331]. As soon as a general particulate flow model is built for low and high solid fractions, the collisions can be a major physical phenomenon, such as in fluidized beds. The fully resolved numerical simulations have to integrate a collision model to prevent unphysical adhesion of particles and try to represent reality. This point is developed in the present section.

#### 8.1.3.1 Existing Models for Particle Interaction

A first approach consists in defining a mass-damping force [22, 318] or an exponential repulsive force [313] that ensures that particles do not collide. These methods do not include a priori the explicit modeling of the influence of the lubrication effects. Several solutions exist to account for lubrication: the first one is based on a local refinement of the flow grid in order to resolve the lubrication effect [332, 333]. Therefore, this approach is very expensive in three dimensions. Another solution consists in adding an equivalent force to the repulsive component in order to reproduce the lubrication force. We have chosen this strategy, based on the work of Breuguem [334].

#### 8.1.3.2 Solid/Solid Collision

A normal solid/solid collision is characterized by the dry coefficient of restitution (in air), $e_d$, defined as the ratio between the velocity after and before the collision. The energy absorbed by the collision is represented by the $e_d$ coefficient. For a solid collision between steel spheres, the collision time is around $10^{-9}$ s. This time is too small to be reproduced by the fluid simulation with a time step chosen according to macroscopic hydrodynamic effects, with $\Delta t$ laying in the range $10^{-6}$–$10^{-2}$ s in most of the fluid mechanics problems involving particles such as a fluidized bed or a sedimentation flow. The solution in which a repulsive force is used to prevent from particle numerical adhesion is known as smooth spheres [1]. The main drawback of this model is that it artificially separate particles or walls as soon as they are distant from one grid cell. In this way, the collision effect is not accurately modeled and the restitution coefficient is generally underestimated.

In order to account explicitly for the restitution coefficient, a damping-mass force is used, instead of a standard smooth sphere repulsive force, according to Eq. (8.32) where $N_c \Delta t$ represents the collision time, which is always overestimated. Preliminary tests have shown that using $N_c = 8$ is a good compromise to get a correct collision model (more details on the numerical implementation are given in [24])

$$\mathbf{F_s} = \frac{m_e\left(\pi^2 + ln(e_d)^2\right)}{[N_c \Delta t]^2}\mathbf{d} - \frac{2m_e ln(e_d)}{[N_c \Delta t]}\dot{\mathbf{d}} \tag{8.32}$$

where $m_e$ is the reduced mass verifying $m_e = \left(m_1^{-1} + m_2^{-1}\right)^{-1}$ for inter-particle collision and $m_e = m$ for particle wall collision, $\boldsymbol{d}$ is the distance vector in each Cartesian direction between the particles or the particle and the wall and $\dot{\mathbf{d}}$ is the time derivative of $\boldsymbol{d}$. Notations $m_1$ and $m_2$ stand for the masses of particle 1 and 2 whereas $m$ denotes the mass of the particle in a particle/wall collision configuration. The restitution coefficient is given by the user depending on the material and particle properties. This input parameter have to be measured experimentally.

#### 8.1.3.3 Lubrication

Many authors propose models that include the lubrication effects [332, 335, 336]. The principle consists in utilizing the analytical development given by [337] or the lower order development provided by [1] of the forces exerted by a viscous fluid between two particles or a particle and a wall. The following formulation of the lubrication force is considered here:

$$\mathbf{F}_l(\epsilon_i, \mathbf{u}_n) = -6\pi \mu_f R \mathbf{u}_n \left[\lambda\left(\epsilon_i\right) - \lambda\left(\epsilon_{al}\right)\right] \tag{8.33}$$

with $\lambda$ being defined for particle-particle interaction as $\lambda_{pp}$ or as $\lambda_{pw}$ for particle-wall interaction

$$\lambda_{pp} = \frac{1}{2\epsilon_i} - \frac{9}{20}\log\left(\epsilon_i\right) - \frac{3}{56}\epsilon_i \log\left(\epsilon_i\right) + 1.346 + \bigcirc\left(\epsilon_i\right) \tag{8.34}$$

$$\lambda_{pw} = \frac{1}{\epsilon_i} - \frac{1}{5}\log\left(\epsilon_i\right) - \frac{1}{21}\epsilon_i \log\left(\epsilon_i\right) + 0.9713 + \bigcirc\left(\epsilon_i\right) \tag{8.35}$$

where $\mathbf{u}_n$ is the normal velocity between the particles or the particle and the wall, and $\epsilon_i = \dfrac{\|\boldsymbol{d}\|}{R}$ is the dimensionless distance between them. The activation distance $\epsilon_{al}$ is given by numerical assumptions explained in [24].

#### 8.1.3.4 Four-Way Coupling

The modeling and simulation of fully resolved particles is intrinsically based on two-way coupling approaches, in which the effect of the fluid on the particles and the modification of the carrier fluid due to the particle motions are explicitly solved. Several reference works exist on this topic [54, 109, 192, 319–321, 324, 325]. In these models, it is assumed that the flow grid is refined enough to undertake the lubrication or collision effects or that the solid concentrations are sufficiently low [331, 338] such that the collision frequency is low too.

As soon as the flow grid and time steps are coarse compared to the lubrication and collision characteristic time and space scales, an explicit modeling of the collision and lubrication forces have to be added to the particle flow model, as discussed for example by [339]. When relevant particle/particle interactions such as turbulent dispersion, transverse lift forces, wall collisions with roughness or inter-particle collisions, are explicitly accounted by the models, the particle model is called four-way coupling. Our finite size particle flow model belongs to this class of approaches, as previously published by [21, 318]. To our opinion, the works of [320, 321, 324, 325] are also four-way coupling based finite size particle models, even if their authors consider them as two-way coupling, as they explicitly account for particle collision. The main originality of our particle collision modeling is the implicit treatment of the collision forces directly in the momentum equations.

Instead of taking into account the particle collisions explicitly during the particle tracking step, it has been chosen to directly plug the resulting interaction forces $\mathbf{F_s}$ and $\mathbf{F_l}$ into the momentum equations. The Lagrangian collision force for a particle $I$ is $\boldsymbol{F}^I_{si} = \mathbf{F_s} + \mathbf{F_l}$. It is projected as a volume force in the momentum equations so as to implicitly prevent particle sticking or penetration during the global one-fluid velocity solving. The two forces $\mathbf{F_s}$ and $\mathbf{F_l}$ are estimated thanks to an extrapolated position of each particle $\boldsymbol{x}^{I,L,n+1}$ at time $(n+1)\Delta t$ with the use of the velocity at the preceding time step. This choice avoids a time splitting error induced by an explicit treatment of the collision, as in the works of [21, 22, 192, 318–321, 324, 325]. This point is the main difference between our implementation and the implementation of the collision force given by [334]. Then, the Lagrangian forces for each particle $I$ are distributed in the Eulerian cells included inside the particle shape by

$$f_{si}(M) = f^I_{si}/(V_p) \tag{8.36}$$

if Eulerian grid point $M$ is inside particle $I$. The particle volume is referred to as $V_p$. The implicit treatment of lubrication and solid-solid collision forces has been validated in [24] against the experiments and correlation of [340] concerning the restitution coefficient of a particle after a collision with a solid wall. By implicit treatment, it is considered that the particle overlapping is avoided directly during the solving of the pressure and velocity field by a time extrapolation of the particle collision force. In reality, this numerical treatment is semi-implicit as the particle position is not updated during the solving of the mass and momentum equations.

### 8.1.4 Parallel Implementation

Our simulation tool has been parallelized for distributed memory architectures with MPI procedures. The parallelism is based on a grid decomposition to which the same simulation algorithm is applied on each processor. Each processor possesses its own data with specific node covering zones used to exchange the solutions at the boundaries of the processor data blocs. On each processor data block, a standard BiCG-Stab II solver is used which is preconditioned thanks to the local available unknowns through a Modified and Incomplete LU (MILU) method. The grid splitting is simply achieved by using a Cartesian cutting in each direction of the global mesh covering all the mesh blocks associated to each processor.

Specific developments have been investigated concerning the tracking of the particles over time in processor discrete space. Indeed, to build the phase function describing the local solid fraction and to update at each time step the physical characteristics of the one-fluid equations, the searching of particles interacting with a solid sphere belonging to a given processor must be undertaken. This type of algorithms has been often studied for many applications as the simulation of granular flows, the molecular dynamics or the universe gravitational forces. The algorithms change depending on the range of the interactions. For short range interactions the existing methods are able to find collisions between millions of particles. The most common algorithms use linked list approach [341, 342], Verlet list [343] or a combination of the two [344]. These algorithms are designed to deal with millions of particles efficiently. In our case they are not straightforwardly required because the number of considered particles is around several thousands and the number of CPU used for the simulations is determined by the needs of the fluid solver and the expected precision of the physical solution. However, brute force algorithm that search all the collisions between all the particles have a complexity of $\mathcal{O}(N_p^2)$ if $N_p$ is the number of particles. The time spent in searching collisions becomes important, i.e. comparable or larger than the solving time of the Navier-Stokes equations, for $N_p \geq 500$ with such a direct approach. A finer implementation using for example octree structures will be required in the future to improve this aspect.

As a consequence, it has been decided to implement a cell-list approach to reduce the CPU time involved by the particle searching. The main idea is to split the physical space in boxes, and to search for collisions between particles only in a given box and the boxes around, as illustrated in Fig. 8.5. Two conditions are necessary, the particles cannot pass through a box during a time step and the length of the boxes cannot be lower than the particle radius. The first condition is satisfied by assuming a CFL condition less than 1. The second condition have to be respected by the definition of the boxes. In our case, the CPU boxes generated by the Cartesian splitting of the global mesh are optimum for the solver with a number of cells of about $30^3$. As the particle considered in our simulations have between 8 and 15 grids cells per radius, this condition is always fulfilled. Therefore, the use of the CPU boxes for the searching algorithm is a simple and efficient solution. In addition, this choice minimizes the communications between CPU.

### 8.1.5 Sum up of the Implemented Eulerian-Lagrangian Algorithm

The global algorithm used to solve one time iteration of fluid-particle interaction with the ITPM, the augmented Lagrangian approach and the Lagrangian VOF particle tracking is summarized in Fig. 8.6. The first part of this work flow is devoted to

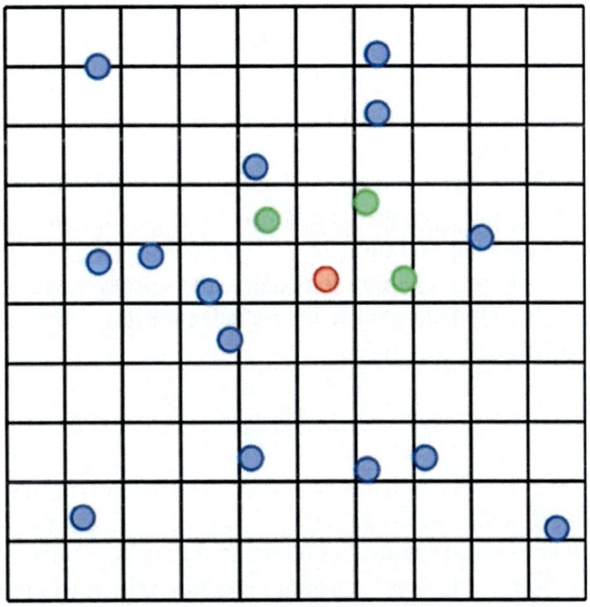

**Fig. 8.5** Searching of particle collision inside a CPU box distribution of $10 \times 10$ in two dimensions—the red particle is detected to interact with green particles whereas no interaction is considered with the blue ones

$$\begin{cases} \tilde{\mathbf{X}}^{n+\frac{1}{2}} = \mathbf{X}^{I,L,n} + \frac{\Delta t}{2}\mathbf{V}^{I,L,n}t \\ \tilde{\mathbf{V}}^{n+\frac{1}{2}} = \frac{3}{2}\mathbf{V}^{I,L,n} - \mathbf{V}^{I,L,n-1} \end{cases}$$

$$\mathbf{F}^n_{si} = \sum_{j \neq i}^{N_p} \mathcal{D}_i \left( \mathbf{F}\left( \tilde{\mathbf{X}}_i^{n+\frac{1}{2}}, \tilde{\mathbf{X}}_j^{n+\frac{1}{2}}, \tilde{\mathbf{V}}_i^{n+\frac{1}{2}}, \tilde{\mathbf{V}}_j^{n+\frac{1}{2}} \right) \right)$$

$$\rho\left(\frac{\mathbf{u}^{*,m} - \mathbf{u}^{*,0}}{\Delta t} + \mathbf{u}^{*,m-1} \cdot \nabla \mathbf{u}^{*,m}\right) - r\nabla(\nabla \cdot \mathbf{u}^{*,m}) = -\nabla p^{*,m-1} + \rho \mathbf{g} + \nabla \cdot [\mu(\nabla \mathbf{u}^{*,m} + \nabla^T \mathbf{u}^{*,m})] + \mathbf{F}^n_{si}$$

$$p^{*,m} = p^{*,m-1} - r\nabla \cdot \mathbf{u}^{*,m}$$

$$\mathbf{V}^{I,L,n+1/2} = \frac{\mathbf{V}^{I,L,n} + \mathbf{V}^{I,L,n+1}}{2}$$

$$\mathbf{X}^{I,L,n+1} = \mathbf{X}^{I,L,n} + \Delta t \mathbf{V}^{I,L,n+1/2}$$

**Fig. 8.6** Work-flow of the four-way coupling with the ITPM associated to augmented Lagrangian approach and Lagrangian VOF particle tracking predicting the position of the particle at time $(n+1)\Delta t$ through an extrapolation of their trajectories. The four-way coupling forcing terms $f_{si}^n$ are deduced from the particle interaction at this tentative position. The mass and momentum equations are then solved in both fluid and solid zones to obtain the pressure and velocities at time $(n+1)\Delta t$. The particles are finally advected in a Lagrangian way with the new velocity field. At the end of each time iteration, Lagrangian positions of the particles are projected on the Eulerian grid to build $C^{n+1}$ and the local densities and viscosities are updated according to $C^{n+1}$ before starting the next physical time iteration. This last procedure allows to penalize the solid behavior according to penalty viscosities in the cells belonging to the particles and also to account for buoyancy effects through the density variations.

## 8.2 Validations

The results presented concerning the fixed arrangements of spheres are taken from [29], the subsection concerning fluidized beds was published in [26, 30] whereas the part devoted to interactions between turbulence and particles is taken from [25].

### 8.2.1 *Monodispersed Arrangements of Spheres*

In the previous section, we have exhibited various successful validations of the force calculation for uniform flows past a fixed isolated sphere. We can now confidently complexity the simulated particulate motion by investigating a flow past fixed particle assemblies which corresponds to a gas-solid flow with high Stokes number. Two way are possible to set up a uniform flow past fixed packed particles:

- by imposing a constant pressure gradient in the domain. In this configuration, the mean fluid velocity $\langle \mathbf{u}_f \rangle = \dfrac{\int_V (1-C)\mathbf{u} dV}{\int_V (1-C) dV}$ evolves to reach a steady state corresponding to the imposed pressure gradient.
- by choosing a desired mean fluid velocity $\langle \mathbf{u}_f \rangle = \mathbf{u}_d$, and so a desired Reynolds number. A source term

$$\mathbf{F}_m = -\rho \frac{\langle \mathbf{u}_f \rangle - \mathbf{u}_d}{\Delta t} + \frac{\sum_i^{N_p} \mathbf{F}_i^D}{V_f} \tag{8.37}$$

is inserted in the momentum conservation equations. It is adjusted until the desired Reynolds number is reached. This is the method used in this work. Its main advantage is to be able to simulate a prescribed Reynolds number without a trial and error procedure unlike to what is require in the pressure gradient technique.

Two ways to distribute the particles are studied in this work: a Face-Centered Cubic arrangement of spheres for both mono and bi-dispersed flows, and random assemblies of spheres only for mono-dispersed flows.

The particulate Reynolds number $Re = \frac{|\langle \mathbf{u}_f \rangle| d}{\nu}$ is used studying the Face-Centered Cubic arrangement of spheres following the work of Massol [345]. On the other hand, in the random assemblies of spheres, another Reynolds number based on the superficial velocity $(1-\alpha_d)|\langle \mathbf{u}_f \rangle|$ is considered to take into account the solid loading in the characterization of the flows [346, 347]. It is given by:

$$Re_S = \frac{(1-\alpha_d)|\langle \mathbf{u}_f \rangle| d}{\nu} \tag{8.38}$$

The relation between the solid volume fraction $\alpha_d$ and the number of particle $N_p$ is:

$$N_p = \frac{6}{\pi}\left(\frac{L}{d}\right)^3 \alpha_d \tag{8.39}$$

As for the uniform flow past an isolated sphere, the grid resolution is fixed by imposing 5 Cartesian cells in the boundary layer as $\Delta x = \frac{d}{5\sqrt{Re_S}}$.

The mean non-dimensional drag force for all the particles is then defined as:

$$F = \frac{|\langle \mathbf{F}^D \rangle|}{3\pi\mu d(1-\alpha_d)|\langle \mathbf{u}_f \rangle|} \tag{8.40}$$

with $\langle \mathbf{F}^D \rangle = \frac{1}{N_p}\sum_i^{N_p} \mathbf{F}_i^D$, $\mathbf{F}_i^D$ being the drag force computed over the $i$th particle.

#### 8.2.1.1 Face-Centred Cubic Periodic Arrangement of Spheres

A Face-Centered Cubic (FCC) array is a cube where three spheres are placed on the faces centers, and one sphere is located on the vertices with periodic boundary conditions, as illustrated in the Fig. 8.7.

This configurations was widely studied [345, 346, 348] to understand and separate the wake effects, observed when spheres are aligned along the flow direction, i.e. streamwise interactions, from the blocking effects, where the spheres are aligned along the direction perpendicular to the flow direction, i.e. lateral interactions. Two regimes, i.e. attached and separated flows, govern the uniform flow past a FCC array of spheres, depending on the Reynolds number and the solid volume fraction. They are considered below.

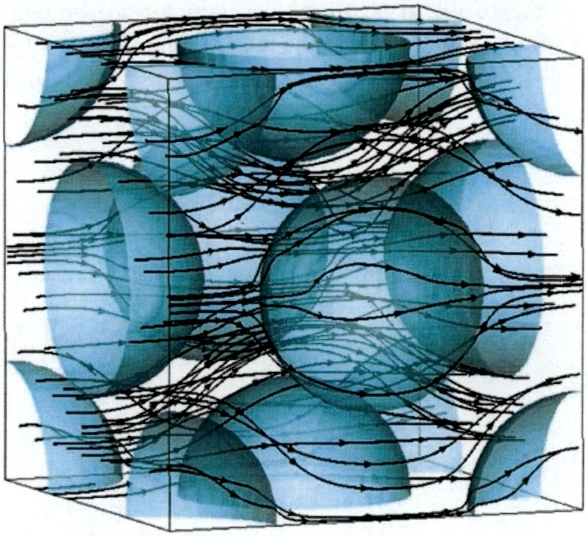

**Fig. 8.7** Streamlines of a steady flow along the x-axis of a face-centred Cubic array of spheres at $Re = 50$ and $\alpha_d = 0.5$. The particle shape is plotted in blue

**Regime 1: attached flows**

For low Reynolds numbers, the boundary layer remains attached to the particles. Unlike the isolated sphere configurations, the boundary layer detachment occurs for higher Reynolds numbers when FCC arrays are considered, due to the blocking effect of the surrounding particles. In fact, the presence of lateral spheres speeds-up the flow between the spheres and blocks the detachment of the boundary layer. Figure 8.8 illustrates this regime for $Re = 50$, $\alpha_d = 0.15$. A very good agreement is observed for the pressure distribution over the spheres in both plans $\beta = 0°$ and $\beta = 45°$ as shown in Fig. 8.8a, b, compared to the results of Massol [345] obtained with body fitted meshes. The same agreement is observed for the axial friction coefficient as illustrated in Fig. 8.8c, d for the plans $\beta = 0°$ and $\beta = 45°$, the spherical coordinates $\alpha$ and $\beta$ are detailed in [29]. It is worthwhile to note that the flow is non-axisymmetric even if all particles see the same flow due to the symmetry of the array and the periodicity of boundary conditions. Note that the pressure coefficient is known at the center of each Lagrangian element. The $C_p$ value given on each $\beta$ plan is that of the nearest Lagrangian element to the given plan.

**Regime 2: separated flows downstream of the spheres**

With increasing Reynolds numbers, a separation of the boundary layer occurs in the downstream hemisphere of the particles. As noticed for an isolated sphere, the recirculating zone length increases with the Reynolds number. The flow confinement due to the presence of the lateral particles is so significant that high speed jet flows are obtained between the particles. These jet flows impact the spheres onto the upstream

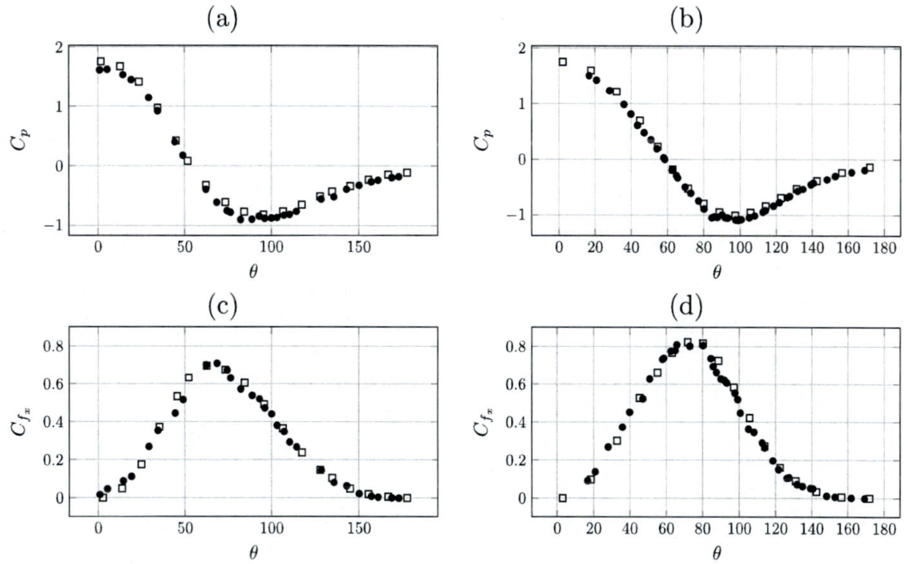


**Fig. 8.8** Pressure coefficient [**a** $\beta = 0$, **b** $\beta = 45$] and axial friction coefficient [**c** $\beta = 0$, **d** $\beta = 45$], for a uniform flow past a FCC at $Re = 50$, $\alpha_d = 0.15$: square for Massol [345] and circles for the present work

hemisphere. As illustrated in Fig. 8.9a, this mechanism induces the development of a fountain effect on the vertical plan $\beta = 0°$, the direct consequence of which is that the maximum pressure is no longer at $\theta = 0°$ but at $\theta = 36°$. Note that the same result was found by Massol [345]. Figure 8.9a, b show again the good agreement of the pressure distribution on the vertical and lateral plans with the body fitted simulations of Massol [345], and the same conclusions is observed in the Fig. 8.9c, d for the axial friction coefficient distribution. It can be noticed that for this Reynolds number of 300, larger differences are observed between our force calculation and body fitted grid results, mostly in the vicinity of the pressure peaks in the $\beta = 0°$ plans.

### Drag force

The non-dimensional drag force $F$ normalized by the isolated sphere non-dimensional drag force (given by Schiller and Naumann [349]) $F_s = \dfrac{1 + 0.15 Re^{0.687}}{(1 - \alpha_d)^2}$, noted as $\dfrac{F}{F_s}$, is compared to existing literature results:

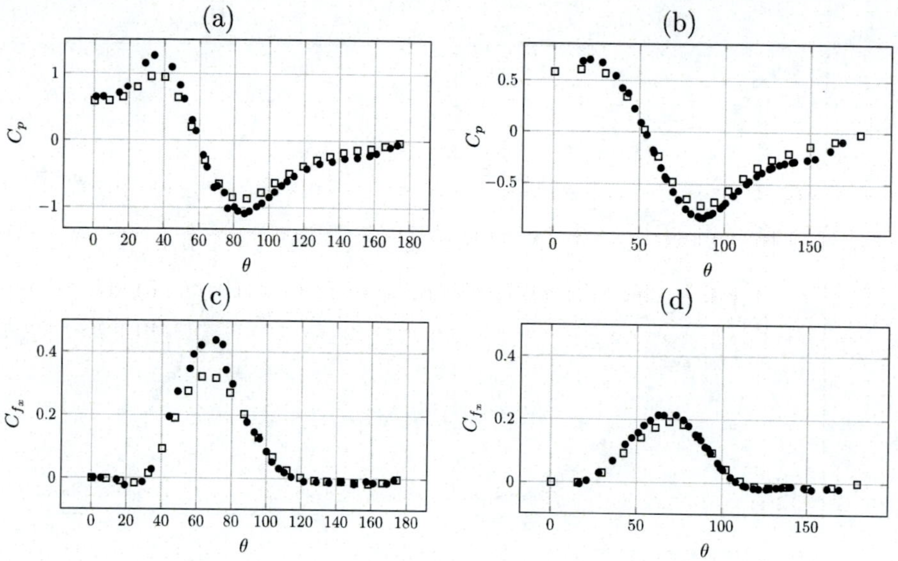


**Fig. 8.9** Pressure coefficient [**a** $\beta = 0$, **b** $\beta = 45$] and axial friction coefficient [**c** $\beta = 0$, **d** $\beta = 45$], for a uniform flow past a FCC at $Re = 300$, $\alpha_d = 0.15$: square for Massol [345] and circle for the present work

- Tenneti et al. [346]:

$$\begin{aligned}F_T &= F_s + F_\phi + F_{\phi,Re}\\ F_\phi &= \frac{5.81\phi}{(1-\alpha_d)^2} + 0.48\frac{\alpha_d^{\frac{1}{3}}}{(1-\alpha_d)^3}\\ F_{\phi,Re} &= (1-\alpha_d)\alpha_d^3 Re\left(0.95 + \frac{0.61\alpha_d^3}{(1-\alpha_d)^2}\right)\end{aligned} \tag{8.41}$$

- Gobin et al. [350], Wen and Yu [351], Ergun [352]:

$$\begin{aligned}F_G &= \begin{cases} F_{WY} & \text{if } \alpha_d \le 0.3\\ \min(F_{WY}, F_E) & \text{otherwise}\end{cases}\\ F_{WY} &= F_s(1-\alpha_d)^{-1.7}\\ F_E &= \frac{150}{18}\frac{\alpha_d}{(1-\alpha_d)^2} + \frac{7}{4}\frac{1}{18}\frac{Re}{(1-\alpha_d)^2}\end{aligned}$$

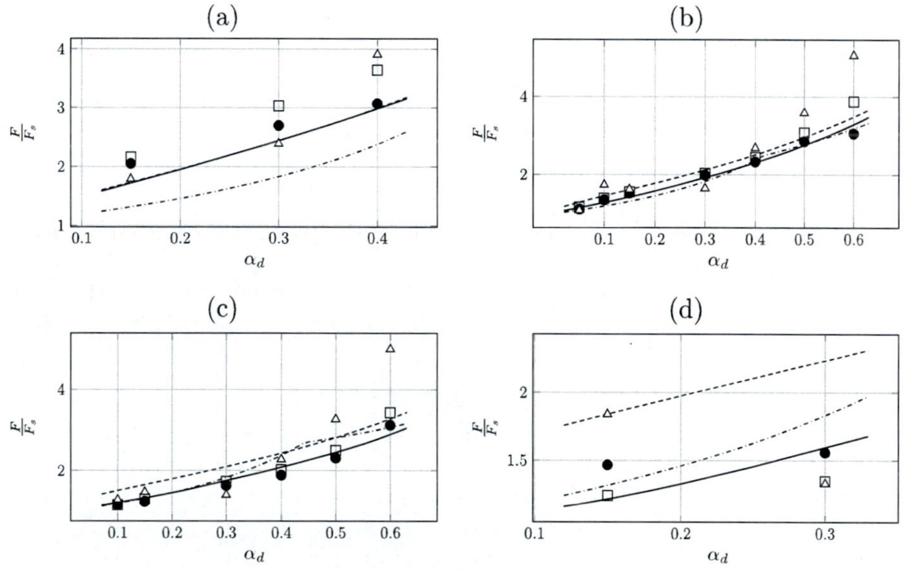


**Fig. 8.10** Drag force for a uniform flow past a FCC, normalized by Schiller and Naumann [349] drag force for a uniform flow past an isolated sphere. Results are presented as a function of the solid volume fraction $\alpha_d$ and Reynolds number **a** $Re = 10$, **b** $Re = 50$, **c** $Re = 100$, **d** $Re = 300$: solid line for Tenneti et al. [346], dash dotted line for Gobin et al. [350], dashed line for Beetstra et al. [347], triangle for Randrianarivelo [353], square for Massol [345] and circle for the present work

- Beetstra et al. [347]:

$$F_B = \frac{10\alpha_d}{(1-\alpha_d)^2} + (1-\alpha_d)^2\left(1+1.5\sqrt{\alpha_d}\right) + \frac{0.413}{(1-\alpha_d)^2}\frac{Re}{24}\,\frac{\frac{1}{(1-\alpha_d)} + 3(1-\alpha_d)\alpha_d + 8.4Re^{-0.343}}{1+10^{3\alpha_d}Re^{-0.5-2\alpha_d}} \tag{8.42}$$

Our FCC results are compared to FCC results of Massol [345] and Randrianarivelo et al. [353]. Laws corresponding to random configurations are also considered, following the conclusion of Massol [345] that FCC arrangement of spheres provides a better agreement to random configurations compared to simple cubic, tetragonal and centered cubic arrangements. These comparisons are presented in Fig. 8.10 at different Reynolds number $Re = 10$ Fig. 8.10a, $Re = 50$ Fig. 8.10b, $Re = 100$ Fig. 8.10c and $Re = 300$ Fig. 8.10d. It can be observed that a nice match is found between our results and those of Massol [345] which was extracted from body fitted simulations and with those of Tenneti et al. [346] extracted from IBM simulations. Reasonable agreement is also observed with Gobin et al. [350], and Beetstra et al. [347] even if larger differences are noticed for $Re = 300$ with Beetstra's results et al. [347] which can be a consequence of insufficient Eulerian mesh resolution at that Reynolds number.

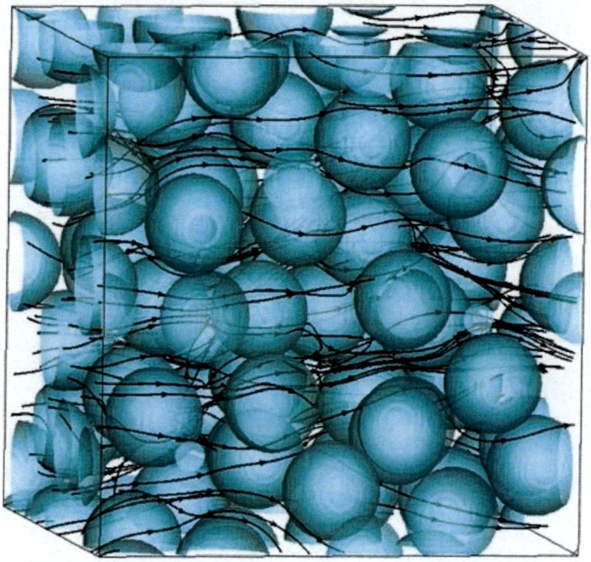

**Fig. 8.11** Random periodic arrangement of spheres, for $Re_S = 50$ and $\alpha_d = 0.3$—the particle shape is plotted in blue while black lines represent streamlines

#### 8.2.1.2 Random Periodic Arrangement of Spheres

A new step is taken in the complexity of the problem by randomly distributing a set of spheres for a given solid volume fraction. An illustration is given in Fig. 8.11 for $Re_S = 50$ and $\alpha_d = 0.3$. The locations of the spheres are set by means of a classical Monte Carlo procedure to distribute them in the domain until the desired compacity is reached, and a hard sphere collisions model is used to prevent spheres from overlapping [354]. To simulate an infinite domain, periodic boundary conditions are specified in all directions.

To be statistically converged, we performed for most cases five Multiple Independent Simulations (MIS), Tenneti et al. [346] conducted the same number of MIS to obtain their correlation for the drag force (due to CPU time requirements, 5 MIS were not conducted for all Reynolds numbers and solid fractions). Consequently, the representative non-dimensional drag force numerically obtained is an average of those calculated over the different realizations (MIS). Simulations with various solid volume fraction $\alpha_d = 0.1, 0.15, 0.2, 0.3$ have been investigated, together with different superficial Reynolds numbers ranging from 20 to 200.

The results presented in Fig. 8.12 demonstrate again a good global agreement of our results with existing correlations of Tenneti et al. [346], Gobin et al. [350] and Beetstra et al. [347] for each solid volume fractions studied here as shown in Fig. 8.12a for $\alpha_d = 0.1$, 8.12b for $\alpha_d = 0.15$, 8.12c for $\alpha_d = 0.2$ and 8.12d for $\alpha_d = 0.3$, all function of the superficial Reynolds number. In all cases, for larger $Re_S$, the correlation of Beetstra et al. [347] is farther from other works.

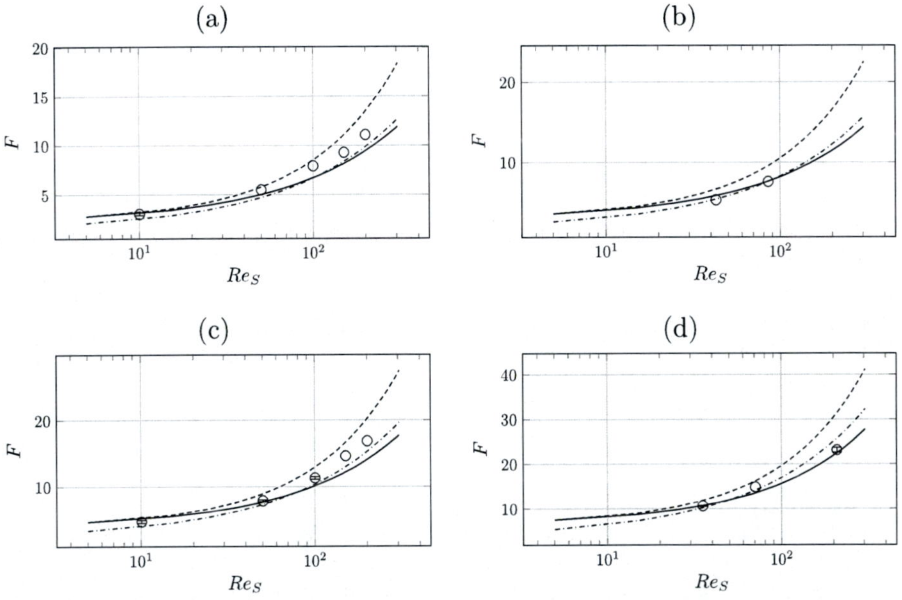


**Fig. 8.12** Normalized drag force for a uniform flow past random packed spheres as function of the superficial Reynolds number $Re_S$ and solid volume fraction **a** $\alpha_d = 0.1$, **b** $\alpha_d = 0.15$, **c** $\alpha_d = 0.2$, **d** $\alpha_d = 0.3$: solid line for Tenneti et al. [346], dash dotted line for Gobin et al. [350], dashed line for Beetstra et al. [347] and circle for the present work. The error bars represent 95% confidence intervals in the estimation of the normalized drag force

The pressure profiles according to $\theta$ are given in Figs. 8.13 and 8.14. They have been obtained with our force calculation and our simulations. It can be noticed that the pressure contribution on the drag force increase with the solid volume fraction in both vertical plan $\beta = 0°$ as illustrated in Fig. 8.13a and lateral plan $\beta = 45°$ see Fig. 8.13b. On the contrary, it seems that the Reynolds number has a small effect on the distribution of the pressure over the particles, and here again for both plans, as illustrated in Fig. 8.14. This result has previously been obtained in the study of Tenneti et al. [346]. One can observe that the same result is obtained for both plans $\beta$.

As can suggest the correlation of Gobin et al. [350], our results (see Fig. 8.15) show that the drag force in a random arrangement of spheres is having the same dependence on the Reynolds number as for an isolated sphere. On the contrary, different behaviors are noticed for Tenneti and Beetstra simulations.

### 8.2.2 Bidisperse Arrangements of Spheres

For a bidisperse arrangement of sphere, i.e. with two types of particle size, we studied a Face-Centred Cubic periodic arrangement for the two species of spheres: the larger particles are distributed in the same configuration as the one previously presented for

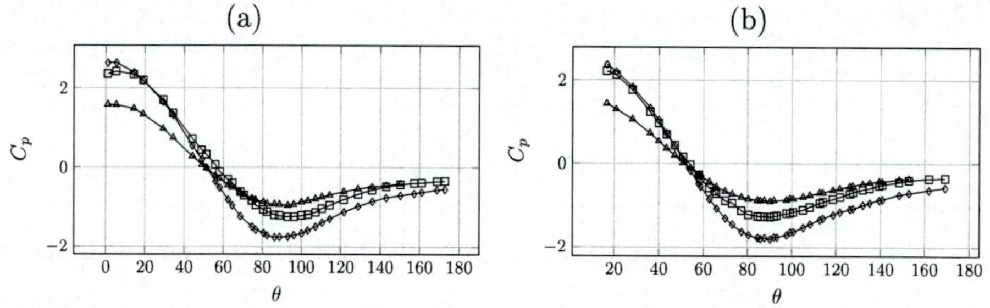


**Fig. 8.13** Pressure coefficient for a uniform flow past random packed spheres at $Re_S = 200$ as function of $\alpha_d$ **a** $\beta = 0$, **b** $\beta = 45$: triangle for $\alpha_d = 0.1$, square for $\alpha_d = 0.2$ and diamond for $\alpha_d = 0.3$

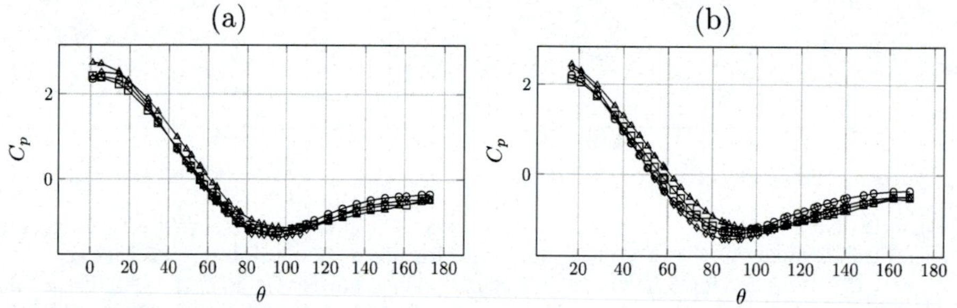


**Fig. 8.14** Pressure coefficient for a uniform flow past random packed spheres at $\alpha_d = 0.2$ as function of $Re_S$ **a** $\beta = 0$, **b** $\beta = 45$: triangle for $Re_S = 50$, square for $Re_S = 100$, diamond for $Re_S = 150$ and circle for $Re_S = 200$

monodisperse arrangements while the smaller particles are positioned at the center of the vertices and at the center of the cubic simulation domain. This geometry of the particle arrangement is illustrated in Fig. 8.16.

Two additional dimensionless parameters, for each species, are now necessary to characterize the flow:

$$x_i = \frac{\alpha_i}{\alpha_d} \quad , \quad y_i = \frac{d_i}{d_s} \tag{8.43}$$

where $d_i$ and $\alpha_i$ are the particle diameter and the solid volume fraction of the specie $i$ respectively. The Sauter mean diameter $d_s$ is given by:

$$d_s = \left[\sum_i^2 \frac{x_i}{d_i}\right]^{-1}$$

The Sauter mean diameter is one of the most important characteristic dimension for the bidisperse particle arrangement that is often used in the literature [347].

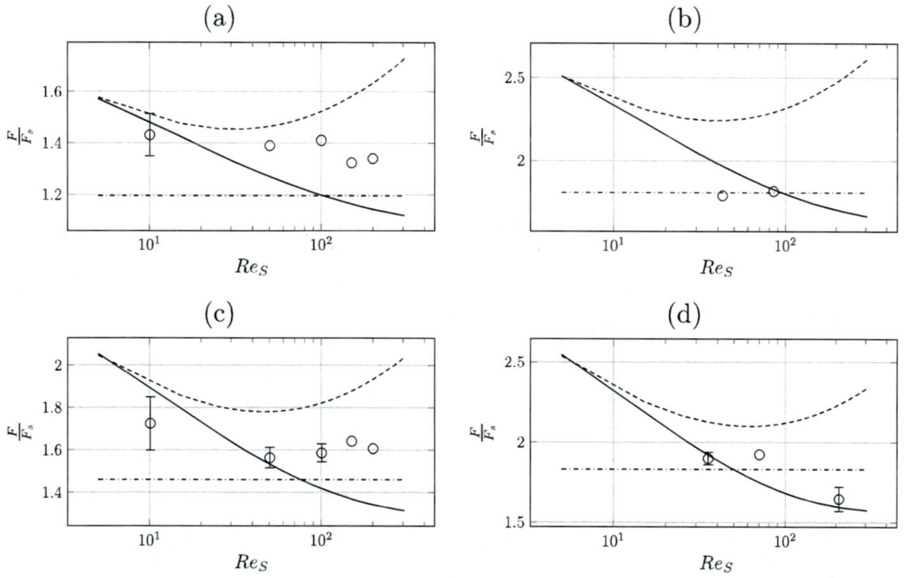


**Fig. 8.15** Drag force for a uniform flow past random packed spheres, normalized by Schiller and Naumann [349] drag force for a uniform flow past an isolated sphere, as function of the superficial Reynolds number $Re_S$ and solid volume fraction **a** $\alpha_d = 0.1$, **b** $\alpha_d = 0.15$, **c** $\alpha_d = 0.2$, **d** $\alpha_d = 0.3$: solid line for Tenneti et al. [346], dash dotted line for Gobin et al. [350], dashed line for Beetstra et al. [347] and circle for the present work. The error bars represent 95% confidence intervals in the estimation of the normalized drag force

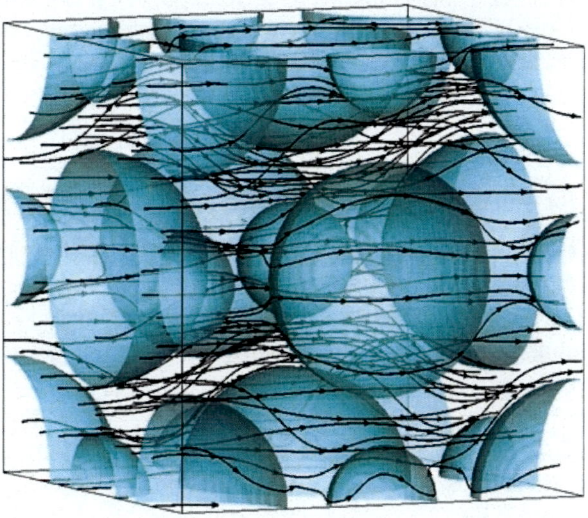

**Fig. 8.16** FCC/FCC bidisperse arrangement of spheres for $Re = 50$ and $\alpha_d = 0.3$—the particle shape is plotted in blue and the black lines are the streamlines

**Table 8.1** Values of the parameters characterizing the bi-disperse flows simulated in this work

| $x_1$ | $x_2$ | $y_1$ | $y_2$ | $d_s$ |
|---|---|---|---|---|
| 0.88889 | 0.11111 | 1.11111 | 0.55556 | 0.0009 m |

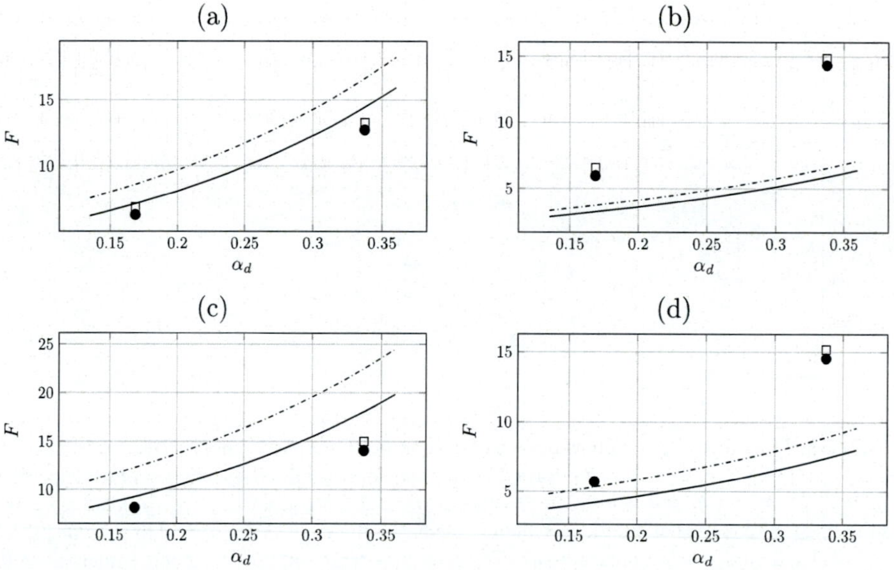


**Fig. 8.17** Normalized drag force for a uniform flow past a FCC/FCC packed spheres as function of the solid volume fraction $\alpha_d$ **a** ($Re = 50, y1$), **b** ($Re = 50, y2$), **c** ($Re = 100, y1$), **d** ($Re = 100, y2$): solid line for Beetstra et al. [347], dash dotted line for Mehrabadi et al. [355], square for Massol [345] and circle for the present work

We have performed simulations of FCC/FCC bidispersed arrangement of spheres, for which $\frac{d_1}{d_2} = 2$, at Reynolds number $Re = 50$ and 100. The solid volume fractions for the larger particles are $\alpha_1 = 0.15$ and 0.3. The corresponding values of $x_i$, $y_i$ and $d_s$ are reported in Table 8.1.

Our force calculation results are given in Fig. 8.17. They are compared to Beetstra et al. [347] correlation that provides the drag force for the $i$th species of particle as follows:

$$(F_B)_i = ((1 - \alpha_d)y_i + \alpha_d y_i^2 + 0.0064(1 - \alpha_d)y_i^3)F_B$$

where $F_B$ is the Beetstra drag force for a mono-dispersed arrangement of spheres given in Eq. (8.42). Mehrabadi et al. [355] took over the Tenneti et al. mono-dispersed work [346] and extend it to bidispersed arrangement of spheres. A correlation was proposed from the PR-DNS data, it is given by:

$$(F_M)_i = ((1 - \alpha_d)y_i + \alpha_d y_i^2)F_T$$

where $F_T$ is the Tenneti drag force for a mono-dispersed arrangement of spheres given in Eq. (8.41). The values of forces obtained with our method are compared to Beetstra [347] and Mehrabadi [355] correlations and to Massol body fitted results [345] in Fig. 8.17. It can be observed that for all Reynolds numbers and solid fractions, our drag forces are in better agreement with Massol simulations than with Beetstra and Mehrabadi laws extracted from their simulations. The drag forces computed on the larger particles ($y_1$) seems to be lower than those in random arrangement case, when the drag forces on the smaller particles ($y_2$) are greater. This can be explained by the fact that in FCC/FCC arrangement of spheres, all the smaller particles are not in flow wake of the greater spheres. It can be assumed that the flow sped up by the lateral larger spheres comes to impact the small particles increasing the FCC drag force compared to the random arrangement case. In this later configuration, some of the small spheres can be behind the larger particles. The difference between FCC/FCC and random configuration forces increases with solid volume fraction (see Fig. 8.17b, d). This discrepancy is less important for the larger spheres of the bidisperse arrangement (see Fig. 8.17).

### *8.2.3 Fluidized Beds*

Al the results presented in this section are taken from [23, 26, 30].

The fluidized bed chosen for the simulations is taken from the experiments of Aguilar Corona [356]. In a cylinder of 0.64 m height and 0.08 m diameter, 2133 pyrex particles are fluidized by the circulation of potassium thiocyanate KSCN at various fluidization velocities (see Fig. 8.18). The diameter of the particles is 0.006 m and the simulated fluidization velocity is 0.12 $ms^{-1}$. The density of the pyrex is $\rho_s = 2230$ $kgm^{-3}$ whereas the density $\rho_l$ and the dynamic viscosity $\mu_l$ of the KSCN are 1400 $kgm^{-3}$ and 3.8 $10^{-3}$ Pas. For the chosen particles in KSCN, the terminal settling velocity is equal to 0.226 $ms^{-1}$, the particle Reynolds number based on this velocity is equal to 530 and the Stokes number is 0.8.

Concerning the DNS of particles, a parallelepiped box of dimensions 8 cm $\times$ 8 cm $\times$ 64 cm is used with a grid mesh containing 160 $\times$ 160 $\times$ 1280 cells. With such a grid resolution, nearly 12 cell length $\Delta x$ per particle diameter is achieved. The other numerical parameters used are a penalty solid viscosity 100 times larger than $\mu_l$ and $\mathcal{K} = 100$. The cylinder is represented numerically with a Darcy penalty method [34]. The method consists in adding a Darcy term $\dfrac{\mu}{K}\boldsymbol{u}$ in the momentum equations. By taking a permeability tending to zero in the cells being outside the cylinder, a null velocity is imposed. A permeability of $10^{-40}$ $m^2$ is used in the simulations in the solid parts outside the cylindrical vein. The permeability is $10^{40}$ $m^2$ elsewhere such that the effect of the Darcy term vanishes in the fluid and particles. The rebound of particles on the walls of the cylindrical tank is treated as a rebound between two particles. The simulations are conducted during a total physical time of 10 s with a numerical time step of $\Delta t = 0.001$ s. They are run on 512 Intel Quad-Core processors of the

**Fig. 8.18** Three-dimensional view of the particles inside the fluidized bed—from left to right, $t = 1, 2, 3, 4, 5, 6, 7, 8, 9$ and $10s$

CICT computing center of Toulouse and CINES computing center of the French Ministry of Research. In the present simulation, 33 millions of grid cells are used in the whole numerical domain, which lead to blocs of 64,000 cells on 512 processors. In this way, the fluidized bed simulation belongs to the range of speed-up 1 for the ITPM, as demonstrated before for a two-dimensional case. In three dimensions, the speed-up equal to one zone is maintained, for a given number of grid points, for a larger number of processors than in two dimensions. The CPU time for simulating $10s$ of physical time is 25 days. The initial distribution of the particle position is randomized in the first 0.47 m of the fluidization cylinder, which corresponds to an average solid fraction of 0.1. For the tensorial penalty of the particles, a reduced diameter of the spheres $d_p - \Delta x$ is considered so as to obtain a correct effective

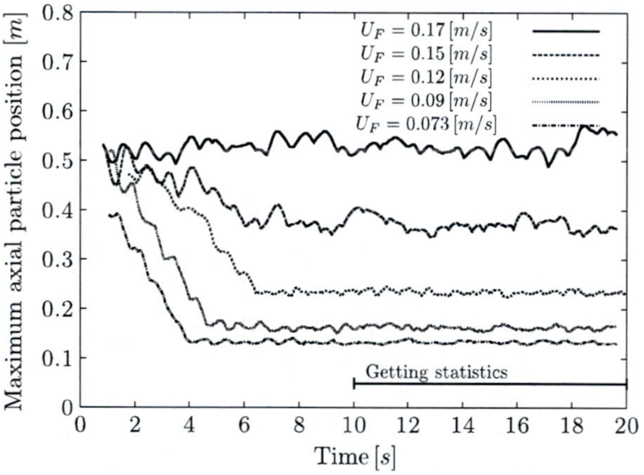


**Fig. 8.19** Time evolution of maximum axial particle position at different fluidization velocities: $U_F = 0.17,\ 0.15,\ 0.12,\ 0.09,\ 0.073\,m/s$ (resp. $\phi_b = 0.11,\ 0.14,\ 0.22,\ 0.31,\ 0.39$)

drag force exerted by the fluid on the particles. Indeed, the projected shape of the spheres is different from the real one as explained in Sect. 5.2. The real particle shape is considered for the definition of the solid density and the treatment of the collisions between the particles.

The fluidization law (relationship between the fluidization velocity and bed solid fraction) is the first step of validation of the numerical model, reflecting the macroscopic balance between buoyancy and drag forces. In order to calculate the bed solid concentration $\phi_b$, the bed height $h_b$ was computed using two different methods, which led to the same result.

First, the bed height was set equal to the time averaged maximum particle position in the axial direction. Figure 8.19 displays the time evolution of this parameter at different fluidization velocities. After a transient period, it oscillates around a steady value for all cases investigated due to particle agitation. The intensity of fluctuations is a decreasing function of the fluidization velocity, with a maximum of the order of 5% for the lowest fluidization velocity.

Second, the bed height was also determined by averaging in time (for 10 s) and space (over the bed volume) a particle phase indicator function $\chi_p(\boldsymbol{x}, t)$. This function is defined on each Eulerian mesh cell, equal to 1 if the node is inside the particle and 0 if not. A vertical profile of the time-section average of the particle phase indicator function (or phase fraction) $\{\chi_p\}_{layer}$ is plotted in Fig. 8.20 at different fluidization velocities. This time-section average is defined as follows: for the axial profiles, we define the average in time, over the particles that cross the disk of diameter $D$ and thickness $\Delta z$, such that $\mathcal{L}(\mathbf{x}, z, \Delta z) = \{1 \text{ if } z < \|\mathbf{x} \cdot \mathbf{e}_z\| < z + \Delta z \text{ else } 0\}$, as:

$$\langle \phi \rangle_{layer} = \frac{\int \sum_{n=1}^{N_p} \phi_n(t) \mathcal{L}(\mathbf{x}_n(t), z, \Delta z) dt}{\int \sum_{n=1}^{N_p} \mathcal{L}(\mathbf{x}_n(t), z, \Delta z) dt} \tag{8.44}$$

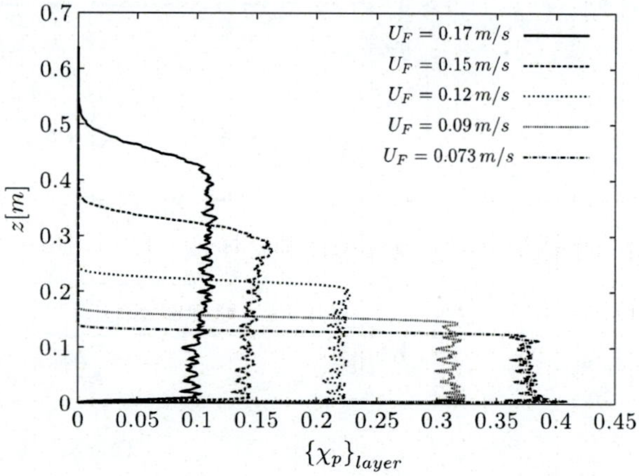


**Fig. 8.20** Axial profile of time-section average of particle phase fraction $\chi_p$

This quantity represents the solid phase fraction averages in disks of diameter $D$ and thickness $\Delta z$. So the integral of $\{\chi_p\}_{layer}$ layer along $z$ is equal to $\phi_b h_b$.

Profiles of Fig. 8.20 show that the phase fraction is rather homogeneous along the bed height but in the freeboard region a gradient of the solid phase fraction develops, becoming stiffer as the fluidization velocity is decreased. The bed height is then computed by applying a linear regression with a high order polynomial interpolation on the phase fraction profile at the interface between the bed and the freeboard region, and the $z$ value of the inflection point of that function is defined as the bed height $h_b$.

Both estimations lead to the same value of $h_b$ for all velocities with difference of a few percent. The bed solid fraction $\phi_b$ is then calculated according to:

$$\phi_b = \frac{2}{3}\frac{n_p d_p^3}{h_b D^2} \tag{8.45}$$

Plotting the fluidization velocity as a function of the bed solid concentration $\phi_b$ gives the fluidization law. Figure 8.21 shows this plot together with the experimental data of [356] and the correlation of [357]:

$$U_F = U_{F0}(1 - \phi_b)^n \tag{8.46}$$

where $n$ is a function of $Re_t$, $U_F$ is the fluidization velocity and $U_{F0}$ is the fluidization velocity leading to particles entrainment. The exponent value best fitting experimental data is $n = 2.41$ ($Re_t = 530$), in quite good agreement with the value predicted by Richardson-Zaki correlation at that particle Reynolds number ($n = 2.39$ for $Re_t > 500$). Experimental value of $U_{F0}$ is found equal to $0.226\,m/s$, and the measured terminal velocity of the particles is $V_t = 0.24\,m/s$. The ratio of $\frac{U_{F0}}{V_t}$ is equal to 0.94, which corresponds to the upper limit of the range of values of this parameter in liquid fluidization. Note also that it is not related to the ratio $\frac{d_p}{D}$ [358]. The agreement between experiments and numerical simulations is quite good, and to the best of

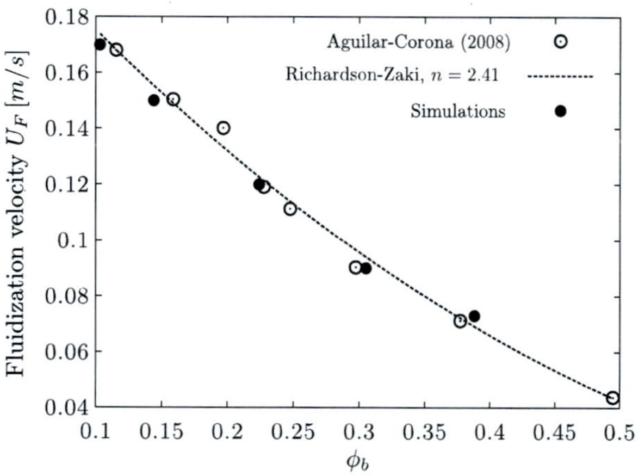


**Fig. 8.21** Fluidization velocity with respect to the bed solid concentration. ⊙: [356], – – –: Richardson's correlation [357] with $n = 2.41$ and $o$: present simulations

our knowledge, this result is the first validation of direct numerical simulation of fluidization law in a full 3-D fluidized bed in that range of particle Reynolds number.

## *8.2.4 Interaction Between Particles and Turbulence*

### 8.2.4.1 Energy Distribution

Based on fully resolved simulations we aim at getting more insight on the collisional regime of finite size particles in a turbulent flow. In order to obtain converged collision statistics over long times, it is necessary to simulated sustained turbulence. To do so, turbulence is sustained by using the linear forcing proposed by [359]. In our implementation the velocity field $\mathbf{u}$ is rescaled to $\mathbf{u}^+$ at each time step as to keep a constant kinetic energy $\mathbf{u}_0^2$ within the simulation domain:

$$\mathbf{u}^+ = \sqrt{\frac{\int_\Omega \mathbf{u}_0^2 dv}{\int_\Omega \mathbf{u}^2 dv}}\mathbf{u} \tag{8.47}$$

A $256^3$ mesh is used for direct numerical simulations enforcing adequate resolution of the Kolmogorov scale $\eta$ (with $\eta/\Delta x = 0.56$). Simulating one large eddy turn-over time takes 10 h CPU on 512 cores of an SGI Altix Ice supercomputer.

In order to obtain the initial turbulent flow, a single-phase turbulence is initialized with a predefined homogeneous isotropic turbulence spectrum [9, 360]. Then, single-phase simulations with sustained turbulence are achieved with the previous forcing scheme until reaching steady state spectrum. The Taylor micro-scale Reynolds number of the turbulent flow computed from this spectrum is $Re_\lambda = 73$. This moderate Reynolds number corresponds to a moderate length scale separation: the ratio of Taylor micro-scale over Kolmogorov scale is $\lambda/\eta = 17$. First, the flow is forced

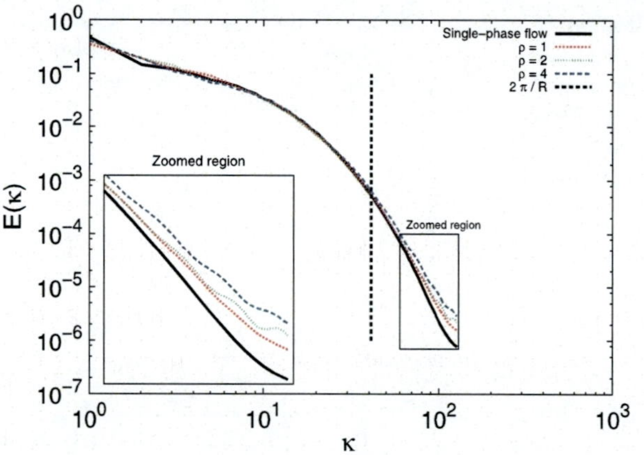


**Fig. 8.22** Energy spectra with and without particles. Solid line for single phase flow DNS. Dotted, dash-dotted and light dashed lines stand for density ratio equal to 1, 2 and 4 respectively. The vertical dash line represents the wavenumber corresponding to the particle diameter. Inset is a close-up for high wavenumbers content of spectra

from its initial condition (analytic spectrum) to reach statistical steady state. Then, 512 particles are seeded and a transient time is considered before starting averages for two-phase flows. The particle radius is $R = 0.59\lambda = 10\eta$. The particle volume fraction is 3%. To obtain converged particle statistics, simulations are carried out over 12 large eddy turn-over times. In the present paper three different simulations are presented for solid to fluid density ratio $\rho$ equal to 1, 2 and 4. No gravity effect is considered.

A study of the spatial spectrum has been carried out to verify that the presence of particles has only a minor influence on the turbulence statistics. In Fig. 8.22, the single phase flow spectrum of energy is compared to two-phase flow simulations for the three density ratios. Only slight modifications are observed at wavenumbers higher than the inverse of the particle radius. The energy spectrum is formed over the entire domain (accounting for the solid body motion of the particles). As shown by [361], the inclusion of the solid domain creates some oscillations on the wavenumbers larger than the inverse of the particle radius. As shown in Fig. 8.22, our simulations stand in a very dilute regime yielding these oscillations to be negligible. Only a weak increase of energy at high wavenumbers is observed. It can be explained by the linear forcing applied in the entire domain.

In order to confirm the weak effect of the particles on the carrying fluid turbulence (the local effect will be discussed later) and overtake the problem of the computation of the spatial spectrum including grid points inside the particles, the temporal Lagrangian spectrum has been computed.

This spectrum is computed from 200,000 Lagrangian fluid elements (point particles following the flow) which are randomly seeded in the fluid domain. Their trajectories are integrated while the turbulent suspension is simulated. Then, the Lagrangian velocity auto-correlation function, $R_L^f = \frac{<\mathbf{V}_i(t_0)\cdot\mathbf{V}_i(t_0-t)>}{<\mathbf{V}_i(t)^2>}$, is computed from those tra-

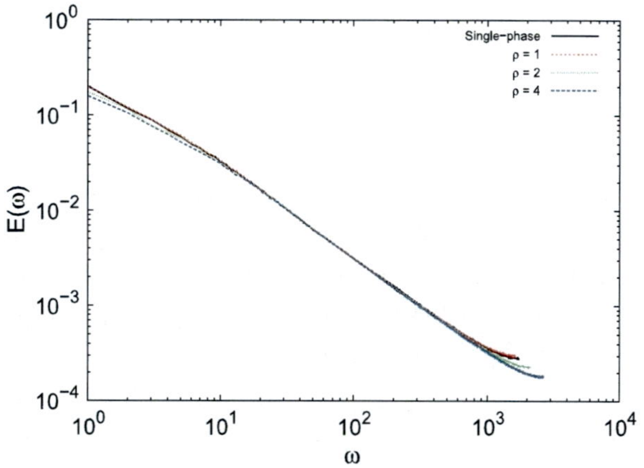


**Fig. 8.23** Temporal spectrum of the single phase simulation and three configurations of two-phase flows. These spectra were computed from tracer particle trajectories. Solid line for single phase flow DNS. Dotted, dash-dotted and light dashed lines stand for density ratio equal to 1, 2 and 4 respectively

jectories to evaluate the energy temporal spectrum $E(\omega) = \frac{1}{2}\mathfrak{F}(R_L^f)\mathfrak{F}^*(R_L^f)$. Where, $\mathfrak{F}$ and $\mathfrak{F}^*$ are the Fourier transform and its conjugate. The Lagrangian spectrum is given in Fig. 8.23. This statistical quantity is really a measure of the turbulence of the carrying flow and is not flawed by accounting simultaneously the fluid and solid in statistics. Based on Fig. 8.23, we can conclude that the carrying flow statistics remain constant for the three considered density ratios.

To characterize the particle response to fluid flow fluctuations, two different Stokes numbers are determined, based on the Kolmogorov time scale and the large eddy turnover time. The Stokes number based on the Kolmogorov time scale is relatively large $St_k = \{26, 52, 104\}$, as a consequence, the particles are inertial relative to the Kolmogorov scales. The preferential concentration of inertial particles in low vorticity regions has been commonly observed and quantified for small particles (pointwise or particles smaller than the Kolmogorov length scale). Particles whose Stokes number based on the Kolmogorov time scale are near unity are experiencing strong preferential concentration. For larger particles (finite size), it is expected that preferential concentration may occur but on a different length scale where particle response time and fluid flow time scale match. For our simulations, the Stokes number based on the large eddy turnover time is less than 10 ($St_E = \{1.5, 3.0, 6.0\}$) and it can be expected that it is for this scale that preferential concentration could appear. During the simulations no clusters formation was observed. Statistics on the radial pair distribution function did not show evidences of preferential concentration. Different explanations can be proposed. The first one is that the simulation time (12 eddy turn-over times) is not long enough for the particles to migrate in accumulation zones. The second explanation is that the physical life-time of the low vorticity regions at large scales is too short to give enough time for the particles to cluster. The segregation and clustering of finite size particles is still an open subject to study.

#### 8.2.4.2 Particle Statistics

The dynamics of particles smaller than the Kolmogorov scale has been studied extensively. Characterizing the relation between the particle statistics (kinetic energy and the velocity autocorrelation time) and the turbulence properties has been established in [362–366]. In Table 8.2 statistics are reported for finite-size particles and scaled by the fluid tracer statistics. As expected the particle kinetic energy, averaged over the $N_p$ particles during $N_t$ time steps $q_p^2 = \frac{1}{2}\frac{\sum_j \sum_i \mathbf{V}_i^2(t_j)}{N_t N_p}$, is reduced when the density ratio increases while the velocity autocorrelation time increases. This reduction is predicted by Tchen-Hinze theory [362, 363]:

$$\left(\frac{q_p^2}{q_f^2}\right)_{T.-H.} = \frac{\frac{T_L^f}{\tau_p} + b^2}{\frac{T_L^f}{\tau_p} + 1} \tag{8.48}$$

where, $b = \frac{3\rho_f}{2\rho_p+\rho_f} = \frac{3}{2\rho+1}$.

However, the particle kinetic energy is less important than the prediction of the Tchen-Hinze theory, $\left(\frac{q_p^2}{q_f^2}\right)_{T.-H.}$. Indeed, Tchen-Hinze theory is based on an estimate of the fluid kinetic energy at the position of particles assuming that their size is much smaller than the Kolmogorov length scale. In our case, the particles are significantly larger than the Kolmogorov length scale therefore they are not experiencing of the full fluid fluctuations (large particles behave like a spatial low pass filter of the flow kinetic energy).

The Lagrangian velocity autocorrelation time $T_L = \int_0^\infty R_L(t)dt = \int_0^\infty \frac{<\mathbf{V}_i(t_0)\cdot\mathbf{V}_i(t_0-t)>}{<\mathbf{V}_i(t)^2>}dt$ is presented in Table 8.2. This autocorrelation time $T_L$ is normalized by the autocorrelation time of the fluid particles $T_L^f$ obtained thanks to the autocorrelation function $R_L^f$ (see Sect. 8.2.4.1). It characterizes the typical time a particle takes to change its velocity direction. The present simulations show that $T_L$ increases as density ratio rises. Similar trend is observed for inertial point-particles simulations.

As can be seen in Table 8.2, whatever the density ratio is, even for neutrally buoyant particles, the normalized autocorrelation time is greater than 1, clearly showing finite size effect.

**Table 8.2** Particle statistics

| $\rho$ | $\frac{q_p^2}{q_f^2}$ | $\left(\frac{q_p^2}{q_f^2}\right)_{T.-H.}$ | $\frac{T_L}{T_L^f}$ |
|---|---|---|---|
| 1 | 0.85 | 1 | 1.31 |
| 2 | 0.79 | 0.5 | 1.45 |
| 4 | 0.74 | 0.2 | 1.65 |

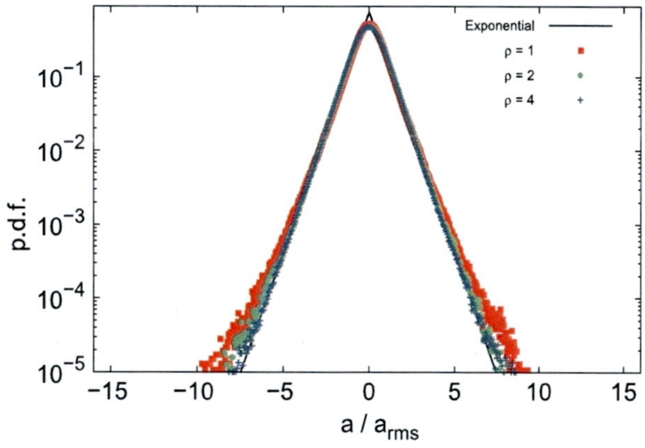


**Fig. 8.24** Probability density function of particle acceleration. Exponential p.d.f. is given as a reference, $\exp^{-|a|^\beta}$, $\beta = 1$. without collision events

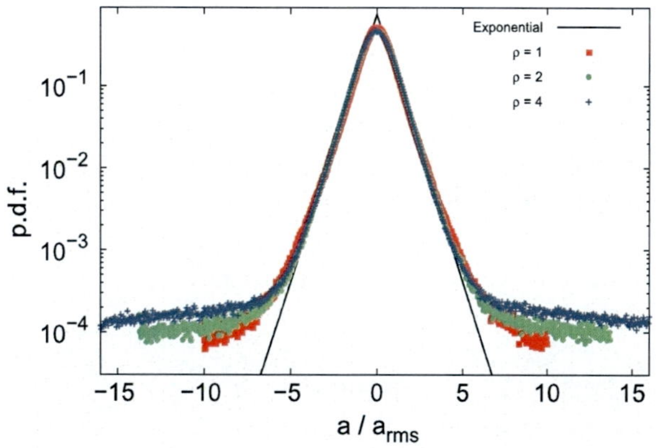


**Fig. 8.25** Probability density function of particle acceleration. Exponential p.d.f. is given as a reference, $\exp^{-|a|^\beta}$, $\beta = 1$. with collision events

Analysing the acceleration statistics provides insights on the dynamic response of particles to the turbulence forcing. It has been shown in literature [367–369] that the probability density function of finite size solid particles acceleration $P(a)$ can be approximated by an exponential law $\exp^{-|a|^\beta}$ where $\beta \leq 1$. As showed by [370] the constant $\beta$ approaches unity for inertial particles (here for high density ratio $\rho$). In Figs. 8.24 and 8.25 two p.d.f. of particle accelerations normalized by their r.m.s. values have been reported for the different simulations. In Fig. 8.24 the p.d.f. has been obtained without considering the acceleration occurring during collisions.

For this case, the p.d.f. closely follows an exponential law with $\beta = 1$ whatever the density ratio. Therefore, this is a measure of acceleration statistics strictly related to the interaction with turbulence in the very dilute regime (no collision).

Concerning, the case where acceleration during collision is taken into account for the p.d.f. acceleration (Fig. 8.25, the same exponential behaviour is observed for low acceleration values. This is no longer the case for larger values. Indeed the tails of the p.d.f. in Fig. 8.25 are shifted towards higher values. This is related to rare but strong collision events on the acceleration.

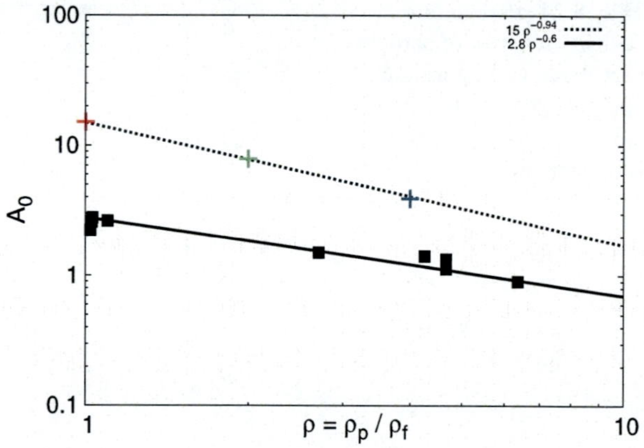


**Fig. 8.26** Dimensionless acceleration r.m.s. compared to density ratio. Present simulations (cross) and experiments of Qureshi [368] (squares) and their respective exponential law

In [368] an experimental study has been conducted on the interaction between finite size soap bubbles and turbulent air channel flow. It was shown that the r.m.s. acceleration scaled by the fluid dissipation rate ($A_0 = a^2_{r.m.s.}/\varepsilon_t^{-3/2}\nu^{1/2}$) decreases as $\rho$ increases. In order to compare to this result the dissipation of the turbulent flow $\varepsilon_f$ is estimated in the present simulations from the spectrum: $2\nu\int_0^\infty \kappa^2 E(\kappa)d\kappa$. In [368] an exponential decrease has been observed from $\rho = 1$ to $\rho = 10$ as $A_0 = 2.8\rho^{-0.6}$. From our simulations the same trend can be observed as $15\rho^{-0.94}$ (Fig. 8.26).

#### 8.2.4.3 Collision Statistics

- Collisions with solid contact only

Understanding the collision dynamics is a key point for modeling suspension flows and predicting coalescence events for drops. In this section, only the solid-solid model has been activated without lubrication.

We have followed the present procedure to form statistics. If we call $\Delta t_{coll}$ the time between two collisions for a given particle, and $\Delta t_{coll,i}$ the averaged of all the $\Delta t_{coll}$ concerning the $i$th particle, the inter-collision time $\tau_c$ for $N_p$ particles is defined as $\tau_c =< \Delta t_{coll,i} >_{N_p}= \frac{1}{N_p}\sum_{i=1}^{N_p} \Delta t_{coll,i}$. Another way to compute $\tau_c$ is to count the number of collision $N_{coll}$ during the total time of analysis $\Delta T$, and for the $N_p$ particles to obtain $\tau_c = \frac{N_p}{2N_{coll}}\Delta T$.

This characteristic time can be estimated theoretically (Eq. 8.49) from the average relative velocity of particles $\langle|w_r|\rangle$ as proposed in [371].

$$\frac{1}{\tau_c} = \frac{1}{2} g_0 n_p^2 \pi (2R)^2 \langle|w_r|\rangle \tag{8.49}$$

When the motion of particles is fully uncorrelated, the relative velocity of particles $\langle|w_r|\rangle$ can be evaluated from the particle kinetic energy $q_p^2$, [372] :

**Table 8.3** Comparison of inter collision times for various density ratios

| $\rho$ | $[\frac{\tau_c^{n,1}}{\tau_k} .. \frac{\tau_c^{n,2}}{\tau_k}]$ based on computations | $\frac{\tau_c^{th}}{\tau_k}$ theory Eq. (8.50) |
|---|---|---|
| 1 | [22.6 .. 24.5] | 21.5 |
| 2 | [19.3 .. 20.8] | 22.4 |
| 4 | [16.2 .. 17.1] | 23.1 |

The reference Kolmogorov time is $\tau_k = 0.0635$ s

$$\frac{1}{\tau_c^{th}} = \frac{1}{2} g_0 n_p^2 \pi (2R)^2 \sqrt{\frac{16}{\pi}\frac{2}{3} q_p^2} \tag{8.50}$$

In Table 8.3 the inter collision time is reported for different density ratios. Two different inter collision times have been considered. The first one $\tau_c^{th}$ is obtained thanks to Eq. 8.50, by extracting the particle kinetic energy $q_p^2$ from simulations with the assumption that $g_0$ is equal to 1 (particle dilute flow) and using the effective collision radius ($1.1R$ because the collision is activated for $\epsilon = 0.1$) and $n_p = \frac{512}{(2\pi)^3}$ the particles number density. The second inter collision time $\tau_c^n$ is directly obtained by estimating the time between two collisions in simulations. This estimate has been achieved with two methods:

$\tau_c^{n,1}$ We average all the times between the end of a collision and the beginning of the next collision for all the particles. In this estimate the solid contact, which is not instantaneous with our soft sphere model, is not taken into account.

$\tau_c^{n,2}$ Is the ratio between the total number of collisions occurring during the simulation and the simulation time.

The first method gives a lower bound of $\tau_c$ while the second one provides an upper bound.

The orders of magnitude of inter collision times are comparable whatever method used. The gap between $\tau_c^{th}$, $\tau_c^{n,1}$, $\tau_c^{n,2}$ is always less than 15%. As expected form Eq. (8.50), $\tau_c^{th}$ is increasing with the density ratio as $q_p$ is decreasing. In contrast, the collision time measured is decreasing with increasing density ratio. One can explain this behaviour by considering that the evaluation of $\tau_c^{th}$ is done under the assumption of $g_0 = 1$.

In order to understand those discrepancies the probability density function of inter-collision times is now considered. In simulations, the p.d.f. is obtained by counting the number of occurrences of each inter collision time $\Delta t_{coll}$ for all the particles. The p.d.f. is divided by the total number of occurrences. From the kinetic theory of granular flows (uncorrelated motion of particles) the p.d.f. follows

$$P(\Delta t_{coll}) = \frac{1}{\tau_c^{th}} \exp\left(-\frac{\Delta t_{coll}}{\tau_c^{th}}\right) \tag{8.51}$$

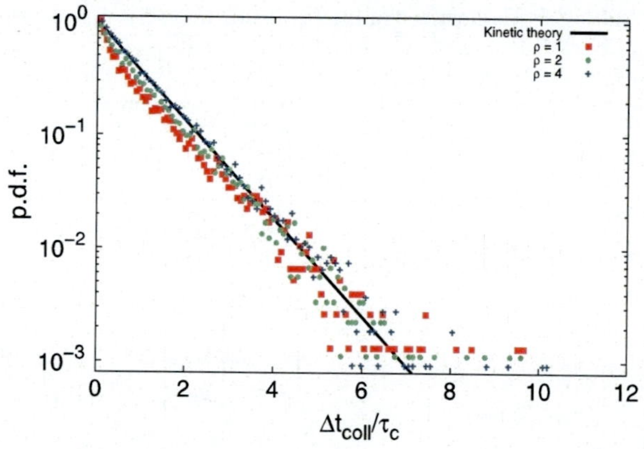


**Fig. 8.27** Inter-collision time probability density function. Solid line corresponds to Eq. (8.51), ($\tau_c = \tau_c^{th}$ for kinetic theory and $\tau_c = \tau_c^{n,1}$ for the measured p.d.f.)

In Fig. 8.27, the probability density function is reported for three density ratios. A comparison is provided between simulations and kinetic theory.

The p.d.f. obtained with simulations follows the kinetic theory for $\Delta t_{coll} < 4\tau_c$. For larger values of $\Delta t_{coll}$, the measured p.d.f. has a large dispersion due to the small number of occurrences that are counted. For this reason, the comparison to the kinetic theory is not representative for $\Delta t_{coll} > 4\tau_c$.

If we focus on the small inter collision times, the comparison between kinetic theory and measured p.d.f. shows a larger difference for $\rho = 1$ than for $\rho = 4$. This is the consequence of the particle motion that follows more easily the flow and comes into contact with slightly more correlated trajectories when $\rho = 1$. This discrepancies for $\rho = 1$ could be inferred to the effect of secondary collisions that will be affected by the lubrication model. A discussion on the secondary collisions is given in Sect. 8.2.

In order to describe more precisely the collision regime, statistics of the velocity orientations when collision occurs are analysed. For uncorrelated motion of particles, the p.d.f. of the angle between two particle velocities at contact is predicted by [365]:

$$f(\theta_{coll}) = \frac{\int_0^\infty \int_0^\infty 8\pi \sin(\theta_{coll}) V_a^2 V_b^2 \sqrt{V_a^2 + V_b^2 - 2V_a V_b \cos(\theta_{coll})}\, f_v(V_a) f_v(V_b) dV_a dV_b}{\int_0^\pi \int_0^\infty \int_0^\infty 8\pi \sin(\theta) V_a^2 V_b^2 \sqrt{V_a^2 + V_b^2 - 2V_a V_b \cos(\theta)}\, f_v(V_a) f_v(V_b) dV_a dV_b d\theta} \tag{8.52}$$

where $f_v$ is the velocity p.d.f..

In Fig. 8.28 the kinetic theory corresponding to Eq. 8.52 for the collision angle p.d.f. is compared to simulation results. A Gaussian profile has been used for $f_v$ corresponding to the behaviour observed in our simulations.

Whatever the density ratio is, the maximum of the measured p.d.f. is obtained for $20° < \theta_{coll} < 40°$, whereas the kinetic theory provides a peak at 108°. This is characterized by a shift of the p.d.f. from large to low angles indicating that particles come into contact with nearly parallel velocities, evidencing a particle to particle tra-

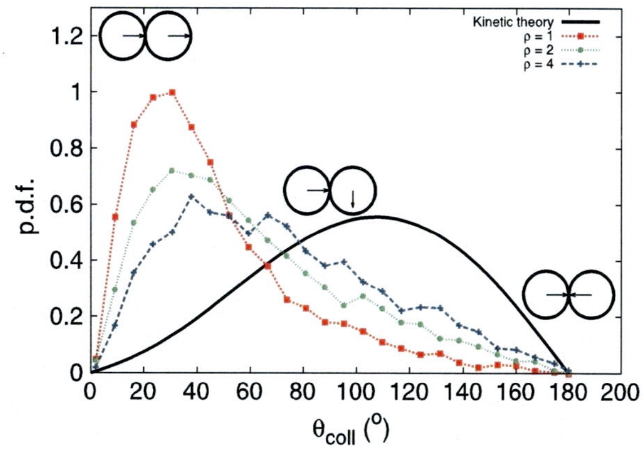


**Fig. 8.28** Probability density function of the angle between particle velocity at contact. Theory for uncorrelated motion is given by Eq. (8.52) (solid line). A sketch of particle velocity at contact is provided for $\theta_{coll} = \{0, 90, 180\}$ degrees

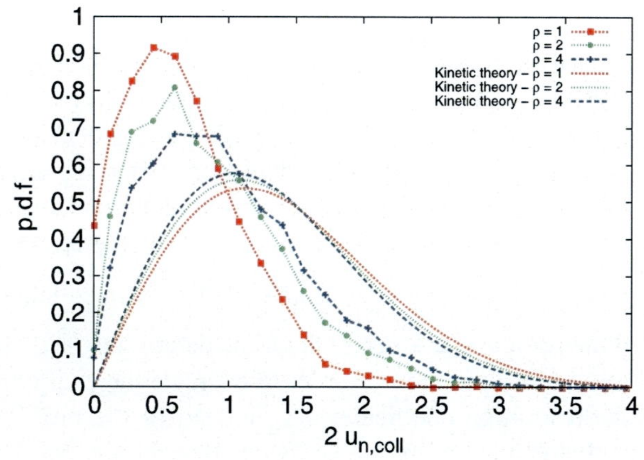


**Fig. 8.29** Probability density function of the relative velocity at contact (theory and simulations)

jectory correlation. This shift decreases for higher density ratio showing that particle inertia yields larger decorrelation between the particles and the carrying fluid flow. From this result, it is obvious that the relative velocity at contact should be lower than the one obtained with the kinetic theory discussed below.

We now compute the relative velocity at contact $w_r = 2V_{r,coll} = \left(\mathbf{V}_p^a - \mathbf{V}_p^b\right) \cdot \mathbf{n}$, where $\mathbf{n}$ is the unit vector connecting the centers of the two particles. For uncorrelated velocity particles with a Gaussian velocity distribution, the p.d.f. of $w_r$ is given by

$$f_{coll}(|w_r|) = \frac{2}{\left(8/3q_p^2\right)^2}|w_r|\exp\left(-\frac{|w_r|}{8/3q_p^2}\right) \tag{8.53}$$

In Fig. 8.29, we compare the p.d.f. of the relative velocity at contact obtained with the kinetic theory and simulation results. In Eq. (8.53), the kinetic energy $q_p^2$ is obtained from the simulations.

As already observed for p.d.f. of the angle, a clear shift of the p.d.f. of velocity is also reported in simulations compared to the kinetic theory. As expected, the relative

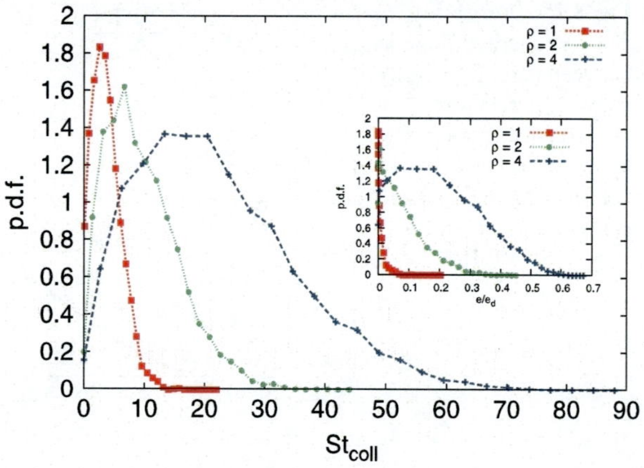


**Fig. 8.30** Probability density function of the impact Stokes number obtained from the relative velocity. Inset plot represents the p.d.f. of the restitution coefficient obtained with the correlation given by [340]

velocity at contact is lower in the simulations than in an uncorrelated motion. Particle inertia increases decorrelation of particle motion and fluid flow yielding a better agreement with kinetic theory for the largest density ratio.

The relative velocity at contact characterizes the energy involved in collisions. When collisions occur in a viscous fluid, part of this energy is dissipated by viscous effect during the fluid drainage. As a result, the velocity after the collision is reduced. It has been shown that the impact Stokes number defined as $St_{coll} = \dfrac{2}{9}\dfrac{RV_{r,coll}\rho_p}{\mu}$ is a dimensionless parameter which can determine the collision regime (from dry contact to viscous rebound). In order to estimate the energy dissipated during the collision the restitution coefficient $e$ is defined as the ratio between the velocity after the collision and before the collision. Based on experiments, the correlation proposed by [340], $\frac{e}{e_d} = \exp\left(-\frac{35}{St_{coll}}\right)$, gives the relation between the restitution coefficient and the Stokes number (the restitution coefficient is scaled by its value for a dry contact $e_d$). In Fig. 8.30 the p.d.f. of impact Stokes numbers is presented as well as the distribution of the effective restitution coefficients based on the p.d.f. of the relative velocity, Fig. 8.29.

It can be seen that for the case $\rho = 1$ most of collisions are in the Stokes range corresponding to no rebound ($e \simeq 0$). Rebound will not occur and particles will stay stuck together. For the case $\rho = 4$ the values of the Stokes numbers cover a wide range corresponding to restitution coefficients lower than 0.5. This means that for the range of physical parameters we simulated, lubrication effects during particle encounters are important and must be accounted for.

- The effect of lubrication on collision statistics

The effect of lubrication on collisions will be studied for cases $\rho = 1$ and $\rho = 4$. The lubrication model is now activated at distance $1.1\ R$ similarly to the solid-solid model in the previous section. The solid-solid collision model is not modified and is activated only for the particles that overlap.

Several physical quantities and statistics will be compared in order to highlight the effect of film drainage on collisions. The kinetic energy of particle agitation, the autocorrelation time, the velocity and acceleration distributions are not drastically modified by including the effect of lubrication. A first difference between simulations with and without lubrication model appears on the averaged distance between particles. For example, the mean distance to the nearest neighbor is decreased by about 5%, [373]. The same result can be observed in the radial distribution function: for small distances the probability to observe a particle increases, meaning that particles tends to stay closer to each other. The activation of lubrication between two particles decreases the relative velocity between them during the encounter yielding a longer time of nearby interaction.

Here we focus on the modification of the collisional regime. In order to analyse this collisional regime it is necessary to make the difference between two different interactions: the encounter, defined as the activation of lubrication model, and the solid-solid collision, defined as the activation of solid-solid model.

The inter-encounter time obtained as the average time between two activations of the lubrication model is $\{21.6, 16.1\}\ \tau_k$ for $\rho = \{1, 4\}$ respectively. The elapsed time between two successive interactions is close to the results obtained for the simulations without lubrication model: $\tau_c^{n,1} = \{22.6, 16.2\}\ \tau_k$. It shows that the encounters between particles do not depend on the lubrication model but only on the flow dynamics which carries the particles towards contact. In Fig. 8.31 the p.d.f. of the inter-encounter and inter-collision times are presented for the simulations with and without lubrication model.

In all cases an exponential decay is observed for large inter collision or inter encounter times. This is resulting from the uncorrelated motion in agreement with the kinetic theory.

When the lubrication is considered, the ratio of collisions occurring at small times increases. As a consequence, the level of the exponential law decreases. For the small inter collision times the exponential law is no more valid. This effect is enhanced for $\rho = 1$, but also present for $\rho = 4$. Because of the increase of small $\Delta t_{coll}$ occurrences, the inter collision time $\tau_c^{n,1}$ is reduced from 22.6 $\tau_k$ to 2.5 $\tau_k$ for neutrally buoyant particles. This is explained by the occurrence of many body solid-solid collisions during the activation of the lubrication model. The lubrication slows down the particles during separation after a collision, increasing solid-solid collision events. These phenomena have been also reported in [374], so-called secondary collisions.

As proposed by [374] the p.d.f. can be split into two contributions. The first one, relative to primary collisions has an exponential tail corresponding to long time collisions: $P(\Delta t) = \frac{\alpha}{\tau_c} \exp\left(-\frac{\Delta t_{coll}}{\tau_c}\right)$. The second one, relative to secondary collisions is obtained as the difference between the overall p.d.f. and the first contribution. $\alpha$ is a parameter representing the gap between the real p.d.f. and the kinetic theory. It is obtained by fitting the p.d.f. for long inter-collision times. The exponential law of the kinetic theory is recovered when $\alpha = 1$.

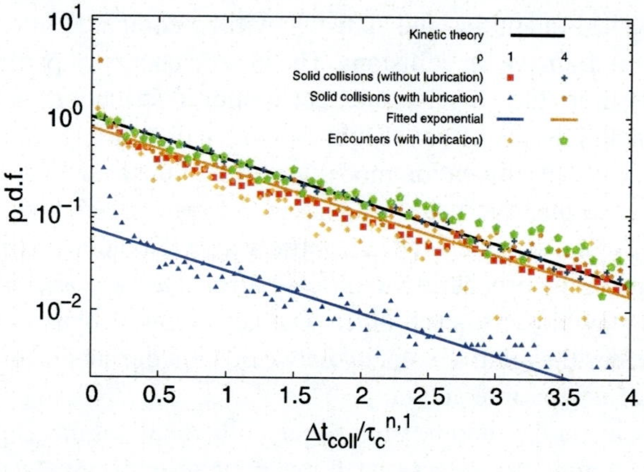


**Fig. 8.31** Inter-collision time probability density function for simulations with and without lubrication model. Fitted exponential law is given for solid-solid collisions for simulations with lubrication model. For comparison purpose the $\Delta t_{coll}$ time is scaled by the inter collision time obtained without lubrication model (see Table 8.3). The kinetic theory (Eq. 8.51) corresponds to solid line

From our simulations we obtain $\alpha$ for the solid-solid collisions with the lubrication model:

- For $\rho = 1, \alpha = 7\%$: only 7 paths over 100 are reproduced by the kinetic theory, the other 93 are free paths between secondary collision. In the simulations of [374], $\alpha$ was ranging between 8 and 11% which agrees with our results.
- For $\rho = 4$: the occurrence of consecutive collisions is less pronounced. The inter-collision time is reduced only from 16.2 $\tau_k$ to 14.6 $\tau_k$. The parameter $\alpha$ is equal to 77%. Here, particle inertia drives the particles towards more energetic collisions and rebounds. The effect of lubrication is weaker and cannot keep the particles together after collision.

# Chapter 9
# Multiscale Euler–Lagrange Coupling

Part of this chapter has been published in [7].

## 9.1 Introduction

In the wide topic of multi-phase flows, atomization is a key process involved in many engineering applications, especially in combustion problems. The detailed numerical simulation of aeronautical combustion chambers has become a major topic in the last years, answering to the need to enhance the engine efficiency and to reduce the polluting agents. Among the most common injectors found in aeronautical engines are the *airblast* injectors. These devices are meant to use the assisted atomization to pulverize the fuel, which is obtained through the injection of a high pressurized liquid through a small fence (planar or annular) sandwiched by airflows. The role of the atomization process is to generate a cloud of fine droplets on a very small distance, in order to maximize the liquid surface and improve the combustion efficiency. The geometry of these injectors can be either planar or annular, both share similar developments of the atomization process.

The physics of atomization involves a wide range of phenomena and a strongly multi-scale dynamics, where many order of magnitude span between larger and smaller structures (see Fig. 9.1). Very different topologies such as drop, ligament, liquid sheet, may appear due to the many instabilities arising [375–379]. The most common approach used in industrial oriented Large Eddy Simulations of reacting flows is to directly inject the spray with assumed characteristics, given by semi-empirical correlations. The liquid being modelled by a dispersed phase approach, no atomization is resolved within the simulation. Examples can be found in [380, 381].



S. Vincent et al., *Small Scale Modeling and Simulation of Incompressible Turbulent Multi-Phase Flow*, CISM International Centre for Mechanical Sciences 607,
https://doi.org/10.1007/978-3-031-09265-7_9

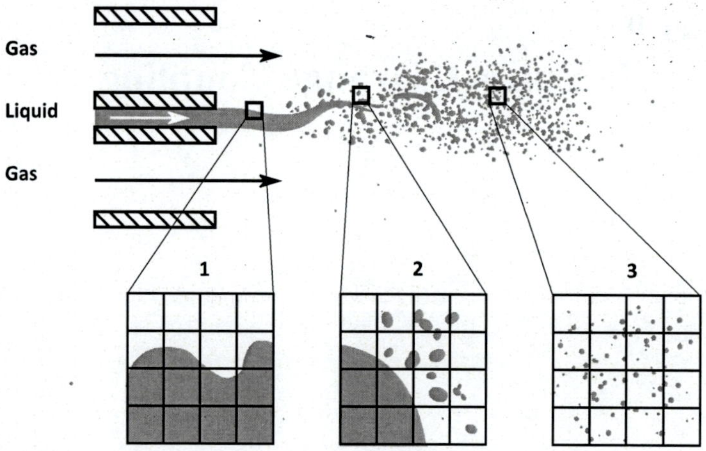


**Fig. 9.1** Representation of the multi-scale aspect of assisted atomization

In more research-oriented simulations, focused on the atomization mechanisms, many numerical methods allow the Direct Numerical Simulation of these complex two phase flows, usually in much more simplified and small scale configurations. The most widely employed methods are the Level-Set and the interface reconstructed VOF methods, where the interface position and topology are tracked as accurately as possible in time. Among their most advanced evolutions are the Coupled Level-Set Method [89], the Conservative Level-Set [382] and the mass conserving Level-Set [383]. Thanks to those sharp-interface-capturing algorithms, some first numerical simulations of atomization have become possible: three-dimensional DNS simulations of liquid jet breakup have been reported in [150, 292, 383, 384], while in [385, 386] a full assisted atomization is simulated in different configurations. A way to efficiently perform such simulation of two phase flows is to use a variable spatial discretization which is adapted to the interface shape. This scenario has been investigated by [387] using a force refinement level set grid method on structured and unstructured meshes, or by [388] with VOF method applied on an octree adaptative mesh refinement to treat the atomization process. Other works focus on the adaptive mesh refinement to maximize the resolution on the interface itself in order to lessen the computational burden, such as in [160, 163, 169, 182, 389, 390]. However, these method have proved ineffective to correctly capture the dynamics of small droplets because they required a drastic refinement of spatial mesh, especially if the cloud of drop disperses uniformly into the computational domain.

As the detailed simulation of all the spatial scales remains very expensive, the particle tracking method appear interesting for the smallest particles. Indeed, these structures are sufficiently small by regard with the smallest computational mesh size and have a spherical shape due to surface tension effects, so that the dispersed phase model assumptions are validated. Following this idea, the problem can be numerically split in two "moments" following the two stages of atomization, in order to adapt the local two-phase modelling. A separated phases model as the VOF or Level-Set could describe the primary atomization, where the two phases are clearly distinguished and the interface in subject to low frequency instabilities, while the dispersed phase model would be better adapted to describe the evolution of the sprays in terms of secondary atomization, transport and eventually evaporation. In [385] (by extension of [387] with a Level Set method), [391, 392] the coupled DNS and Lagrangian method technique has been successfully applied to the atomization of jets in cross-flow configuration. In [386] (coupled to a VOF method [388] ) and in [393] a similar approach involving VOF and CLSVOF methods were developed to resolve assisted atomization. Both of them tracked point particle liquid particles with a drag force to couple Eulerian and Lagrangian quantities, this methodology seeming accurate enough to capture both large and small inclusion. In other approaches, the generation of the spray is not direct but it is instead treated by a more statistical approach, as in [394, 395], in order to avoid the use of cell sizes small enough to resolve the smallest drops. However, in most of those approaches the transition from Eulerian to Lagrangian formulation is still controversial as well as the treatment of medium inclusions regarding to the grid size.

Indeed, if one consider a medium liquid drop, i.e. a quasi-spherical inclusion whose radius is equivalent to a few mesh cell, the Eulerian formulation is not able to correctly predict the dynamics. Actually, between 6 or 10 mesh points per radius are necessary to described the evolution of such drop to avoid disappearance of liquid inclusion due numerical diffusion of level-set method, or less accurate interface representation of the piecewise linear reconstruction of VOF method. Moreover, the Lagrangian tracking could not be applied, the assumption of dispersed phase is not fulfilled any more. Hence, a third modelling scale have to be introduced so that the spatial gap between the two models is filled, as it seems inappropriate.

In order to lessen these problems, the present work proposes an approach to model this third scale between pure Eulerian and Lagrangian representations: medium drops are supposed to have a rigid spherical shape, and they are followed by rigid translation while their mass and momentum properties are projected on the Eulerian field. This formulation results in a more conservative representation of the initial characteristics of the droplet. The hypothesis of spherical shape allows an efficient coupling with an AMR technique, where allowing a progressive mesh coarsening around the droplet makes the transition between purely Eulerian and Lagrangian formulations consistent. The reciprocal transformation has been adopted to treat the collision between Lagrangian particle with the interface of a large liquid inclusion. A CLSVOF method is used (from [89, 153]) to accurately predict the interface shape with the Level-Set description and to ensure local mass conservation with the VOF formulation. An adaptive mesh refinement (given by the PARAMESH package [165]) is dynamically performed around the resolved CLSVOF interface throughout the computation, so that jump conditions are accurately satisfied. The smallest particles are tracked with a Lagrangian formulation and a classical drag force, while medium ones are treated using the mixed Eulerian-Lagrangian method previously described. The whole procedure has been paralleized to allow relatively large computational domains. The numerical code is an evolution of the one described in [155, 160].

This paper is organized as follows: Sect. 9.2 details the governing equation of two-phase flow, Sect. 9.3 describes the numerical procedure for large scale inclusions (CLSVOF method, pressure solver, adaptative mesh refinement). Section 9.4 details the original conservative treatment for the medium inclusions as well as the more classical Lagrangian particle tracking for the small drops. In this section, the Euler to Lagrange transition criteria for the inclusions (size, shape) are presented; the recombination of a particle with a large scale interface is described as well. Finally, Sect. 9.5 shows some academic and realistic two-phase problems and show the capability of the present method to described multi-scale problems, in particular the assisted atomization of a planar liquid sheet.

## 9.2 Governing Equations

The numerical methodology proposed here aims to solve two-phase flows under the incompressibility hypothesis. Hence we consider the Navier-Stokes equations in each phase under free divergence velocity field constraint:

$$\frac{\partial \boldsymbol{u}}{\partial t} + (\boldsymbol{u} \cdot \nabla)\boldsymbol{u} - \frac{1}{\rho}\nabla \cdot \boldsymbol{T} = \boldsymbol{f}, \tag{9.1}$$

$$\nabla \cdot \boldsymbol{u} = 0, \tag{9.2}$$

where $\boldsymbol{u} = \left[u_i, u_j, u_k\right]^T$ represents the velocity vector field, $p$ the hydrodynamic pressure, $\rho$ the fluid density, $\mu$ the fluid dynamic viscosity and $\boldsymbol{f} = \left[f_i, f_j, f_k\right]^T$ the external body forces like gravity.

$$\begin{aligned} \boldsymbol{T} &= -p\mathbf{I} + \boldsymbol{D} \\ \boldsymbol{D} &= \mu\left(\nabla\boldsymbol{u} + (\nabla\boldsymbol{u})^T\right) \end{aligned} \tag{9.3}$$

Those equations are solved in both liquid and gases phases. Within this paper, the liquid and gazes properties will be subscribed with "l" and "g" respectively. Across the interface, the no slip condition imposes the continuity of velocity field $\boldsymbol{u}_l = \boldsymbol{u}_g$, while jump conditions are imposed by viscous constraints equilibrium and surface tension effects ($\sigma$ being the liquid-gas surface tension coefficient):

$$\begin{aligned} [p] - \boldsymbol{n} \cdot \left[\mu\left(\nabla\boldsymbol{u} + (\nabla\boldsymbol{u})^T\right)\right] \cdot \boldsymbol{n} &= \sigma\kappa, \\ \mathbf{t} \cdot \left[\mu\left(\nabla\boldsymbol{u} + (\nabla\boldsymbol{u})^T\right)\right] \cdot \boldsymbol{n} &= 0, \end{aligned} \tag{9.4}$$

where $\kappa$ is the interface curvature, $\mathbf{t}$ and $\boldsymbol{n}$ the tangential and normal unit vectors oriented from gas to liquid and $[.] = (.)_l - (.)_g$ any jump across the interface.

## 9.3 Resolved Liquid Structures—Eulerian Modelling

### 9.3.1 Interface Tracking

An interface tracking method is employed to solve the primary atomization region, up to the formation of the droplets. The coupled Level-Set/Volume-Of-Fluid (CLSVOF) formulation is used in reason of its accurate description of the interface shape, given by the Level-Set function, and the crucial mass conservation properties given by the VOF. The chosen implementation of the CLSVOF follows [89] and the improved version from [153], with some further modifications described in [155]. The Level-Set function $\phi(\boldsymbol{x}, t)$ gives implicitly the interface position, as it is defined as a signed distance to the interface. The evolution of $\phi$ follows a linear advection equation driven by the free-divergent fluid velocity:

$$\frac{\partial \phi}{\partial t} + \boldsymbol{u} \cdot \nabla \phi = 0 \tag{9.5}$$

Physical properties of each fluid is given by the sign of $\phi$:

$$\alpha(\boldsymbol{x}, t) = \begin{cases} \alpha_l & \text{if} \quad \phi > 0 \\ \alpha_g & \text{elsewhere.} \end{cases} \tag{9.6}$$

The regularity of the Level-Set function allows us a straightforward computation of the geometrical properties at the interface, namely normal vector and curvature:

$$\boldsymbol{n} = \frac{\nabla \phi}{||\nabla \phi||}, \qquad \kappa = \nabla \cdot \left( \frac{\nabla \phi}{||\nabla \phi||} \right) \tag{9.7}$$

The VOF function $C$ represents the ratio of liquid volume inside a computational cell over the total cell volume

$$C_i = C(\boldsymbol{x}_i) = \frac{V_{l,i}}{V_i} \tag{9.8}$$

where $\boldsymbol{x}_i$ is a cell centered and $V_i = V_{l,i} + V_{g,i}$. This function is driven by the same linear passive advection equation as the Level-Set (9.5):

$$\frac{\partial C}{\partial t} + \boldsymbol{u} \cdot \nabla C = 0. \tag{9.9}$$

### 9.3.2 Temporal Integration

Let $(\boldsymbol{u}^n, p^n, \phi^n, C^n)$ the solution at a given time $t^n$, $n$ being the temporal subscript. The necessary steps to integrate the flow equations can be regrouped into two moments. At fist, the interface motion given by the two Eqs. (9.5) and (9.9) is solved; subsequently, the incompressible flow Eq. (9.2) is solved using the updated interface position. The CLSVOF steps can be summarized as:

- The Level-Set function is used to evaluate the surface representing the PLIC reconstructed interface by solving a least-square problem on the local nine-point stencil (as described in [89]; a local stencil correction has been introduced to better resolve some difficult configurations, such as when two interfaces are present in the base stencil).
- Semi-Lagrangian fluxes are computed at each cell face by geometric computation given by [396].
- The Level-Set and VOF functions are integrated simultaneously using the previously computed fluxes: Eqs. (9.5) and (9.9) are solved alternatively in each space direction with a second order conservative split scheme;
- The updated VOF value is used to offset the value of the Level-Set ("enforce"), in order to assure local mass conservation.

Once $C^{n+1}$ and $\phi^{n+1}$ are obtained, they are used to obtain $\rho^{n+1}$ and $\mu^{n+1}$ through the ghost fluid method as in [160]. Then, the flow equations are solved by a Chorin projection method:

- Equation (9.2) is computed without pressure term, using a fully explicit second order Adams-Bashforth scheme (or 2th and 3th order Runge-Kutta), in order to predict the velocity field $\boldsymbol{u}^*$:

$$\boldsymbol{u}^* = 1.5\ \Delta t \mathbf{F}^n - 0.5\ \Delta t \mathbf{F}^{n-1} \tag{9.10}$$

$$\mathbf{F}^n = -(\boldsymbol{u}^n \cdot \nabla)\boldsymbol{u}^n + \frac{1}{\rho^{n+1}} \nabla \cdot \left(\mu^{n+1} \left(\nabla \boldsymbol{u}^n + (\nabla \boldsymbol{u}^n)^T\right)\right) + \boldsymbol{f}^n, \tag{9.11}$$

  The viscous term is evaluated using a spatial second order centered scheme and the convective one with a 5th order WENO scheme (see again [160] for the explicit discretization).
- The pressure is evaluated by solving the elliptic equation

$$\nabla \cdot \left(\frac{1}{\rho^{n+1}} \nabla p^{n+1}\right) = -\frac{\nabla \cdot \boldsymbol{u}^*}{\Delta t} \tag{9.12}$$

  This equation is solved using a Krylov BiCG-Stab preconditionned by a multilevel algorithm built on the AMR tree. The interested reader is referred to [160] for more details.

- Finally, the free divergent velocity field is obtained by correcting the predicted one:

$$u^{n+1} = u^* - \Delta t \frac{1}{\rho^{n+1}} \nabla p^{n+1} \tag{9.13}$$

### 9.3.3 Adaptive Mesh Refinement

The dynamic adaptation of the mesh resolution (AMR) is performed through the open source parallel PARAMESH package. The whole numerical procedure is detailed in [160]. Here is a summary. The computational domain is recursively decomposed in an arbitrary number of blocks, resulting into an oct-tree data structure. Each block is a Cartesian domain with a user-defined fixed number of cell in the three directions but with a variable grid spacing (Fig. 9.2). The maximal and minimal grid spacing is fixed at the beginning of the simulation. In case of need, during checkpoints they can be altered with a small effort. The refinement ratio from one level to another is limited to 2, so that any neighbour of a given cell can't be more than two times larger or smaller than its size. The refinement process is automatically performed following one (or more) user-defined criterion. Within this work, the refinement is forced to be at the maximum level around the interface, where the discontinuities are located, while the mesh is allowed to de-refine into the single phase regions. The block numbering and parallel repartition is performed using a Morton type algorithm and the data transfer from a coarser grid to a finest one (or reciprocal) is done through second-order interpolation process. The velocity field is subject to second-order divergence-free interpolation from [397], the scalar fields to conservative second-order Q1 interpolation. Any interested reader is referred to the PARAMESH documentation or study of [165, 398] for more details.

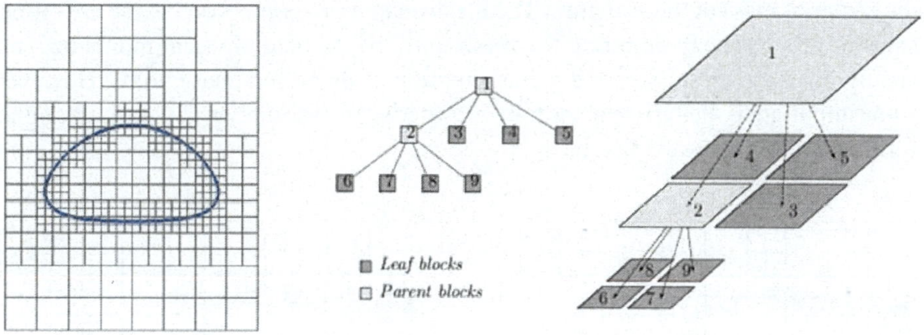


**Fig. 9.2** Left: illustration of the refinement around the interface on a two dimensional rising bubble test case. Right: example of an octree block structure with Morton numbering

## 9.4 Multi-scale Approach

As mentioned in Sect. 9.3.3, the refinement process is automatically performed around the interface in order to correctly resolve the small liquid scales. However, in realistic long-time simulations the atomization process is expected to generate droplets of dimension comparable to some mesh cell sizes ($L \sim \Delta x$), which can not be correctly resolved by the CLSVOF (or any other Eulerian interface-tracking method) because of the insufficient resolution. Worse, those small structures force a useless local grid refinement so that, in numbers, they may defeat the purpose of the AMR by refining almost the whole computational domain. A second point is that the interface tracking method can not track the formation of very small liquid inclusions of size inferior to the finest mesh cell, which are formed when liquid structures like membranes or ligaments are locally under resolved.

A relevant solution to both problems is to couple the mesh refinement strategy with a dispersed phase approach for the mall liquid inclusions. In this approach, a small droplet is assumed to be a stable rigid sphere, in the sense that the gas shearing force is considered to be no more strong enough to overcome the surface tension and thus induce a secondary atomization. The characteristics of the droplet are then given by its center of mass, its radius and velocity. Resolving the dispersed phase by a Lagrangian modelling, the droplet motion can be tracked by advecting its center of mass with a drag-force model on a very coarse mesh, for a fraction of the Eulerian computational cost. This approach allows the resolution of the primary atomization up to the droplet formation by the CLSVOF and the most refined mesh, while the droplets can be tracked far outside the primary atomization region by the Lagrangian algorithm on a cheaper de-refined mesh. Figure 9.3 shows the principles of this approach.

The most basic transition involves the removal of the liquid phase of the droplet, resolved by the CLSVOF on a small number of cells ($L \approx \Delta x$), from the Eulerian field and the injection of a volume-less point particle in the position corresponding to the center of mass of the structure. The mesh adaptation is supposed in the next time-steps to progressively de-refine the mesh until the particle physical radius become smaller than the mesh size, and the assumption of dispersed phase holds. However, some important issues arises in the transition between the Eulerian and the Lagrangian formulation if care is not taken:

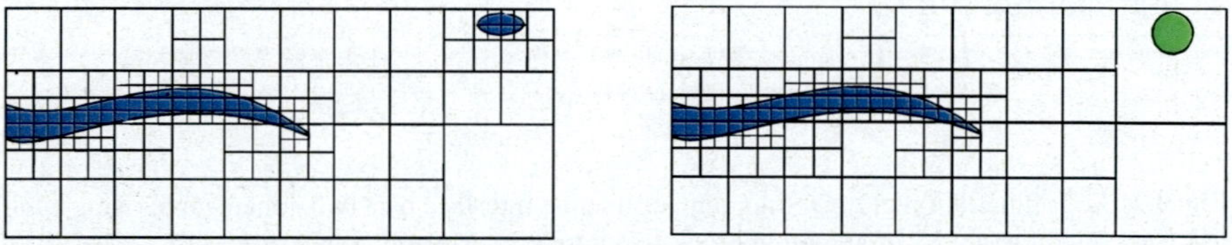

**Fig. 9.3** Left: illustration of the refinement around a (under-)resolved droplet. Right: illustration of the coupled effect of AMR and point-particle Lagrangian tracking

- First, the direct transformation on the refined grid artificially changes the local flow topology, as the domaine of droplet influence being reduced from the initial Eulerian volume to the cell containing the center of mass only. Moreover, the source term is very stiff.
- Second, the drag-force model requires an undisturbed velocity to compute the correct dynamics of the droplet. This velocity is not directly available from the cell containing the point particle, nor the surrounding cells. The resulting droplet acceleration is in consequence incorrect.
- Third, if a larger resolved liquid structure is present near the droplet, the mesh coarsening is not allowed until the droplet or the structure leaves the block. The radius to cell size ration remain in this case unchanged.

If the first problem can be addressed by integrating the source terms over a set volume of influence, the second is trickier. One possible solution is given in [399], where the actual influence of the droplet on the underlying Eulerian field is explicitly subtracted in order to obtain the undisturbed velocity by its own definition. For the droplet whose size is lesser than the mesh cell, the Lagrangian approximation can be considered as correct, as only the cell velocity is affected.

The approach proposed in this paper proposes the separate treatment of the liquid inclusions in two main scenarios, depending on the to-be-converted particle size:

1. the droplet size is lesser than the local mesh cell size. This kind of droplet will be called "small inclusion" within this paper;
2. the droplet size is greater than the local mesh size. This kind of droplet will be called "medium inclusion" within this paper.

In the first case, the dispersed phase approach hypothesis are considered as satisfied, so that a classical treatment is done, either in one or two ways coupling[1]; in the second, a kind of penalty method has been developed to allow the more complex interaction of the medium droplets and the underlying flow.

In the following paragraphs both algorithms are detailed. The proposed multi-scale approach includes the Lagrangian to Eulerian inverse conversion. This is needed because in the primary atomization process the droplets co-exist with larger structures: this may lead to coalescence processes, in which the particles collide a resolved liquid body. The collision is handled differently for the small and medium particles. In the former situation, the droplet is deleted, and its relatively small mass and momentum are added to the local cell; in the latter, the collision is naturally resolved by the Eulerian solver by projecting the droplet back to the Eulerian field on the fine mesh.

[1] In the application foreseen in this paper, the one way approach is considered as sufficient, even if the small droplets can in practice have almost the size of the cell.

## 9.4.1 Treatment of Medium Structures

Handling the medium droplets involves three main items: their detection and subsequent conversion to Lagrangian droplets, their advection and their potential liquid collision.

### 9.4.1.1 Detection and Transformation

The first step of the multi-scale algorithm is to detect the liquid inclusion to be transformed. A tag-propagation algorithm is employed. As illustrated on Fig. 9.4, each liquid inclusion is detected by propagation of a tag over each liquid-cell, i.e. a cell where $C_i > 0.5$. At first, each liquid-cell is tagged with an unique number. Then, the smallest tag is iteratively propagated among each neighbouring liquid-cells. When the tag is entirely propagated among all liquid cells which constitute the liquid inclusion, the tag is then extended to all surrounding cells where $0 < C_i < 0.5$. Considering a liquid inclusion $l$, one can evaluate the volume of this inclusion $\nu_l$, its center of mass $\boldsymbol{x}_l$ and average velocity $\boldsymbol{v}_l$ over all cells $i_l$ which contains the liquid inclusion through the VOF function:

$$
\begin{aligned}
\nu_l &= V_{cell} \sum_{i \in i_l} C_i \\
\boldsymbol{x}_l &= V_{cell} \frac{1}{\nu_l} \sum_{i \in i_l} \boldsymbol{x}_i C_i \\
\boldsymbol{v}_l &= V_{cell} \frac{1}{\nu_l} \sum_{i \in i_l} \boldsymbol{v}_i C_i
\end{aligned}
\tag{9.14}
$$

where $\boldsymbol{v}_i$ is the eulerian velocity estimated at the center of cell $i$ and $V_{cell}$ the volume of a mesh-cell.

In a typical atomization process, the flow is highly disturbed due to many instabilities that may occurs (like Kelvin-Helmholtz, Rayleigh-Taylor, Plateau-Rayleigh ...). Hence, it is quite difficult to determine a physical parameter (like Weber number) to predict the stability of any drop and allow the transition between Eulerian and Lagrangian resolution. Hence, a possible transition is determined here on geometrical considerations. Let $N_l$ be the total number of cell contained in a liquid inclusion $l$. For very small inclusions where $N_l < 2^m$, with $m$ the dimension of the considered problem ($m = 2$ for bidimensional problems, $m = 3$ for three dimensional cases), the inclusion is automatically treated in a Lagrangian point particle way, without any other arguments. For medium inclusion $2^m \leq N_l \leq 6^m$, a sphericity criterion is used, as illustrated on Fig. 9.5.

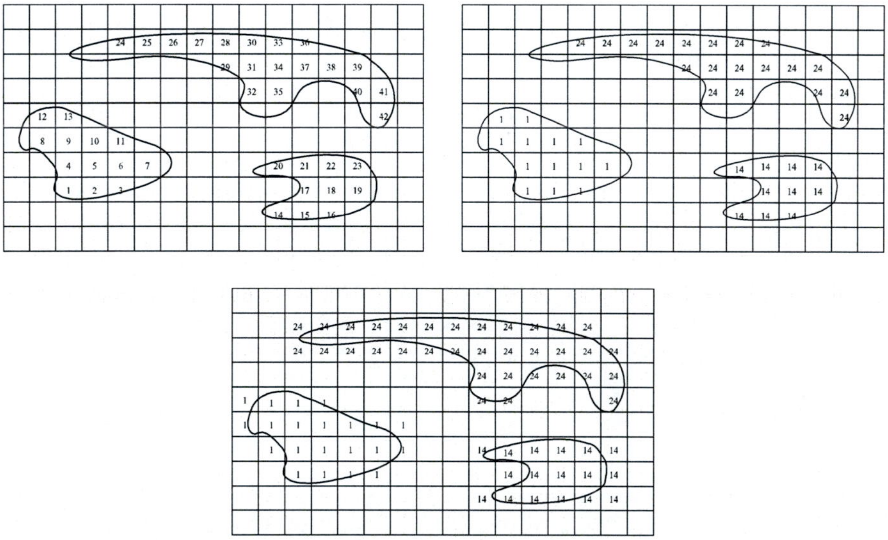

**Fig. 9.4** Illustration of liquid inclusion detection. Each "liquid cell" ($C_i > 0.5$) is tagged with a unique number (top left). The smallest tag is propagated among each neighboring liquid cell (top right). Finally, this tag si propagated to each cell where $0 < C_i < 0.5$ (bottom).eps

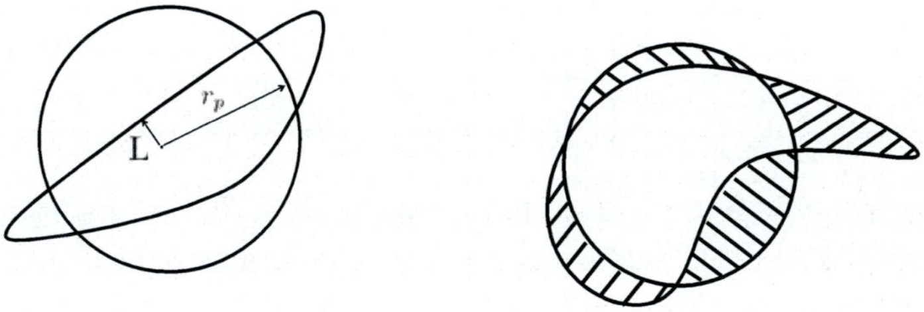


**Fig. 9.5** Sphericity criterion for medium inclusion. A sphere of equivalent volume is superimposed on the considered inclusion. The inclusion is becomes a Lagrangian particle if the difference in length $r_p - L$ and volume (dashed zone) are sufficiently small

To this way, a sphere of equivalent volume is superimposed on the center of mass of the liquid inclusion, for which the radius is

$$r_p = \left(\frac{v_l}{\pi}\right)^{1/m} \left(\frac{3}{4}\right)^{(m-2)/m} \tag{9.15}$$

and the difference of caracteristic length $\delta_l$ and volume $\delta_V$ (dashed part on the right of Fig. 9.5 are computed:

$$\begin{cases} \delta_l = r_p - L \\ \delta_V = V_{cell} \sum_{i \in i_l} |C_i - \chi_i| \end{cases} \tag{9.16}$$

where $\chi_i = 1$ if $||\boldsymbol{x}_i - \boldsymbol{x}_l|| \leq r_p$, 0 otherwise and $L$ is the smallest length between the center of mass and the interface, i.e. $L = \phi(\boldsymbol{x}_l)$. The inclusion is assumed as quasi-spherical if

$$\frac{\delta_l}{r_p} < \alpha \quad \frac{\delta_V}{\nu_l} < \beta. \tag{9.17}$$

After some numerical test, $\alpha$ and $\beta$ have been set between 0.5 and 1 (these values can be increased to force more conversions). If those two criterion are satisfied, the inclusion becomes a Lagrangian particle of volume $\nu_l$, radius $r_p$ and initial velocity $\boldsymbol{v}_p = \boldsymbol{v}_l$ at the location $\boldsymbol{x}_p = \boldsymbol{x}_l$. If they are not satisfied, the inclusion is still treated by the Eulerian CLSVOF resolution until a possible stabilization or fragmentation occurs.

#### 9.4.1.2 Tracking

An original tracking method has been developed for the medium drop, in a way of a penalty method. Indeed, the idea of this method is to project the mass and momentum particle on the Eulerian field in the same way than Eq. (9.6). Hence, for a given particle:

$$(\rho^{n+1}, \mu^{n+1}, \boldsymbol{u}^n)_i = \begin{cases} (\rho_l, \mu_l, \boldsymbol{v}_p) & \text{if} \quad |\boldsymbol{x}_i - \boldsymbol{x}_p| \leq r_p \\ (\rho_g, \mu_g, \boldsymbol{u}_i^n) & \text{otherwise} \end{cases} \tag{9.18}$$

Then, Eqs. (9.11) and (9.12) are solved on the Eulerian mesh with the new values of density, viscosity and velocity updated with respect to particle positions. The pressure inside drops is corrected by Laplace relation:

$$p_i^{n+1} \rightarrow p_i^{n+1} + (\text{ndim} - 1)\frac{\sigma}{r_p} \quad \text{if} \quad |\boldsymbol{x}_i - \boldsymbol{x}_p| \leq r_p \tag{9.19}$$

with ndim $= 2$ in two dimensions and ndim $= 3$ in three. Afterwards, Eq. (9.13) is solved with the corrected pressure. Then, the new particle velocity is obtained by averaging the Eulerian velocity inside the considered drop

$$\boldsymbol{v}_p = \frac{1}{\nu_p}\sum_i V_{cell,i} C_i \boldsymbol{u}_i^{n+1} \quad \text{if} \quad |\boldsymbol{x}_i - \boldsymbol{x}_p| \leq r_p \tag{9.20}$$

and finally Eq. (9.28) is solved with this new particle velocity. This formulation take implicitly into account the drag force on the particle and lead to a more conservative form than the method presented into previous section, while allowing the de-refinement process as the particle moves away from the interface.

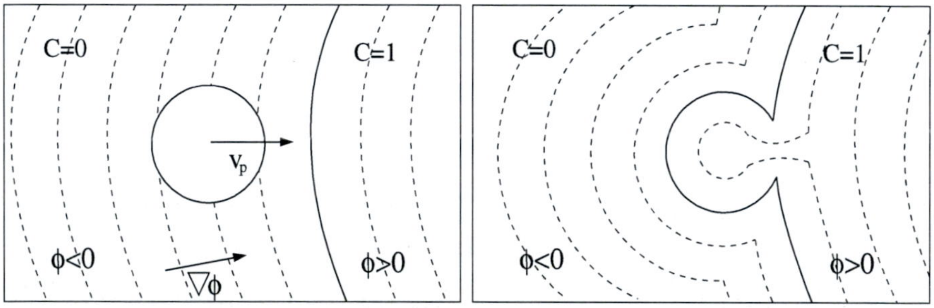


**Fig. 9.6** Illustration of the re-impact algorithm. Fat continuous line: interface ($\phi = 0,\ C = 0.5$), discontinuous line: iso-value of $\phi$. Left: before collision, right: after collision. When a particle is closed to the interface moving throw it, the re-impact is allowed: both VOF and level set functions are reconstruct into the drop, and the Lagrangian particle is removed

#### 9.4.1.3 Drop-Interface Collision

In the present work, a re-impact algorithm is implemented to take into account of a possible collision between particles and the liquid-gas interface. The re-impact is allowed if a given particle is moving throw the interface and if this particle is sufficiently closed to it (as illustrated of Fig. 9.6), i.e.:

$$\boldsymbol{v}_p \cdot \nabla\phi > 0,\ |\phi_p| < r_p + \Delta_{x,min} \tag{9.21}$$

where $\phi_p$ is the interpolation of $\phi$ at a location $\boldsymbol{x} = \boldsymbol{x}_p$ and $\Delta_{x,min}$ the grid size. If those criterion are satisfied, both Level Set and VOF functions are reconstruct inside the drop:

$$\phi_i^n = r_p - ||\boldsymbol{x}_p - \boldsymbol{x}_i||,\ C_i^n = 1 \tag{9.22}$$

for all cells i inside the drop (i.e.$||\boldsymbol{x}_p - \boldsymbol{x}_i|| < r_p$). To the drop periphery, the Level Set function is fixed to 0 and the difference between the drop volume and the VOF newly added is equally distributed to the VOF function on all cells at the drop periphery, to ensure mass conservation. Afterwards, the velocity inside de drop is imposed on the Eulerian field with momentum conservation consideration:

$$\boldsymbol{u}_i^n \rightarrow \frac{C_i^n \rho_l \boldsymbol{v}_p + (1 - C_i^n)\rho_g u_i^n}{C_i^n \rho_l + (1 - C_i^n)\rho_g} \tag{9.23}$$

which is exactly equal to $\boldsymbol{v}_p$ for full cells (i.e. $C_i^n = 1$) and is a mass weighted for mixed cells or small droplet ($r_p \leq \Delta_{x,min}$). Finally, the redistance algorithm is used to regularize the function $\phi^n$ into the gas phase.

### 9.4.2 *Small Droplets*

The detection of droplets smaller than the finest mesh size is performed by detecting any cells characterized by a value of VOF $0 < C_i < 0.5$ not attached to any larger structure, i.e. surrounded by cells with $0 < C_i < 0.5$. In this case the Level-Set function does not see any liquid mass, so that for the CLSVOF this liquid would be effectively lost. For these small Lagrangian droplet (i.e. $r_p \leq \Delta_{x,min}$), a classical drag coefficient for spherical point particle is used, from Schiller and Neumann:

$$C_D(R_{e,p}) = \begin{cases} 24R_{e,p}^{-1}(1 + 0.15R_{e,p}^{0.687}) & \text{if} \quad R_{e,p} < 1000 \\ 0.424 & \text{otherwise} \end{cases} \tag{9.24}$$

with

$$R_{e,p} = \frac{2r_p\rho_g||\boldsymbol{v}_p - \boldsymbol{v}_{p,e}||}{\mu_g} \tag{9.25}$$

and where $v_{p,e}$ the Eulerian velocity interpolated at the particle location. These values give a relaxation time in the form

$$\tau_p = \frac{4\rho_l}{3\rho_g}\frac{2r_p}{C_D||\boldsymbol{v}_p - \boldsymbol{v}_{p,e}||}. \tag{9.26}$$

Then, the particle velocity follows the equation

$$\frac{d\boldsymbol{v}_p}{dt} = \tau_p^{-1}||\boldsymbol{v}_p - \boldsymbol{v}_{p,e}|| \tag{9.27}$$

and its position

$$\frac{d\boldsymbol{x}_p}{dt} = \boldsymbol{v}_p. \tag{9.28}$$

Equations (9.27) and (9.28) are integrated using a classical second order Runge-Kutta scheme. The potential collision of these droplets with an interface can happen in two different ways. In the first case, the droplet is considered as small even on the finest mesh, so that its mass and momentum can be directly injected into the local cell. In the second, the particle approaching the interface crosses one or more mesh refinement jumps: once its size exceeds the cell size, it is detected by the medium droplets search algorithm and switches in this category, so that it is projected on the Eulerian field before the actual impact (see Sect. 9.4.1.3).

## 9.5 Results and Validation

In this section, the capabilities of the above developed method are presented. At first, dedicated test cases are provided to validate each step of the global algorithm; then, a more complex practical case consisting of a sheared liquid sheet atomization is shown.

### 9.5.1 *Drop in a Uniform Flow*

This test case is devoted to the numerical verification of the medium droplet treatment. The depicted configuration consists of a spherical droplet sheared by an air flow. This is the typical scenario expected in the atomization droplet generation, where liquid structures are sheared and disintegrated, and the resulting droplets accelerated by drag forces. This situation corresponds to the one described in Sect. 9.4.1, where the droplet spans over some cell size diameters and point-particles drag-force models cannot be employed. This test compares the solution given by a pure CLSVOF resolution versus the presented medium-droplet advection algorithm.

#### 9.5.1.1 Drop Initially at Rest

In this first example, a three dimensional liquid drop at rest ($\rho_l = 2$ kgm$^{-3}$, $\mu_l = 0.001$ kg.m$^{-1}s^{-1}$) is immersed into an uniform gas stream ($\rho_g = 1$ kgm$^{-3}$, $\mu_g = 0.001$ kgm$^{-1}s^{-1}$) with constant velocity in the $x$ direction $\boldsymbol{u}(t = 0) = [U_0, 0, 0]^T$ with $U_0 = 0.1$ ms$^{-1}$ and a surface tension $\sigma = 0.001$ Nm$^{-1}$. The computational domain consists of a square box of size $L = 0.25$ m with slip conditions on the walls tangent to the gas stream and respectively uniform inlet and outlet on the two walls orthogonal to the stream. This lead to following non dimensional numbers, based on the drop radius $R = 0.05$ m:

$$W_e = \frac{\rho_l U_0^2 R}{\sigma} = 1, \qquad R_e = \frac{\rho_l U_0 R}{\mu_l} = 100 \tag{9.29}$$

This set of parameters allow us to avoid Stokes flow and ensure that the drop remain stable and quasi spherical. Indeed, [400] show in an asymptotical study that considering a drop moving in a gaseous medium, its distortion follows:

$$h = \frac{a^2}{b^2} = 1 - \frac{9}{32} W_e + \frac{63}{640} W_e^2 \tag{9.30}$$

where $a$ is the diameter of the drop in the cross section and $b$ the drop diameter into the flow direction. In this case, $h \approx 0.82$, indicating that the shape is not stable in time. In order to avoid droplet deformation during its motion, the inlet gas stream in injected with a time ramp $U_{inj} = \min(U_0, U_0 \times t/10)$ so that the acceleration is smooth and the shape stable. An increase of the surface tension coefficient to reduce the Weber number has not been possible, as the spurious currents developing on very poorly resolved meshes did not allow a correct representation of the droplet acceleration (incidentally, this problem supports the scope of the method proposed in this paper).

The test consists into a study of the acceleration of the droplet up to a physical time of $t_{end} = 2$ s done with the CLSVOF with increasing refined meshes. The coarsest mesh is composed by 32 (M32) cells, which correspond to 7 cells per droplet diameter. The droplet can still be considered (but barely) as resolved; in atomization processes a lot of droplets are expected to be less resolved than that. The more refined meshes are given by 64 and 128 cells (M64 and M128). The medium Lagrangian droplet computation is performed on the coarsest (M32) of the meshes. The particle is initialized as in the pure Eulerian case, but the multi-scale algorithm is allowed to detect and transform at the very first time-step: the Level-Set and VOF liquid fields are removed and a Lagrangian point particle is added with the radius equal to R = 0.05 m. Figures 9.7 and 9.8 show a direct comparison between the mesh converging Eulerian results and the Lagrange medium droplet formulation. The visualization clearly shows that the final position of the droplet ($x(t = t_{end})$ is shifted forward when a more accurate resolution is performed. The M32 mesh droplet is clearly late in comparison with the reference result of the M128 mesh. Conversely, the Lagrangian result (on M32 mesh) shows a good agreement with the most accurate M64 and M128 meshes. The velocity field obtained with the mixed formulation is close to the one obtained with the Eulerian, whereas a classical Lagrangian drag-force model fails to correctly reproduced the flow topology.

The position and velocity of the droplet have been measured, their values plotted against time in Fig. 9.9. In the CLSVOF computation, the position of the center of mass and resulting velocity of the droplet have been evaluated as

$$\boldsymbol{x}_{drop} = \frac{\int_{\Omega_l} \boldsymbol{x} \, dv}{\int_{\Omega_l} dv} = \frac{\sum_i C_i \boldsymbol{x}_i}{\sum_i C_i} \tag{9.31}$$

$$\boldsymbol{v}_{drop} = \frac{\int_{\Omega_l} \boldsymbol{u} \, dv}{\int_{\Omega_l} dv} = \frac{\sum_i C_i \boldsymbol{u}_i}{\sum_i C_i} \tag{9.32}$$

In the Eulerian/Lagragian case, the point particle coordinate is directly available, as well as the advection velocity computed by (9.20) within the medium droplet algorithm. A Lagrangian computation with a very coarse mesh (M16) has been plotted as well. The plots clearly show how the Lagrangian medium droplet modelling

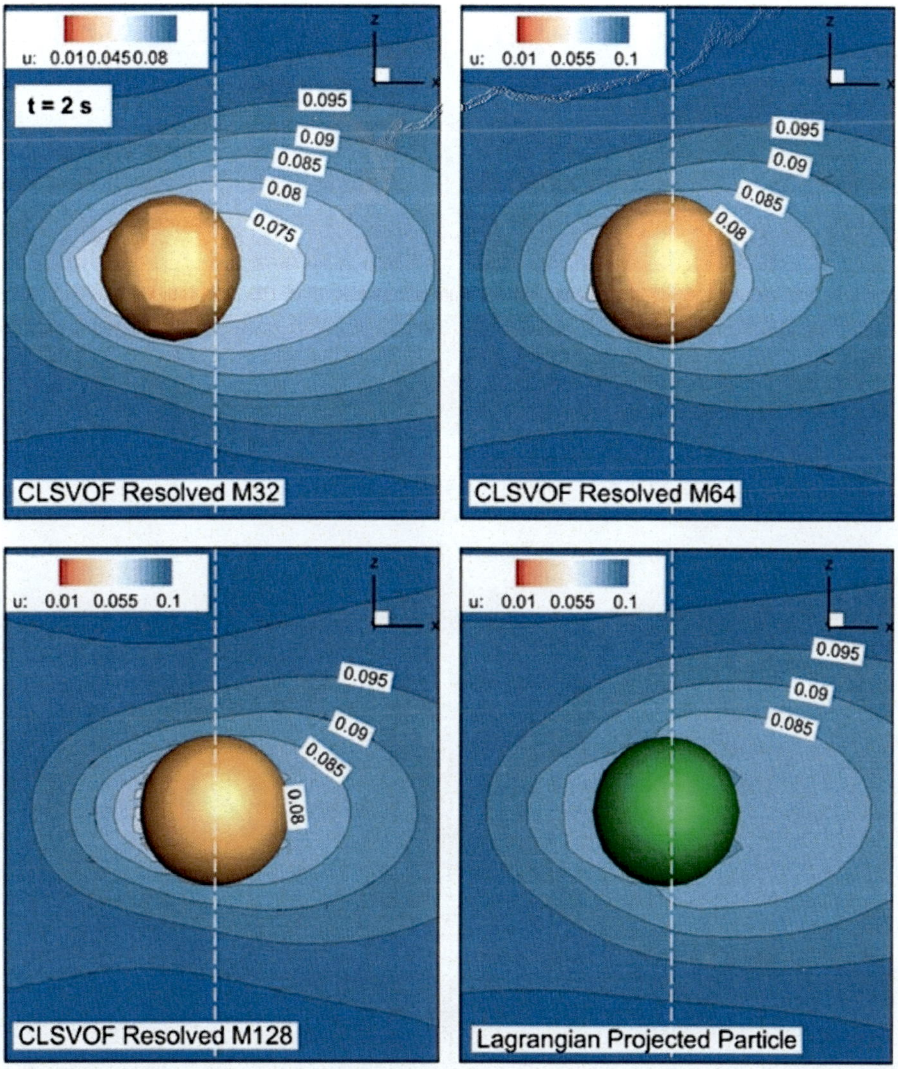


**Fig. 9.7** Advection of the liquid droplet at $t_{end} = 2$ s. Three CLSVOF results with increasing mesh refinement (M32, M64, M128) and the result of the multi-scale computation. The colour field is the x component of the velocity $\boldsymbol{u}$. The vertical dotted white line represents the x position of the droplet center of mass, as calculated with the CLSVOF M128 (the finest) mesh at $t = t_{end}$

is able to reach almost the same accuracy as the Eulerian modelling with a mesh size twice as refined. It is also visible how the inaccurate prediction of the increasing velocity for the M32 CLSVOF solution is affected by an initial plateau between $t = 0$ s and $t = 0 = 0.5$ s: the initial acceleration being delayed, the following evolution is shifted towards lower values.

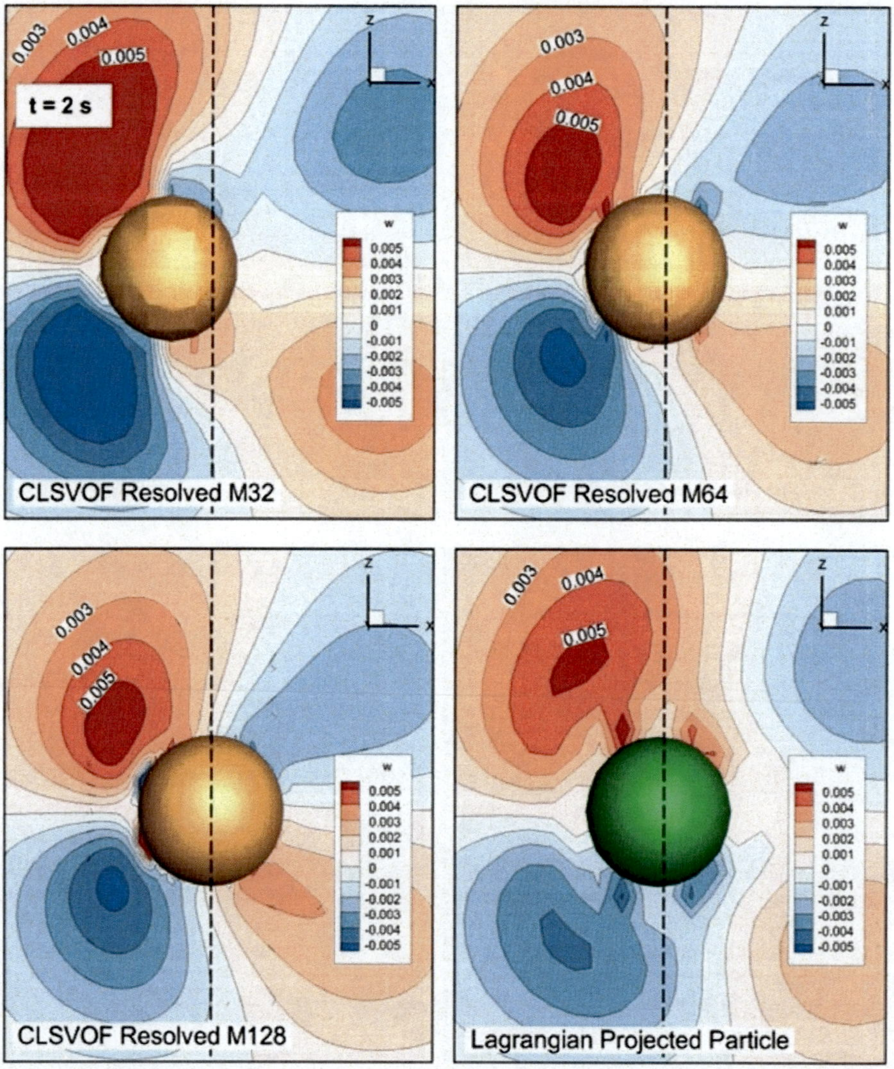


**Fig. 9.8** Advection of the liquid droplet at $t_{end} = 2$ s. Three CLSVOF results with increasing mesh refinement (M32, M64, M128) and the result of the multi-scale computation. The colour field is the z component of the velocity $\boldsymbol{u}$. The vertical dotted white line represents the x position of the droplet center of mass, as calculated with the CLSVOF M128 (the finest) mesh at $t = t_{end}$

#### 9.5.1.2 Drop with Transversal Initial Velocity

In this second test, the same configuration as the previous one has been retained. The difference lays in the initial condition of the droplet. Instead of being at rest, an initial uniform transversal velocity in the y direction ($\boldsymbol{v}_{drop}(t = 0) = [0, V_0, 0]^T$, with $V_0 = 0.1\ \mathrm{ms}^{-1}$) is imposed at t = 0 s. This initial set-up is meant to reproduce the "ejection" of the droplets following the longitudinal oscillations of an atomizing

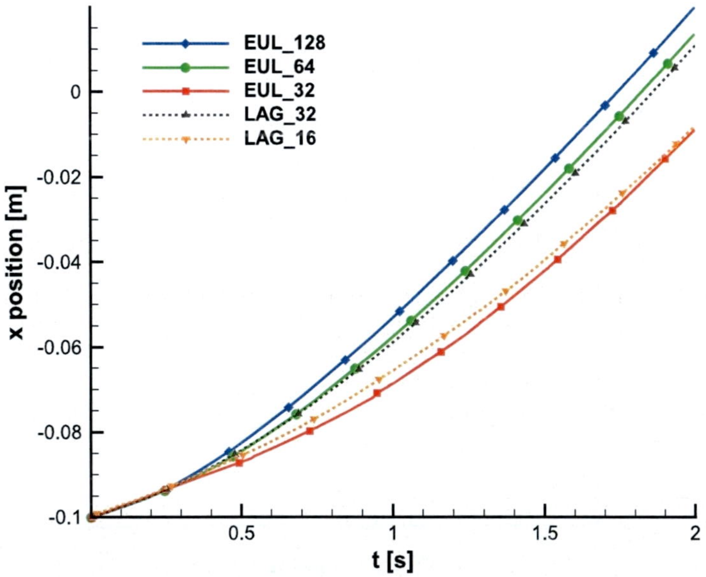


(a)

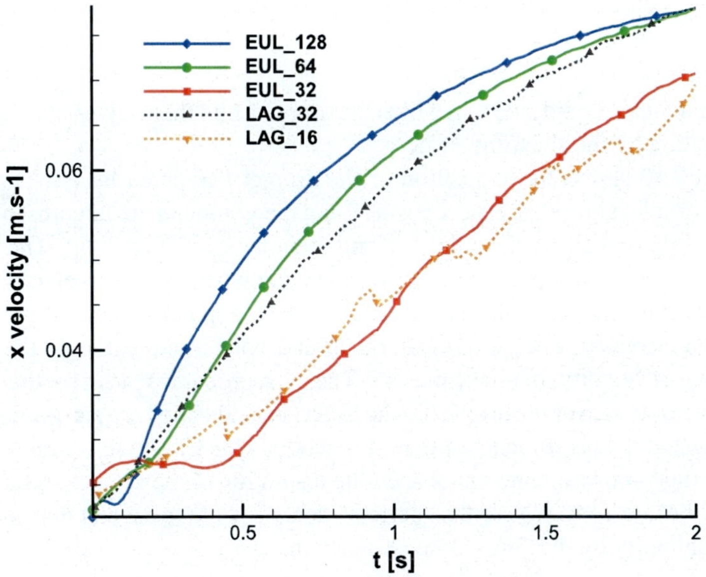


(b)

**Fig. 9.9** **a** $x(t = t_{end})$ position of the droplet, **b** $x$ velocity component of the droplet versus time. Continuous lines: Eulerian computation, dotted lines: Eulerian/Lagragian

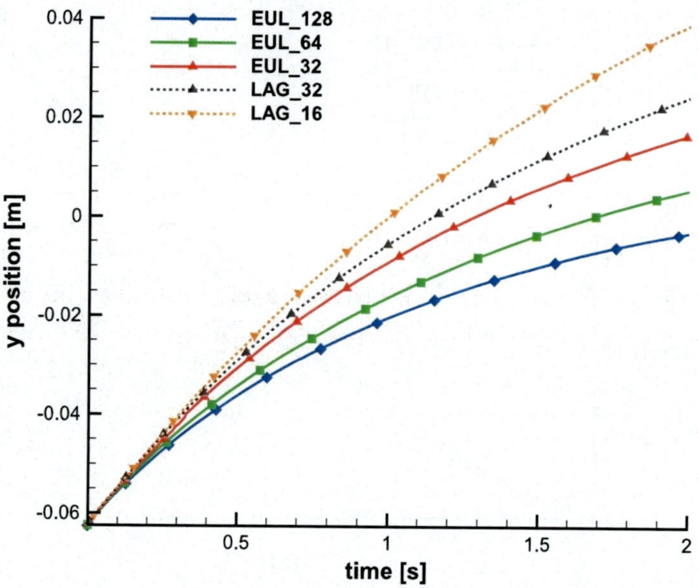


**Fig. 9.10** $y(t = t_{end})$ position of the droplet, continuous lines: Eulerian computation, dotted lines: Eulerian/Lagragian

planar or annular liquid sheet, in which the droplets find themselves in a "cross-flow" configuration by the shearing air flow.

Figure 9.10 shows the y position of the droplet computed by (9.31) and by the medium droplet algorithm. The x position and the x velocity are not presented as they superpose quite well on the results of the previous case (Fig. 9.9a). The transverse velocity is not as well predicted as for the streamwise one. The velocity decreases over time as result of the viscous effects, the reduction being more important as the resolution increases. The Lagrangian medium droplet treatment underestimates this decreasing of velocity for all the meshes. The most probable reason for this behaviour is that the projection of the droplet on the Eulerian field (9.18) is a zero-order injection (the projected field is somehow "aliased"), while in the CLSVOF solution the Ghost-Fluid method allows a much more accurate resolution of the viscous jumps. It would probably be interesting to use the "perfect" geometry of the droplet to better evaluate the discontinuity by the Ghost-Fluid algorithm.

### 9.5.2 Drop-Free Surface Collision

In this section, the numerical procedure is experienced with a drop colliding a free surface at rest by comparing Eulerian and mixed Eulerian/Lagrangian computation. A water drop of radius $r = 0.176$ m ($\rho_l = 1000$ kgm$^{-3}$, $\mu_l = 10^{-3}$ kgm$^{-1}$s$^{-1}$) at rest is immersed in air ($\rho_g = 1.225$ kgm$^{-3}$, $\mu_g = 10^{-5}$ kgm$^{-1}s^{-1}$), into a cubic

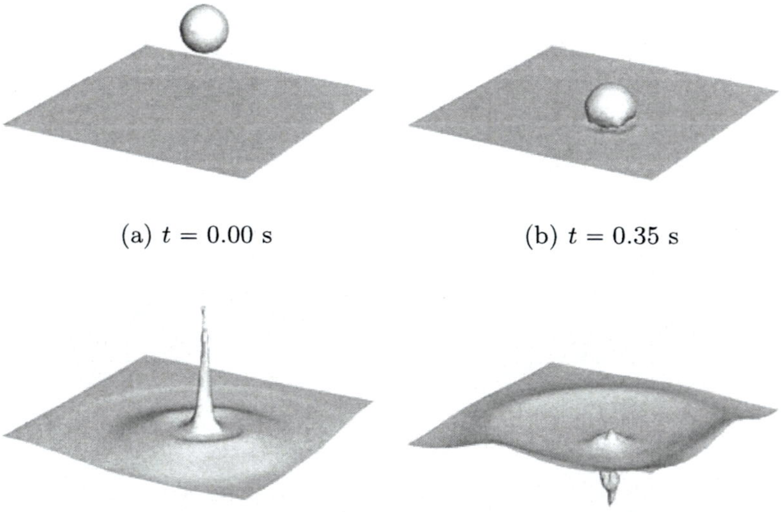

(a) $t = 0.00$ s (b) $t = 0.35$ s

(c) $t = 1.25$ s (d) $t = 1.75$ s

**Fig. 9.11** Results of the CLSVOF computation of a droplet splashing on a liquid free surface, resolution $64 \times 64 \times 64$

domain of size $L = 1$ m. The droplet is suspended over a free liquid surface placed at $y = 0.25$ m; the initial position of the droplet is $\boldsymbol{x}_d = [0.25, 0.25, 0.75]$ m (the offset from the center of the domain is to avoid the particle crossing an interface between mesh blocks). A gravity field is imposed as $\mathbf{g} = [0, -9.81, 0]$ ms$^{-1}$. The droplet is let to accelerate under the gravity effect, until impact on the surface. This test is meant to test both the conversion and transport of medium drops algorithm as well as the particle re-impact.

As for the previous example, the domain is discretized with increasingly fine meshes for both the pure Eulerian and the Eulerian/Lagrangian medium particle model. The meshes are M16, M32, M64. For the Lagrangian computation, the algorithm detects and transforms the droplet at the very first time-step. No slip conditions are applied everywhere.

Figure 9.11 shows the results of the fall and impact of the droplet up to $t_{end} = 2$ s, given by pure CLSVOF computation with the finest M64 mesh. At $t \simeq 0.35$ s the droplet impacts the surface (the exact time depends on the mesh size). Circular surface waves develop after the impact, and a first geyser forms and reach its peak at around $t \simeq 1.3$ s. The height of the rising geyser can give a good measure about the exchange of momentum between the falling droplet and the surface.

Figure 9.12 illustrates a direct comparison between the CLSVOF and the multi-scale resolution of the impact realized on the same M32 mesh (the finest mesh in which the droplet can still be transformed into Lagrangian medium particle). The results show a very good agreement between the two methods: in particular, the time of the impact is almost the same, and the dynamics of the surface waves and the

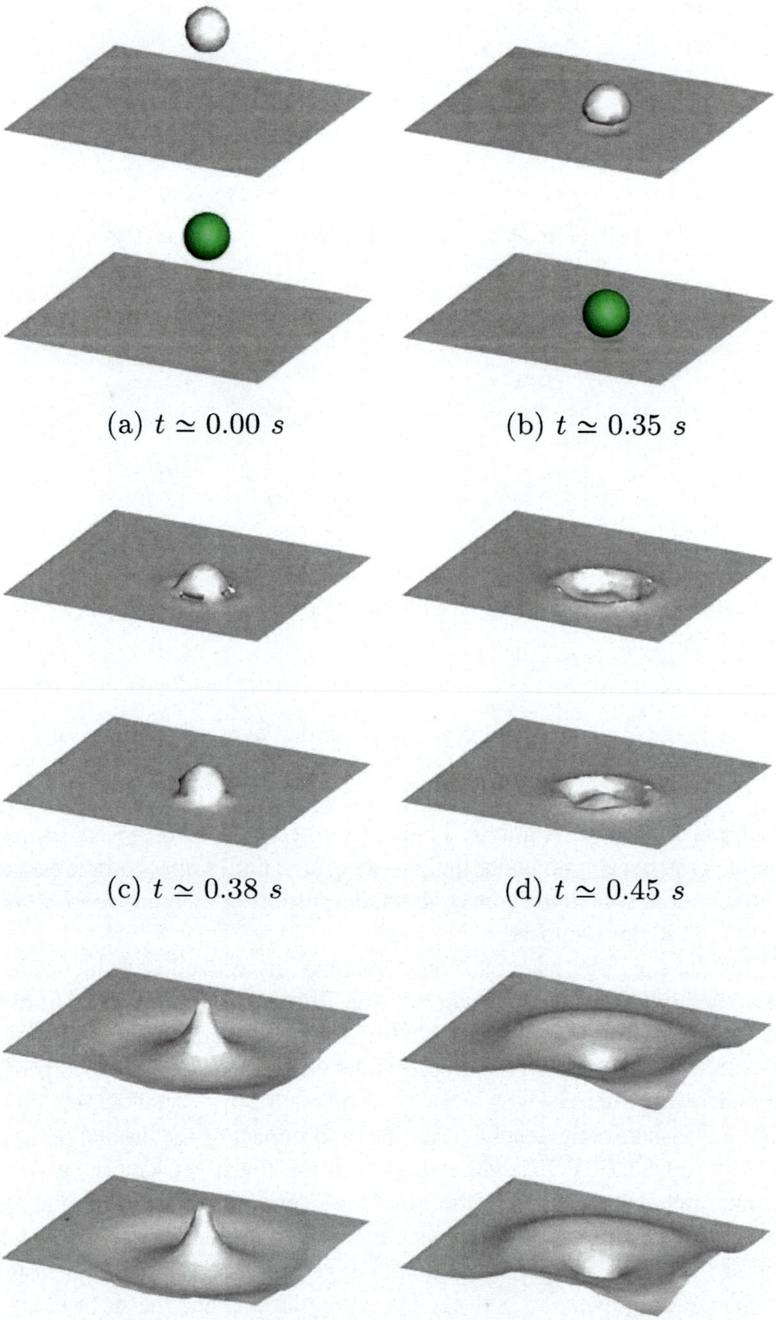


**Fig. 9.12** Visual comparison between the CLSVOF (upper image) and the multi-scale resolution (lower image) of a droplet splashing on a liquid free surface, resolution 32 × 32 × 32

rising geyser are almost identical. This means that a good droplet tracking has been performed, and that the Lagrangian to Eulerian conversion is conservative in mass and momentum.

Figure 9.13 shows the maximum and minimum of the interface position, tracking the geyser and the sink after the impact.

It can clearly be seen how the Lagrangian solution is better than the pure Eulerian one as the drop becomes less resolved, being more conservative in momentum. A small overshoot can be distinguished in the well, possibly due to the droplet keeping rigid until the activation of the re-impact algorithm, while the resolved drop can start deforming a little sooner. Tests with a point-particle drag-force model have clearly shown how the classical dispersed phase approach is not adapted in this configuration.

### *9.5.3 Assisted Atomization of a Liquid Sheet*

The whole multi-scale algorithm has been applied to the simulation of the primary atomization of a liquid sheet sheared by two parallel air streams. In this configuration, the sheet-shape is unstable. Hydrodynamic instabilities are the source of the atomization mechanisms and determine the primary break-up characteristics. During the atomization process, two different moments can be distinguished [375–378]. In the primary atomization the sheet becomes subject to longitudinal instabilities, which are the results of the shearing effect: Kelvin-Helmholtz instabilities perturb the plane sheet, starting a sinusoidal streamwise oscillation all along the sheet. Then, fully three dimensional instabilities generate transverse modulations. The sheet breaks into smaller liquid packs, ligaments and bag-like structures. This continuous fragmentation ends with the formation of a polydisperse spray of droplets. The considered numerical configuration aims to reproduce a simple atomization device as the one experimentally investigated in [401]. The injector, depicted in Fig. 9.14a, consists of a NACA-shaped injector immersed into a channel gas flow, discharging the liquid by a rectangular fence 300 $\mu$m thick (visible in the middle of the device nozzle).

Different geometrical configurations as well as liquid and gas flow regimes are considered by the authors. In this work a liquid sheet maintaining the same non dimensional parameters has been simulated. A ratio of 100 has been imposed between the liquid and gas density, as a ratio of 1000 has shown to produce numerical instabilities leading to non-convergence in presence of strong shearing flows. The chosen physical parameters of the simulation are typical of a "stretched ligament" break-up, where the primary atomization manifests in the form of membranes and longitudinal ligaments formation. The momentum ratio involving the "stretched ligament" break-up are $0.5 < M < 4$ (from [378, 402]). The parameters are summarized in Tables 9.1 and 9.2.

The simulation domain, with reference to Fig. 9.14b, is a rectangular box of $L_x = 40.96$ mm, $L_y = 10.24$ mm, $L_z = 5.12$ mm. AMR blocks of $16 \times 16 \times 8$ cells have been used; a total of 5 levels of refinement give an equivalent fine mesh of $1024 \times 256 \times 128$, which means $\Delta x = 40$ $\mu$m. Inflow condition are imposed on the left

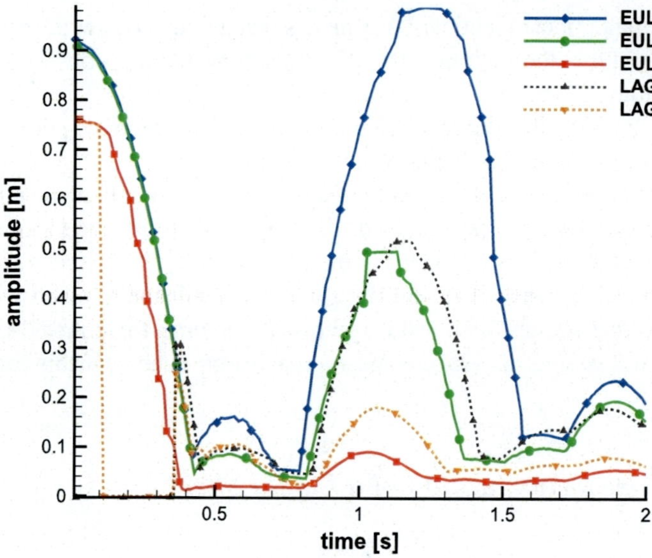


(a) Maximal amplitude

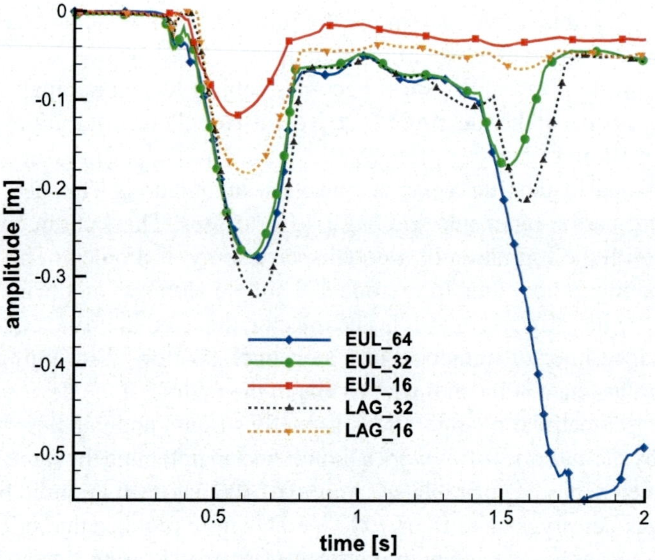


(b) Minimal amplitude

**Fig. 9.13** Maximal and minimal positions of the interface in the $y$ direction (along the trajectory of the droplet). The maximal amplitude shows the primary geyser after the impact, the minimal the depth of the depression/imprint/well

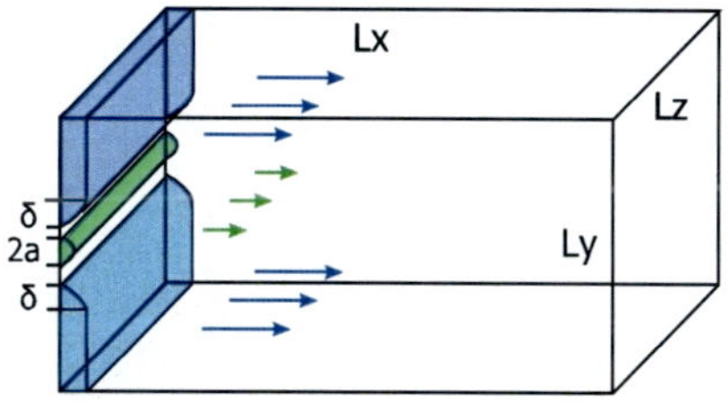


(a) (b)

**Fig. 9.14** **a** Actual liquid sheet injector device. **b** Numerical computation set-up

**Table 9.1** Dimensionless parameters of the 3D atomization simulation

| Non-dimensional parameters | | |
|---|---|---|
| Gas Reynolds | $Re_g = \frac{\rho_g u_g \delta}{\mu_g}$ | 1115 |
| Liquid Reynolds | $Re_l = \frac{\rho_l u_l a}{\mu_l}$ | 104 |
| Weber | $We = \frac{\rho_g (u_g - u_l)^2 a}{\sigma}$ | 17.42 |
| Momentum ratio | $M = \frac{\rho_g u_g^2}{\rho_l u_l^2}$ | 0.91 |

**Table 9.2** Geometric parameters of the 3D atomization simulation

| Geometrical parameters | | |
|---|---|---|
| Sheet thickness | $2a$ | 300 μm |
| Boundary layer thickness | $\delta$ | 270 μm |

face, slip conditions on the horizontal and vertical faces and outflow on the right face. The droplet detection algorithm is capped at a maximum droplet size of seven $\delta x$, giving a possible diameter range of [0:280] μm. Figure 9.15 shows the initial destabilization of the liquid sheet. The longitudinal instability starts to develop. Hints of the transverse instabilities are visible as well. In Fig. 9.16a the instabilities are further developing, and liquid packs resulting from the first sheet break-up have detached from the sheet main body. Ligaments have begun to from under the three-dimensional instabilities effects. Droplets begin to detach from the ligaments and the membranes, their size already varying from the smallest to the largest detectable by the algorithm. Figure 9.16b shows the sheet atomization established regime, with the continuous generation of liquid blobs and small droplets. Many large liquid structures keep stable up to the exit of the numerical domain. A variety of droplet diameters are found by the detection algorithm.

Figure 9.17 zooms to the mechanism of ligament break-up: the ligament is stretched by a Rayleigh-Plateau instability up to the detachment of a satellite droplet, which is immediately transformed into a Lagrangian point particle. It is visible how

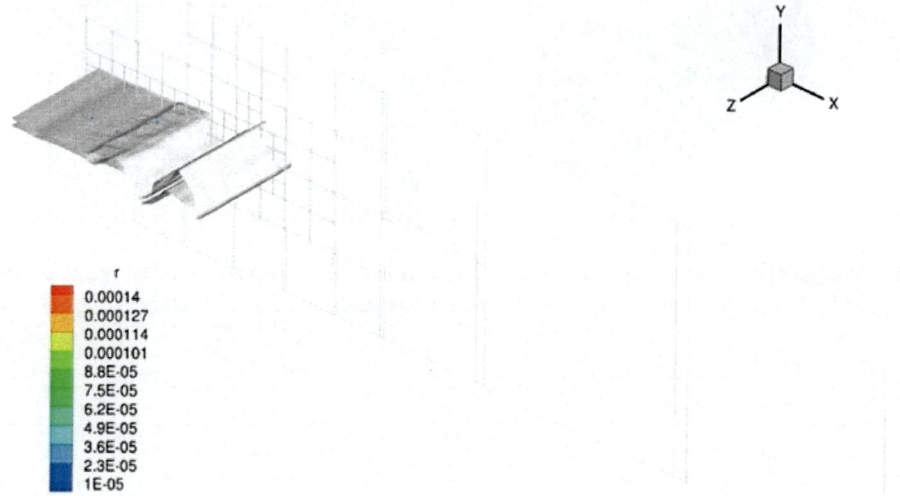


**Fig. 9.15** Results of the planar liquid sheet atomization: initial destabilization, longitudinal oscillation and birth of transver instabilities

the droplet moves to a less refined zone, far from the resolved sheet, as it is expected from the multi-scale algorithm.

A comparison of some large scale features of the atomization process has been made with respect to the experimental results found in [401], in terms of visual aspect, oscillation frequency and mean break-up length. A visualization of the behaviour of the sheet is given in Fig. 9.18. The simulation correctly captures the growing longitudinal instability of the sheet, as well as the transversal modulation. The thin ligaments formed by the primary atomization at the break-up point are fully captured by the CLSVOF eulerian solution. A first wave of droplets is formed at this location. The liquid detached from the main sheet body undergoes further deformation and break-up into a fully three-dimensional flow: it can be seen in both images that large liquid parcels coexist with smaller stable droplets. The oscillation frequency of the simulation is quite higher than the measured one, while the breakup length is underestimated. This seems to be a recurrent problem in two phase DNS simulations of liquid sheet atomization [58, 155, 403, 404].

A preliminary analysis has been performed on the droplet spray, the results compared to the experience by laser diffraction system allowing measurement of spray droplet size distributions. The statistics involve the droplets in the whole numerical domain; their characteristics are averaged in time, the samplings taken with a frequency corresponding to an average convection time in order to avoid repeated registrations of the same droplet. The particles of diameter inferior to the finest mesh cell (40 $\mu$m) have been discarded, as no physics is involved in their creation. Their mass can still be kept in account for liquid mass flows if needed.

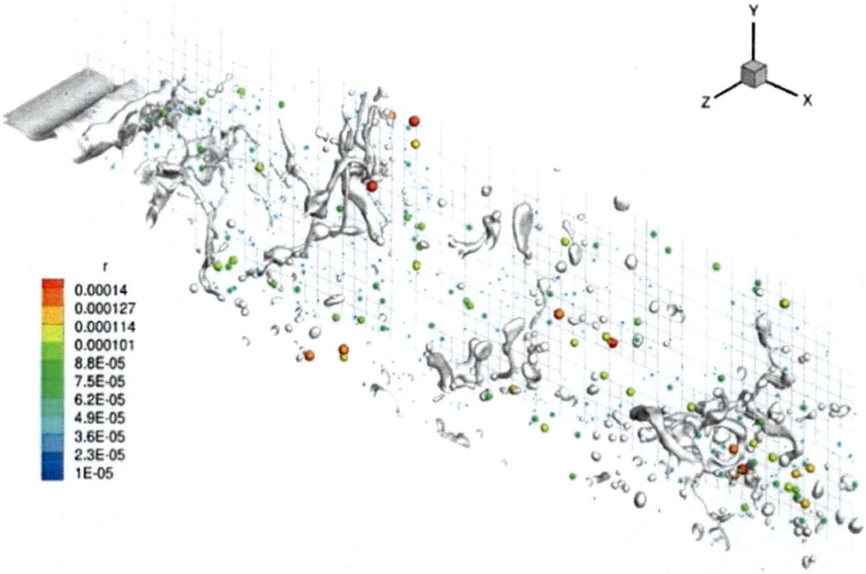


(a)

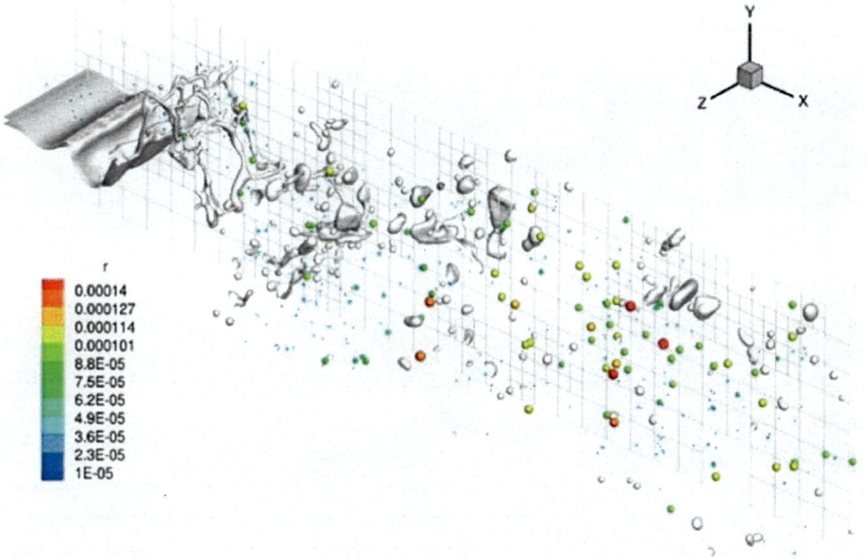


(b)

**Fig. 9.16** Results of the planar liquid sheet atomization. **a** Fully developed atomization, **b** later time. Coloured spheres: Lagrangian particles (both small and medium ones), actual size, coloured by radius

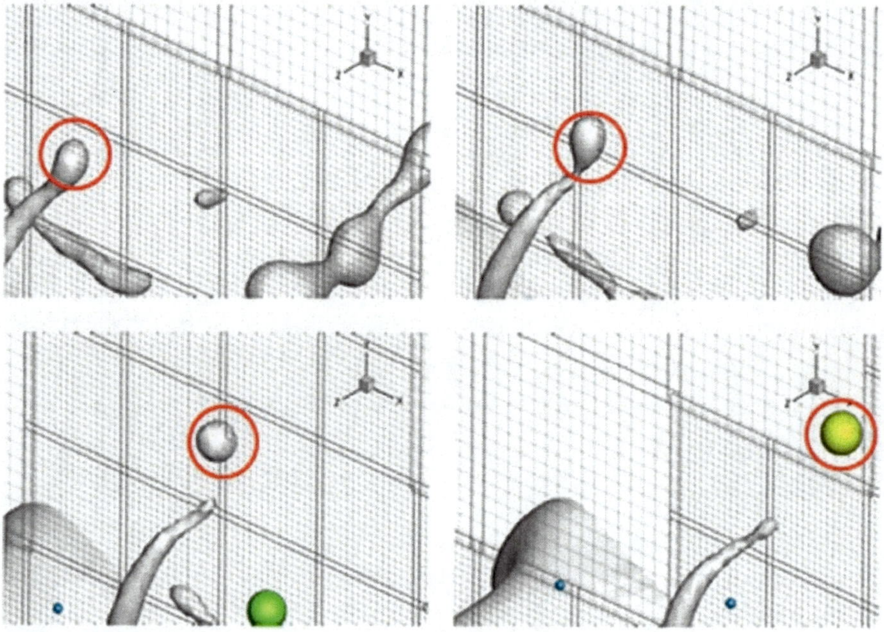

Fig. 9.17 Capture of the generation of a droplet: ligament stretching, break-up and Eulerian to Lagrangian transition

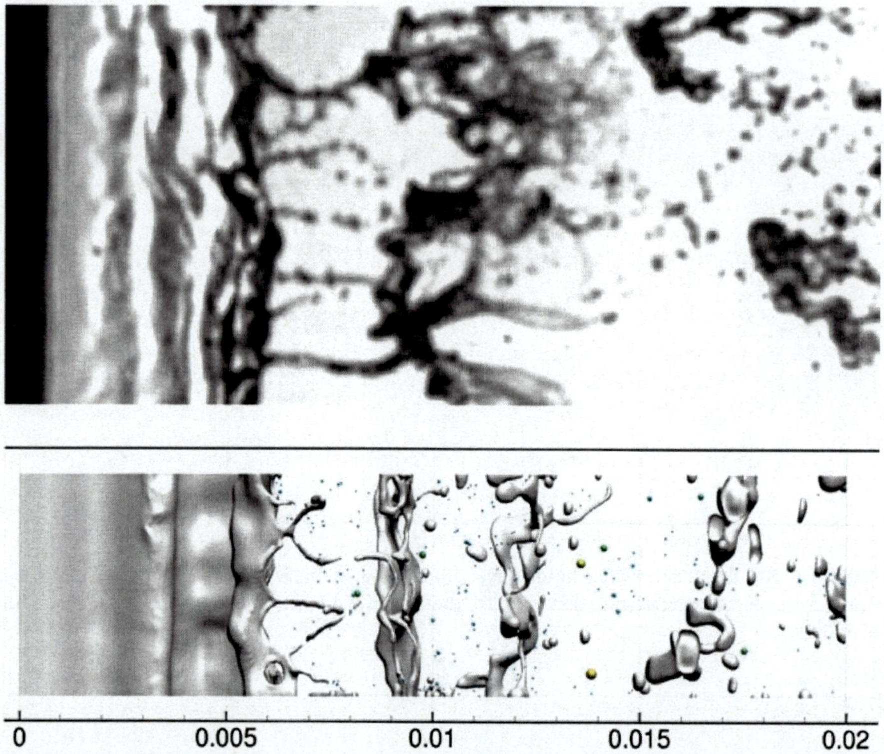


Fig. 9.18 Snapshots of both experimental visualization and simulation, arbitrary time step

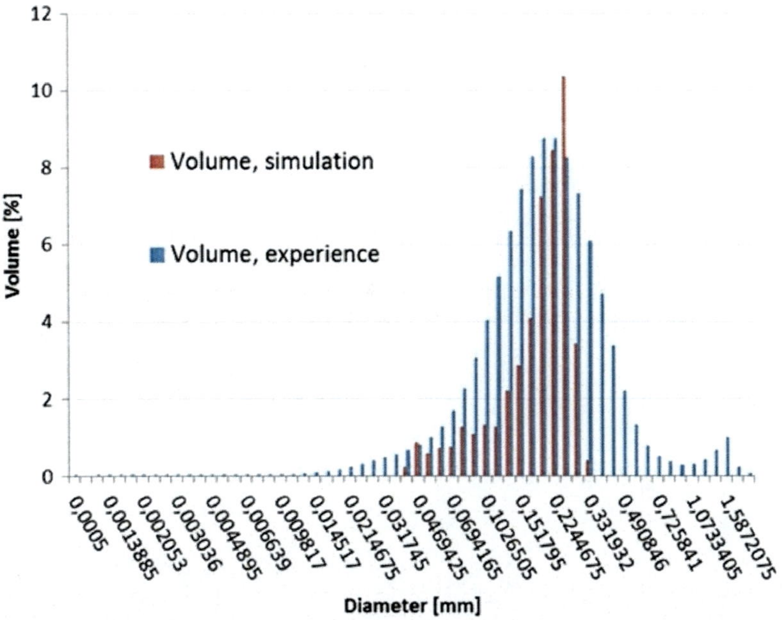


**Fig. 9.19** Droplet volume distribution obtained from the simulation dispersed phase, comparison with the experience

A total of 18457 droplets ($40 < D < 320$ μm) have been registered. Figure 9.19 shows the corresponding distribution in volume. The simulation seems able to capture the predominant diameters around 255 μ, but no information has been extracted from structures larger than 320 in diameter, so that the distribution is truncated after this value. It would seem that the evolution of the distribution with decreasing diameters is underestimated, a possible consequence of the considered volume of interest close to the injector and the absence of a secondary atomization model.

#### 3.2.1.3 Temporal Discretization

An high order temporal resolution is necessary as well for the correct evolution of the numerical interface. The time derivative $\frac{\partial \phi}{\partial t}$ is discretized by a third order Runge-Kutta scheme from [144]. If the spatial discretization of the advection term is written as $-(\boldsymbol{u}.\nabla)\phi = L(\phi)$, the Eq. (3.64) can be written as

$$\frac{\partial \phi}{\partial t} = L(\phi) \tag{3.84}$$

and an explicit Euler step would look like

$$\phi^{n+1} = \phi^n + \Delta t L(\phi^n) \tag{3.85}$$

stable under the CFL condition $\Delta t_{Euler} < \min\left(\frac{\Delta x}{|u|}, \frac{\Delta y}{|v|}\right)$. The Runge Kutta scheme can be seen as a combination of several Euler like "guesses" of the solution:

$$\begin{cases} \phi^{(i)} = \sum_{k=0}^{i-1} \left(\alpha_{ik}\, \phi^{(k)} + \Delta t\ \beta_{ik}\ L(\phi^{(k)})\right) \\ \phi^{(0)} = \phi^n, \qquad \phi^{(m)} = \phi^{n+1}, \qquad i = 1, ..., m \end{cases} \tag{3.86}$$

This scheme is stable under the following CFL condition:

$$\begin{aligned} &\Delta t < c\Delta t_{Euler} \\ &\text{with } c = \min_{i,k} \frac{\alpha_{i,k}}{\beta_{i,k}} \\ &\text{and } \alpha_{i,k} \geq 0\,,\, \beta_{i,k} \geq 0 \end{aligned} \tag{3.87}$$

The coefficients of the third order TVD-RK can be found in [59], and give the final temporal discretization

$$\begin{cases} \phi^1 = \phi^n + \Delta t L(\phi^n) \\ \phi^2 = \frac{3}{4}\phi^n + \frac{1}{4}\left(\phi^1 + \Delta t L(\phi^1)\right) \\ \phi^{n+1} = \frac{1}{3}\phi^n + \frac{2}{3}\left(\phi^2 + \Delta t L(\phi^2)\right) \end{cases} \tag{3.88}$$

This allows a stability CFL condition of $c = 1$ in Eq. (3.87).

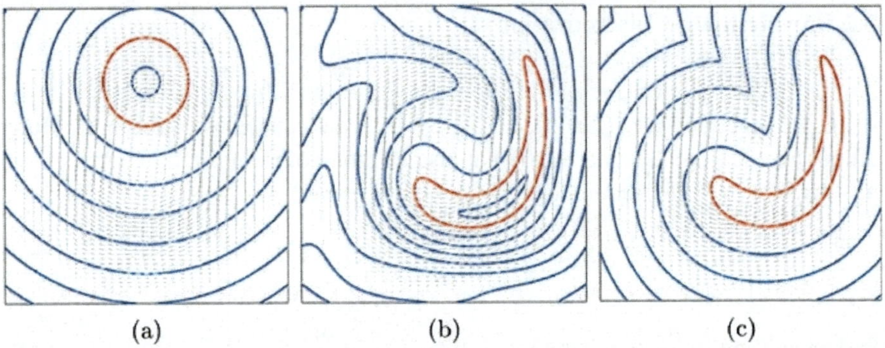

Fig. 3.35 Example of evolution of a Level Set function. **a** Initial condition. **b** After some iterations, the contour lines are no more iso distance lines. **c** With redistance, the contour lines are parallel again

#### 3.2.1.4 The Reinitialization

During the advection of the Level Set, if the velocity field does not impose a rigid translation or rotation, the different contour lines are advected differently, as the local velocity gradient is not uniform: they are no more parallel and they no more represent a constant distance from the interface, $\|\nabla\phi\| \neq 1$ (Fig. 3.35a, b). If the position of the actual interface is not affected by this problem, the characteristic of distance is necessary to the computation of the normal vector and the curvature at the interface. The redistance algorithm has to modify the advected Level Set function, without altering the zero contour, in order to reimpose the $\|\nabla\phi\| = 1$, or the distance property. In [145], Chopp introduced the idea that the level-set could be reinitialised or redistanced, in order to remain a signed distance function. An easely way to do this, is to use a kind of plotting algorithm to extract the $\phi = 0$ iso contour and then geometrically evaluate the distance from each grid point to the interface. Unfortunatly, this procedure is extremely slow as it scales with $N^3$ is $N$ is the number of grid points.

#### 3.2.1.5 The Algorithm

[72] developed an algorithm for the Level Set reinitialization problem. The main idea is that the zero contour line is correct, and the others should be corrected in function of it. The steady state solution of the following Hamilton Jacobi equation will give a signed distance function:

$$\begin{cases} \dfrac{\partial\phi}{\partial\tau} = \text{sign}\,(\phi_0(\boldsymbol{x}, t))\,(1 - \|\nabla\phi\|) \\ \phi(\boldsymbol{x}, \tau = 0) = \phi_0(\boldsymbol{x}, t) \end{cases} \tag{3.89}$$

# Bibliography


1. C.T. Crowe, M. Sommerfeld, Y. Tsuji, *Multiphase flows with droplets and particles* (CRC, 1998)
2. C.T. Crowe, *Multiphase flow handbook* (CRC Press, Taylor and Francis, 2005)
3. C. Marchioli, S. Vincent, Special issue on finite-size particles, drops and bubbles in fluid flows: advances in modelling and simulations. Acta Mech. **230**, 381–386 (2019)
4. S. Vincent, C. Le Bot, F. Sarret, E. Meillot, J.-P. Caltagirone, L. Bianchi, Penalty and eulerian-lagrangian vof methods for impact and solidification of metal droplets plasma spray process. Comput. Fluids **113**, 32–41 (2015)
5. C. Caruyer, S. Vincent, E. Meillot, J.-P. Caltagirone, Modeling the first instant of the interaction between a liquid and a plasma jet with a compressible approach. Surf. Coating Tech. **205**, 974–979 (2010)
6. D. Zuzio, J.-L. Estivalezes, An efficient block parallel AMR method for two phase flow simulations. Comput. Fluids **44**, 339–357 (2011)
7. D. Zuzio, J.-L. Estivalèzes, B. Dipierro, An improved multiscale Eulerian-Lagrangian method for simulation of atomization process. Notes Numer. Fluid Mech. Multidisciplinary Design **176**, 285–301 (2018)
8. P. Trontin, S. Vincent, J.-L. Estivalezes, J.-P. Caltagirone, Detailed comparisons of front-capturing methods for turbulent two-phase flow simulations. Int. J. Numer. Meth. Fluids **56**, 1543–1549 (2008)
9. P. Trontin, S. Vincent, J.-L. Estivalezes, J.-P. Caltagirone, Direct numerical simulation of a freely decaying turbulent interfacial flow. Int. J. Multiphase Flow **36**, 891–907 (2010)
10. D. Zuzio, J.-L. Estivalezes, P. Villedieu, G. Blanchard, Numerical simulation of primary and secondary atomization. C. R. Mécanique **341**, 15–25 (2013)
11. D. Lacanette, A. Gosset, S. Vincent, J.-M. Buchlin, E. Arquis, Macroscopic analysis of gas-jet wiping: numerical simulation and experimental approach. Phys. Fluid **18**(042103), 1–15 (2006)
12. J. Delteil, S. Vincent, A. Erriguible, P. Subra-Paternault, Numerical investigations in Rayleigh breakup of round liquid jets with VOF methods. Comput. Fluids **50**, 10–23 (2011)
13. R. Denèfle, S. Mimouni, J.-P. Caltagirone, S. Vincent, Multifield hybrid approach for two-phase flow modeling—part 1: adiabatic flows. Comput. Fluids **113**, 106–111 (2015)
14. J. Larocque, S. Vincent, P. Lubin, D. Lacanette, J.-P. Caltagirone, Parametric study of LES subgrid terms in a turbulent phase separation flow. Int. J. Heat Fluid Flow **31**, 536–544 (2010)


 
S. Vincent et al., *Small Scale Modeling and Simulation of Incompressible Turbulent Multi-Phase Flow*, CISM International Centre for Mechanical Sciences 607,
https://doi.org/10.1007/978-3-031-09265-7

15. S. Vincent, D. Lacanette, J. Larocque, A. Toutant, P. Lubin, P. Sagaut, Direct numerical simulation of phase separation and a priori two-phase les filtering. Comput. Fluids **37**, 898–906 (2008)
16. S. Vincent, M. Tavares, S. Fleau, J.-L. Estivaelzes, S. Mimouni, and M. Ould-Rouiss, A priori filtering and les modeling of two-phase flows. application to phase separation. Comput. Fluids **176**, 245–259 (2018)
17. B. Lafaurie, C. Nardone, R. Scardovelli, S. Zaleski, G. Zanetti, Modelling merging and fragmentation in multiphase flows with surfer. J. Comput. Phys. **113**, 134–147 (1994)
18. R. Scardovelli, S. Zaleski, Direct numerical simulation of free-surface and interfacial flow. Annual Rev. Fluid Mech. **31**, 567–603 (1999)
19. P. Liovic, D. Lakehal, Multi-physics treatment in the vicinity of arbitrarily deformable gas-liquid interfaces. J. Comput. Phys. **222**, 504–535 (2007)
20. P. Liovic, D. Lakehal, Interface-turbulence interactions in large-scale bubbling processes. Int. J. Heat Fluid Flow **28**, 127–144 (2006)
21. T.N. Randrianarivelo, S. Vincent, O. Simonin, J.-P. Caltagirone, A DNS approach dedicated to the analysis of fluidized beds. Fluid Mech. Appl. **81**, 207–214 (2007)
22. C. Corre, J.-L. Estivalezes, S. Vincent, O. Simonin, S. Glockner, Simulation of a fluidized bed using a hybrid Eulerian-Lagrangian method for particle tracking. Notes Numer. Fluid Mech. Multidisciplinary Des. **110**, 103–110 (2010)
23. S. Vincent, J.C. Brändle de Motta, A. Sarthou, J.-L. Estivalezes, O. Simonin, E. Climent, A Lagrangian VOF tensorial penalty method for the DNS of resolved particle-laden flows. J. Comput. Phys, **256**, 582–614 (2014)
24. J. C. Brändle de Motta, W.P. Breugem, B. Gazanion, J.-L. Estivalezes, S. Vincent, E. Climent, Numerical modelling of finite-size particle collisions in a viscous fluid. Phys. Fluids **25**(083302), 1–9 (2013)
25. J.C. Brändle de Motta, J.-L. Estivalezes, E. Climent, S. Vincent, Local dissipation properties and collision dynamics in a sustained homogeneous turbulent suspension composed of finite size particles. Int. J. Mult. Flow **85**, 369–379 (2016)
26. comparison with experimental data, J.C. Brändle de Motta, A. Ozel, M. Abbas, P. Fède, O. Masbernat, S. Vincent, J.-L. Estivalezes, O. Simonin, Particle resolved direct numerical simulation of a liquid-solid fluidized bed. Int. J. Mult. Flow **89**, 228–240 (2017)
27. J. C. Brändle de Motta, P. Costa, J. J. Derksen, C. Peng, L.-P. Wang, W.-P. Breugem, J. L. Estivalezes, S. Vincent, E. Climent, P. Fède, P. Barbaresco, N. Renon, Assessment of numerical methods for fully resolved simulations of particle-laden turbulent flows. Comput. Fluids. **179**, 1–14 (2019)
28. M.A. Chadil, S. Vincent, J.-L. Estivalezes, Novel method to compute drag force and heat transfers for motions around spheres. Thermo. Interf. Fluid Mech. **1**, 1–23 (2018)
29. M.A. Chadil, S. Vincent, J.-L. Estivalezes, Accurate estimate of drag forces using particle-resolved direct numerical simulations. Acta Mech. **230**, 569–595 (2019)
30. E.I. Thiam, E. Masi, E. Climent, O. Simonin, S. Vincent, Particle-resolved numerical simulations of the gas-solid heat transfer in arrays of random motionless particles. Acta Mech. **230**, 541–567 (2019)
31. A. Chadil, S. Vincent, J.-L. Estivalezes, Improvement of the viscous penalty method for particle-resolved simulations. Open J. Fluid. Dyn. **9**, 168–192 (2019)
32. B. Maury, Direct simulations of 2D fluid-particle flows in biperiodic domains. J. Comput. Phys. **156**, 325–351 (1999)
33. E. Bänsch, Finite element discretization of the navier-stokes equations with a free capillary surface. Numer. Math. **88**, 203–235 (2001)
34. K. Khadra, P. Angot, S. Parneix, J.P. Caltagirone, Fictitious domain approach for numerical modelling of Navier-Stokes equations. Int. J. Numer. Meth. Fluids **34**, 651–684 (2000)
35. I. Kataoka, Local instant formulation of two-phase flow. Int. J. Multiphase Flow **12**, 745–758 (1986)
36. J.M. Delhaye, Jump conditions and entropy sources in two-phase systems. Local instant formulation. Int. J. Multiphase Flow **1**, 395–409 (1974)

37. G. Pianet, S. Vincent, J. Leboi, J.P. Caltagirone, M. Anderhuber, Simulating compressible gas bubbles with a smooth volume tracking 1-Fluid method. Int. J. Mult. Flow **36**, 273–283 (2010)
38. P. Sagaut, *Large eddy simulation for incompressible flows—an introduction* (Springer Verlag, 1998)
39. J. Smagorinsky, General circulation experiments with the primitive equations. i: the basic experiments. Month. Weath. Rev. **91**, 99–165 (1963)
40. J. Bardina, J.H. Ferziger, W.C. Reynolds, Improved subgrid scale models based on large eddy simulation. AIAA Paper **80**, 1357 (1980)
41. E. Labourasse, D. Lacanette, A. Toutant, P. Lubin, S. Vincent, O. Lebaigue, J.-P. Caltagirone, P. Sagaut, Towards large eddy simulation of isothermal two-phase flows: governing equations and a priori tests. Int. J. Multiphase Flow **33**, 1–39 (2007)
42. A. Toutant, E. Labourasse, O. Lebaigue, O. Simonin, DNS of the interaction between a deformable buoyant bubble and a spatially decaying turbulence: a priori tests for les two-phase flow modelling. Comput. Fluids **37**, 877–886 (2008)
43. J. Chesnel, Reveillon J., F.X. Demoulin, T. Menard, Subgrid modeling of liquid atomization, in *6th international conference on multiphase flow*, Leipzig, Germany, 2007
44. J.U. Brackbill, D.B. Kothe, C. Zemach, A continuum method for modeling surface tension. J. Comput. Phys. **100**, 335–354 (1992)
45. R. Fedkiw, T. Aslam, B. Merriman, S. Osher, A non-oscillatory eulerian approach to interfaces in multimaterials flows (the ghost fluid method). J. Comput. Phys. **152**, 457–492 (1999)
46. S. Patankar, *Numerical heat transfer and fluid flow* (CRC Press, 1980)
47. F.H. Harlow, J.E. Welsh, Numerical calculation of time-dependent viscous incompressible flow with free surface. Phys. Fluids **8**, 2182–2189 (1965)
48. A.J. Chorin, Numerical solution of the Navier-Stokes equations. Math. Comput. **22**, 745–762 (1968)
49. R. Temam, Sur l'approximation de la solution des équations de navier-stokes par la méthode des pas fractionnaires. Arch. Rational Mech. Anal. **33**, 377–385 (1969)
50. J.L. Guermond, P. Minev, J. Shen, An overview of projection methods for incompressible flows. Comput. Methods Appl. Mech. Eng. **195**, 6011–6045 (2006)
51. M.L. Minion, D.L. Brown, R. Cortez, Accurate projection methods for the incompressible navier-stokes equations. J. Comput. Phys. **168**, 464–499 (2001)
52. M. Fortin, R. Glowinski, *Augmented Lagrangian: application to the numerical solution of boundary value problems* (North-Holland, Amsterdam, 1983)
53. S. Vincent, J.-P. Caltagirone, P. Lubin, N. Randrianarivelo, An adaptative augmented Lagrangian method for three-dimensional multi-material flows. Comput. Fluids **33**, 1273–1289 (2004)
54. S. Vincent, T. Randrianarivelo, G. Pianet, J.P. Caltagirone, Local penalty methods for flows interacting with moving solids at high Reynolds numbers. Comput. Fluids **36**, 902–913 (2007)
55. S. Vincent, J. Larocque, P. Lubin, J.-P. Caltagirone, G. Pianet, Adaptive augmented lagrangian techniques for simulating unsteady multiphase flows, in *6th international conference on multiphase flow*, vol. 203, Leipzig, Germany, July 9–13, 2007
56. S. Vincent, A. Sarthou, J.-P. Caltagirone, F. Sonilhac, P. Février, C. Mignot, G. Pianet, Augmented Lagrangian and penalty methods for the simulation of two-phase flows interacting with moving solids. Application to hydroplaning flows interacting with real tire tread patterns. J. Comput. Phys. **230**, 956–983 (2011)
57. W. Bizid, A. Etcheverlepo, S. Vincent, J.P. Caltagirone, D. Monfort, M. Hassine, Augmented Lagrangian and penalty methods for the simulation of two-phase flows interacting with moving solids. Application to hydroplaning flows interacting with real tire tread patterns. Notes Numer. Mech. Multidisciplinary Des. **125**, 13–22 (2014)
58. F. Couderc. *Développement d'un code de calcul pour la simulation d'écoulements de fluides non miscibles. Application à la désintégration assistée d'un jet liquide par un courant gazeux.* Ph.D. thesis, ENSAE, 2007

59. CW. Shu, Essentially non-oscillatory and weighted essentially non-oscillatory schemes for hyperbolic conservation laws. Technical report, ICASE—NASA report, 1997
60. M. Kang, R.P. Fedkiw, X.D. Liu, A boundary condition capturing method for multiphase incompressible flow. J. Sci. Comput. **15**, 323–360 (2000)
61. O. Tatebe, *MGCG method, a robust and highly parallel iterative method.* Ph.D. thesis, University of Tokyo, 1996
62. U. Trottenberg, C. W. Oosterlee, A. Schuller, *Multigrid* (Academic Press, 2001)
63. J.W. Cahn, J.E. Hilliard, Free energy of a nonuniform system. I. interfacial energy. J. Chem. Phys. **28**, 258 (1958)
64. B.J. Daly, A technique for introducing surface tension effects in hydrodynamic calculations. JCP **4**, 97–117 (1969)
65. S.O. Unverdi, G. Tryggvason, A front-tracking method for viscous, incompressible, multifluid flows. J. Comput. Phys. **100**, 25–37 (1992)
66. V.T. Nguyen, J. Peraire, B.C. Khoo, P.O. Persson, A discontinuous galerkin front tracking method for two phase flows with surface tension. Comput. Fluids **39**, 1–14 (2010)
67. C. Hirt, J.L. Cook, T.D. Butler, A lagrangian method for calculating the dynamics of an incompressible fluid with free surface. J. Comput. Phys. **5**(1), 103–124 (1970)
68. F. Duarte, R. Gormaz, S. Natesan, Arbitrary Lagrangian-Eulerian method for navier-stokes equations with moving boundaries. Comput. Methods Appl. Mech. Eng. **193**, 4819–4836 (2004)
69. B. Ramaswamy, M. Kawahara, Arbitrary Lagrangian-Eulerian finite element method for unsteady, convective, incompressible viscous free surface fluid flow. Int. J. Numer. Methods Fluids **7**(10), 953–984 (1987)
70. S. Osher, J. Sethian, Fronts propagating with curvature dependant speed? Algorithms based on hamilton-jacobi formulations. J. Comput. Phys. **79**, 12–49 (1988)
71. J.A. Sethian, *Level-set methods and fast marching methods* (Cambridge University Press, 1999)
72. M. Sussman, P. Smereka, S. Osher, A level set approach for computing solutions to incompressible two-phase flow. J. Comput. Phys. **114**, 146–159 (1994)
73. C. Hirt, B.D. Nichols, Volume of fluids (VOF) method for the dynamics of free boundaries. J. Comput. Phys. **39**, 201–225 (1981)
74. W.F. Noh, P. Woodward, SLIC/simple line interface calculation, in *Some methods of resolution of free surface problems*, vol. 59, eds. by A.I. Van De Vooren, P.J. Zandbergen. Lecture notes in physics, Berlin, Springer Verlag, 1976
75. W.J. Rider, D.B. Kothe, Reconstructing volume tracking. J. Comput. Phys. **141**, 112–152 (1998)
76. D. Gueyffier, J. Li, A. Nadim, R. Scardovelli, S. Zaleski, Volume of fluid interface tracking with smoothed surface stress methods for three-dimensional flows. J. Comput. Phys. **152**(2), 423–456 (1999)
77. R. Scardovelli, S. Zaleski, Interface reconstruction with least-square fit and split Eulerian-Lagrangian advection. Int. J. Numer. Methods Fluids **41**(3), 251–274 (2003)
78. B.J. Daly, Numerical study of two fluid Rayleigh-Taylor instability. Phys. Fluids **10**, 297–307 (1967)
79. B.J. Daly, Numerical sudy of the effect of the surface tension on interface instability. Phys. Fluids **12**, 1340–1354 (1969)
80. J. Glimm, O. McBryani, A computational model for interfaces. Adv. Appl. Math. **6**, 422–435 (1985)
81. J. Glimm, J.W. Grove, X.L. Li, K. Shyue, Y. Zeng, Q. Zhang, Three dimensional front tracking. SIAM J. Sci. Comput. **19**(3), 703–727 (1998)
82. S. Shin, D. Juric, Modeling three-dimensional multiphase flow using a level contour reconstruction method for front tracking without connectivity. J. Comput. Phys. **180**, 427–470 (2002)
83. R. DeBar, A method in two-D Eulerian hydrodynamics. Technical report UCID-19683, Lawrence Livermore National Laboratory, 1974

84. J.S. Saltzman, E.G. Puckett, A 3-D adaptive mesh refinement algorithm for multimaterial gas dynamics. Phys. D **60**, 84–104 (1992)
85. B.J. Parker, D.L. Youngs, *Two and three dimensional eulerian simulation of fluid flowwith material interfaces* (Technical report, UK Atomic Weapons Establishment, Aldermaston, 1992)
86. E.G. Puckett, A.S. Almgren, J.B. Bell, D.L. Marcus, W.J. Rider, A high order projection method for tracking fluid interfaces in variable density incompressible flows. J. Comput. Phys. **100**, 269–282 (1997)
87. J.E. Pilliod, *An analysis of piecewise linear interface reconstruction algorithms for volume of fluid methods*. Ph.D. thesis, University of California, Davis, 1992
88. E. Aulisa, S. Manservisi, R. Scardovelli, S. Zaleski, Interface reconstruction with least-square fit and split advection in three-dimensional cartesian geometry. J. Comput. Phys. **225**, 2301–2319 (2007)
89. M. Sussman, E.G. Puckett, A coupled level set and volume of fluid method for computing 3D and axisymmetric incompressible two-phase flow. J. Comput. Phys. **162**, 301–337 (2000)
90. T. Menard, *Développement d'une méthode Level Set pour le suivi d'interface. Application de la rupture de jet liquide*. Ph.D. thesis, Faculté de sciences de l'université de Rouen, 2007
91. S.P. Van Der Pijl, A. Segal, C. Vuik, P. Wesseling, A mass-conserving level-set method for modelling of multi-phase flows. Int. J. Numer. Methods Fluids **47**, 339–361 (2005)
92. D. Enright, R. Fedkiw, J. Ferziger, I. Mitchell, A hybrid particle level-set method for improved interface capturing. J. Comput. Phys. **183**, 83–116 (2002)
93. V. Mihalef, M. Sussman, D. Metaxas, The marker level-set method: a new approach to computing accurate interfacial dynamics. Not published, 2007
94. P. Trontin, S. Vincent, J.L. Estivalezes, J.P. Caltagirone, A subgrid computation of the curvature by a particle/level-set, applicaiton to a front-tracking/ghost-fluid method for incompressible flows. J. Comput. Phys. 6990–7010 (2012)
95. J.P. Morris, Simulating surface tension with smoothed particle hydrodynamics. Int. J. Numer. Methods Fluids **33**(3), 333–353 (2000)
96. J.U. Brackbill, D.B. Kothe, C. Zemach, A continuum method for modeling surface tension. J. Comput. Phys. **100**(2), 335–354 (1992)
97. C.S. Peskin, Numerical analysis of blood flow in the heart. J. Comput. Phys. **25**, 220–252 (1977)
98. R.J. LeVeque, Z. Li, The immersed interface method for elliptic equations with discontinuous coefficients and singular sources. SIAM J. Numer. Anal. **31**, 1019–1044 (1994)
99. R.J. LeVeque, Z. Li, Immersed interface method for stokes flow with elastic boundaries or surface tension. SIAM J. Comput. Phys. **18**, 709–735 (1997)
100. R. Fedkiw, T. Aslam, S. Xu, The ghost fluid method for deflagration and detonation discontinuities. J. Comput. Phys. **154**, 393–427 (1999)
101. R. Fedkiw, S. Xu, The ghost fluid method for for viscous flows, in *Progress in numerical solutions of partial diferential equations*, ed. by M. Hafez (Arcachon, France, 1998)
102. M. Kang, R. Fedkiw, X.D. Liu, A second order accurate symmetrical discretization of the poisson equation in irregular domains. JCP **176**, 205–227 (2001)
103. R. Guillaument, S. Vincent, J. Caltagirone, An original algorithm for vof based method to handle wetting effect in multiphase flow simulation. Mech. Res. Commun. **63**, 26–32 (2015)
104. S. Popinet, Numerical models of surface tension. Ann. Rev. Fluid Mech. **50**, 49–75 (2018)
105. S.T. Zalesak, Fully multidimensionnal flux-corrected transport algorithms for fluids. J. Comput. Phys. **31**, 335–362 (1979)
106. P. Trontin, S. Vincent, J.-L. Estivalezes, J.-P. Caltagirone, A subgrid computation of the curvature by a particle/level-set method. Application to a front-tracking/ghost-fluid method for incompressible flows. J. Comput. Phys. **231**, 6990–7010 (2012)
107. M. Sussman, E. Fatemi, An efficient, interface-preserving level set redistancing algorithm and its application to interfacial incompressible fluid flow. SIAM J. Sci. Comput. **20**, 1165–1191 (1999)

108. J.B. Bell, P. Colella, H.M. Glaz, A second-order projection method for the incompressible navier-stokes equations. J. Comput. Phys. **85**, 257–283 (1989)
109. J.-B. Ritz, J.P. Caltagirone, A numerical continuous model for the hydrodynamics of fluid particle systems. Int. J. Numer. Meth. Fluids **30**, 1067–1090 (1999)
110. D. Bhaga, M.E. Weber, Bubbles in viscous liquids: shapes, wakes and velocities. J. Fluid Mech. **105**, 61–85 (1981)
111. J. Hua, J. Lou, Numerical simulation of bubble rising in viscous liquid. J. Comput. Phys. **222**, 769–795 (2007)
112. S. Osher, R. Fedkiw, *Level set method and dynamic implicit surfaces*, vol. 153 (Springer Applied Mathematical Sciences, 2003)
113. S. Osher, R. Fedkiw, Level set methods: an overview and some recent results. J. Comput. Phys. **169**, 463–502 (2001)
114. E. Aulisa, S. Manservisi, R. Scardovelli, A mixed markers and volume-of-fluid method for the reconstruction and advection of interfaces in two-phase and free-boundary flows. J. Comput. Phys **188**, 611–639 (2003)
115. T. Hahn, Cuba—a library for multidimensional numerical integration. Comput. Phys. Commun. **168**, 78–95 (2005)
116. S.J. Cummins, M.M. Francois, D.B. Kothe, Estimating curvature from volume fractions. Comp. Struct. **83**, 425–434 (2005)
117. S. Bnà, S. Manservisi, R. Scardovelli, P. Yecko, S. Zaleski, Numerical integration of implicit functions for the initialization of the VOF function. Comput. Fluids **113**, 42–52 (2015)
118. S. Bnà, S. Manservisi, R. Scardovelli, P. Yecko, S. Zaleski, VoFi—a library to initialize the volume fraction scalar field. Comput. Phys. Commun. **200**, 291–299 (2016)
119. R.P. Brent, *Algorithms for minimization without derivatives* (Prentice-Hall, Englewood Cliffs, NJ, USA, 1973)
120. E. Polak, G. Ribiere, Note sur la convergence de méthodes de directions conjuguées. ESAIM Math. Modell Numer. Anal. Modél Math Anal Numér **3**, 35–43 (1969)
121. B. Van Leer, Towards the ultimate conservative difference scheme II. Monotonicity and conservation combined in a second order scheme. J. Comput. Phys. **14**, 361–370 (1974)
122. J.P. Boris, D.L. Book, Flux-corrected transport, I. SHASTA, a fluid transport algorithm that works. J. Comput. Phys. **11**, 38–69 (1973)
123. E. Berberović, N.P. van Hinsberg, S. Jakirlić, I.V. Roisman, C. Tropea, Drop impact onto a liquid layer of finite thickness: dynamics of the cavity evolution. Phys. Rev. E **79**(036306), 1–15 (2009)
124. F. Xiao, S. Ii, C. Chen, Revisit to the THINC scheme: a simple algebraic VOF algorithm. J. Comput. Phys **230**, 7086–7092 (2011)
125. W.F. Noh, P. Woodward, SLIC (simple line interface calculation), in *Lecture notes in physics*, eds. by A.I. van Dooren, P.J. Zandbergen (Springer, New York, 1976)
126. M. Rudman, Volume-tracking methods for interfacial flow calculations. Int. J. Numer. Meth. Fluids **24**, 671–691 (1997)
127. J.E. Pilliod, E.G. Puckett, Second-order accurate volume-of-fluid algorithms for tracking material interfaces. J. Comput. Phys. **199**, 465–502 (2004)
128. S. Popinet, S. Zaleski, A front tracking algorithm for accurate representation of surface tension. Int. J. Numer. Methods Fluids **30**, 775–793 (1999)
129. G. Bornia, A. Cervone, S. Manservisi, R. Scardovelli, S. Zaleski, On the properties and limitations of the height function method in two-dimensional Cartesian geometry. J. Comput. Phys **230**, 851–862 (2011)
130. S.V. Diwakar, S.K. Das, T. Sundararajan, A quadratic spline based interface (QUASI) reconstruction algorithm for accurate tracking of two-phase flows. J. Comput. Phys **228**, 9107–9130 (2009)
131. D. Gueyffier, J. Li, A. Nadim, R. Scardovelli, S. Zaleski, Volume-of-fluid interface tracking with smoothed surface stress methods for three-dimensional flows. J. Comput. Phys **152**, 423–456 (1999)

132. R. Scardovelli, S. Zaleski, Analytical relations connecting linear interfaces and volume fractions in rectangular grids. J. Comput. Phys **164**, 228–237 (2000)
133. G.D. Weymouth, D.K.-P. Yue, Conservative volume-of-fluid method for free-surface simulations on Cartesian-grids. J. Comput. Phys. **229**, 2853–2865 (2010)
134. J. López, J. Hernandez, P. Gómez, F. Faura, A volume of fluid method based on multidimensional advection and spline interface reconstruction. J. Comput. Phys. **195**, 718–742 (2004)
135. D.J.E. Harvie, D.F. Fletcher, A new volume of fluid advection algorithm: the stream scheme. J. Comput. Phys. **162**, 1–32 (2000)
136. S. Tanguy, *Développement d'une methode de suivi d'interface Application aux écoulements diphasiques*. Ph.D. thesis, Faculté des sciences de l'université de Rouen, 2004
137. S. Tanguy, A. Berlemont, Application of a level set method for simulation of droplet collisions. Int. J. Multiphase Flow **31**, 1015–1035 (2005)
138. Z.J. Wang, Spectral (finite) volume method for conservation laws on unstructured grids, basic formulation. J. Comput. Phys. **178**, 210–251 (2002)
139. Y. Liu, M. Vinokur, Z.J. Wang, Spectral difference method for unstructured grids: basic formulation. J. Comput. Phys. **216**, 780–801 (2006)
140. S. Osher, C.-W. Shu, High-order essentially non-oscillatory schemes for Hamilton-Jacobi equations. SIAM J. Numer. Anal. **28**(4), 907–922 (1991)
141. C.W. Shu, G.S. Jiang, Efficient implementation of weighted ENO schemes. J. Comput. Phys. **126**, 202–228 (1996)
142. C.W. Shu, S. Osher, Efficient implementation of essentially non-oscillatory shock capturing schemes. J. Comput. Phys. **83**, 32–78 (1989)
143. R. Fedkiw, B. Merriman, S. Osher, Simplified upwind discretisation for systmes of hyperbolic conservation laws containing advection equations. JCP **157**(1), 302–326 (2000)
144. S. Gottlieb, C.W. Shu, Total variation diminishing runge-kutta schemes. Math. Comput. **67**(221), 73–85 (1998)
145. D. Chopp, Computing minimal surfaces with level-set curvature flows. JCP **106**, 77–91 (1993)
146. G.S. Jiang, D. Peng, Weighted eno schemes for Hamilton-Jacobi equations. SIAM J. Sci. Comput. **21**(6), 2126–2143 (2000)
147. D. Peng, B. Merriman, S. Osher, H.K. Zhao, M. Kang, A PDE based fast local level-set. J. Comput. Phys. **155**, 410–438 (1999)
148. E. Olsson, G. Kreiss, A conservative level set method for two phase flow. J. Comput. Phys. **210**(1), 225–246 (2005)
149. E. Olsson, G. Kreissand, S. Zahedi, A conservative level set method for two phase flow II. J. Comput. Phys. **225**(1), 785–807 (2007)
150. J.O. McCaslin, O. Desjardin, An accurate conservative level set/ghost fluid method for simulating turbulent atomization. J. Comput. Phys. **227**(18), 8395–8416 (2008)
151. O. Desjardin, V. Moureau, H. Pitsch, A localized re-initialization equation for the conservative level set method. J. Comput. Phys. **262**, 408–426 (2014)
152. M. Sussman, A second order coupled level set and volume-of-fluid method for computing growth and collapse of vapour bubbles. J. Comput. Phys. **187**, 110–136 (2003)
153. T. Menard, S. Tanguy, A. Berlemont, Coupling level set/VOF/ghost fluid methods: validation and application to 3D simulation of the primary break-up of a liquid jet. Int. J. Multiphase Flow **33**(5), 510–524 (2007)
154. G. Strang, On the construction and comparison of difference schemes. SIAM J. Numer. Anal. 506 (1968)
155. A. Sarthou, D. Zuzio, J.-L. Estivalèzes, Multiscale Euler-Lagrange method for parallel simulation of atomization induced by air-blast planar injectors, in *21st AIAA computational fluid dynamics conference* (2013)
156. G. Tryggvason, R. Scardovelli, S. Zaleski, *Direct numerical simulations of gas-liquid multiphase flows* (Cambridge University Press, 2011)
157. A. Sarthou, S. Vincent, J.-P. Caltagirone, A second-order curvilinear to cartesian transformation of immersed interfaces and boundaries. Application to fictitious domains and multiphase flows. Comput. Fluids **46**, 422–428 (2011)

158. C. Peskin, The immersed boundary method. Acta Numer. **11**, 1–39 (2002)
159. R. McDermott, S.B. Pope, The parabolic edge reconstruction method (perm) for Lagrangian particle advection. J. Comput. Phys. **227**, 5447–5491 (2008)
160. D. Zuzio, J.-L. Estivalèzes, An efficient block parallel amr method for two phase flow simulations. Comput. Fluids **44**, 339–357 (2011)
161. M.J. Berger, *Adaptive mesh refinement for hyperbolic partial differential equations*. Ph.D. thesis, Stanford University, 1982
162. D. DeZeeuw, K.G. Powell, An adaptative cartesian mesh for the Euler equations. J. Comput, Phys. (1992)
163. M. Sussman, A.S. Almgren, J.B. Bell, P. Colella, L.H. Howell, M.L. Welcome, An adaptive level set approach for incompressible two-phase flows. J. Comput. Phys. **148**, 81–124 (1999)
164. F. Couderc, J.L. Estivalèzes, Sharp interface capturing method based on level-set method for incompressible flow with surface tension, in *ASME annual summer meeting FEDSM2003-45151*, 2003
165. P. MacNeice, K.M. Olson, C. Mobarry, R. de Fainchtein, C.P. Paramesh, A parallel adaptive mesh refinement community toolkit. Comput. Phys. Commun. **126** (2000)
166. D.F. Martin, P. Colella, D. Graves, A cell-centered adaptive projection method for the incompressible navier-stokes equations in three dimensions. J. Comput. Phys. 1863–1886 (2007)
167. J.D. Brown, L.L. Lowe, Multigrid elliptic equation solver with adaptive mesh refinement. J. Comput. Phys. **209**(2), 582–598 (2005)
168. D. Bai, A. Brandt, Local mesh refinement multilevel techniques. SIAM J. Sci. Stat. Comput. **8**(2), 582–598 (1987)
169. M. Sussman, A parallelized, adaptive algorithm for multiphase flows in general geometries. Comput. Struct. **83**(6–7), 435–444 (2005)
170. S.F. McCormick, *Multigrid methods* (Society for Industrial and Applied Mathematics, 1987)
171. D.F. Martin, K.L. Cartwright, Solving Poisson equation using adaptive mesh refinement. J. Comput, Phys. (1996)
172. D.F. Martin. *An adaptive cell-centered projection method for the incompressible Navier-Stokes equations*. Ph.D. thesis, BSME University of Florida, 1991
173. Y. Yoshida, M. Sussman, M. Ohta, Numerical simulation of bubble motion by a coupled level set/VOF method, in *Proceedings of the 35th annual meeting of the society of chemical engineers*, 2002
174. S. Schaffer, A semicoarsening multigrid method for elliptic partial differential equations with highly discontinuous and anisotropic coefficients. SIAM J. Sci. Comput. **20**, 228–242 (1998)
175. C. Wagner, Introduction to algebraic multigrid. *Course notes of an algebraic multigrid course at the University of Heidelberg*, 1998/99
176. H.A. Van Der Vorst, Bi-cstab: a fast and smoothly converging variant of bi-cg for the solution of nonsymmetric linear systems. SIAM J. Sci. Comput. **12**(2), 631–644 (1992)
177. R. Teigland, I.K. Eliassen, A multiblock/multilevel mesh refinement procedure for CFD computations. Int. J. Numer. Methods Fluids **36**, 519–538 (2001)
178. M. Oevermann, G.M. Berger, A projection method for two-phase incompressible flow with surface tension and sharp interface resolution. Zuse Institute Berlin—report, 17, 2000
179. S. Hysing, S. Turek, D. Kuzmin, N. Parolini, E. Burman, S. Ganesan, L. Tobiska, Quantitative benchmark computations of two-dimensional bubble dynamics. MOX-report, 23, 2008
180. R. Clift, J.R. Grace, M.E. Weber, *Bubbles, drops and particles* (Academic Press, 1978)
181. A. Prosperetti, Motion of two superposed viscous fluids. Phys. Fluids **24**, 1217–1223 (1981)
182. S. Popinet, An accurate adaptive solver for surface tension driven interfacial flows. J. Comput. Phys. **228**, 5838–5866 (2009)
183. M.R. Hestenes, Multiplier and gradient methods. J. Optimization Theory and Applications **4**, 330–320 (1969)
184. H. Uzawa, *Iterative method for concave programming*, in *Studies in linear and nonlinear programming*, eds. by K.J. Arrow, L. Hurwicz, H. Uzawa (Stanford University Press, 1958)
185. K. Goda, A multistep technique with implicit difference schemes for calculating two- or three-dimensional cavity flows. J. Comput. Phys. **30**, 76–95 (1979)

186. J.-P. Caltagirone, J. Breil, A vector projection method for solving the Navier-Stokes equations. Comptes Rendus de l'Académie des Sciences Paris, Série IIb **327**, 1179–1184 (1999)
187. J.-P. Caltagirone, S. Vincent, Tensorial penalisation method for solving Navier-Stokes equations. Comptes Rendus de l'Académie des Sciences Paris, Série IIb **329**, 607–613 (2001)
188. I. Ramière, P. Angot, M. Belliard, A fictitious domain approach with spread interface for elliptic problems with general boundary conditions. Comput. Meth. Applied Mech. Eng. **196**, 766–781 (2007)
189. I. Ramière, P. Angot, M. Belliard, A general fictitious domain method with immersed jumps and multilevel nested structured meshes. J. Comput. Phys. **225**, 1347–1387 (2007)
190. A. Sarthou, S. Vincent, J.-P. Caltagirone, P. Angot, Eulerian-Lagrangian grid coupling and penalty methods for the simulation of multiphase flows interacting with complex objects. Int. J. Numer. Meth. Fluid **56**, 1093–1099 (2008)
191. R. Leveque, Z. Li, The immersed interface method for elliptic equations with discontinuous coefficients and singular sources. SIAM J. Numer. Anal. **31**, 1019–1044 (1994)
192. T.N. Randrianarivelo, G. Pianet, S. Vincent, J.-P. Caltagirone, Numerical modelling of the solid particle motion using a new penalty method. Int. J. Numer. Meth. Fluids **47**, 1245–1251 (2005)
193. S. Vincent, J.-P. Caltagirone, Efficient solving method for unsteady incompressible interfacial flow problems. Int. J. Numer. Meth. Fluid **30**, 795–811 (1999)
194. A. Sarthou, *Fictitious domain methods for high order elliptic equations and Navier-Stokes equations: application to fluid-structure coupling*. Ph.D. thesis, Université de Bordeaux, 2009
195. J.-P. Caltagirone, S. Vincent, A kinematics scalar projection method (KSP) for incompressible flows with variable density. Open J. Fluid Dyn. **5**, 171–182 (2015)
196. D. Lacanette, S. Vincent, A. Sarthou, P. Malaurent, J.-P. Caltagirone, Efficient solving method for unsteady incompressible interfacial flow problems. Int. J. Heat Mass Transf. **52**, 2528–2542 (2009)
197. S. Vincent, J.-P. Caltagirone, A one-cell local multigrid method for solving unsteady incompressible multiphase flows. J. Comput. Phys. **163**, 172–215 (1999)
198. N. Mordant, J.-F. Pinton, Velocity measurements of settling sphere. Eur. Phys. J. B **18**, 343–352 (2000)
199. D.L. Youngs, Time-dependent multimaterial flow with large fluid distortion, in *Numerical methods for fluid dynamics*, eds. by K.W. Morton, M.J. Baines (Academic, New-York, 1982)
200. J.F. Richardson, W.N. Zaki, Sedimentation and fluidization. Part 1. Trans. Inst. Chem. Eng. **32**, 35–53 (1954)
201. D.J. Tritton, Experiments on the flow past a circular cylinder at low reynolds numbers. J. Fluid Mech. **6**, 547–567 (1959)
202. J. Kim, D. Kim, H. Choi, An immersed-boundary finite-volume method for simulations of flow in complex geometries. J. Comput. Phys. **171**, 132–150 (2001)
203. D.V. Le, B.C. Khoo, J. Peraire, An immersed interface method for viscous incompressible flows involving rigid and flexible boundaries. J. Comput. Phys. **220**, 109–138 (2006)
204. R.D. Moser, J. Kim, N.N. Mansour, Direct numerical simulation of turbulent channel flow up to re = 590. Phys. Fluids **11**, 943–945 (1999)
205. G. Medic, G. Daenincky, J.A. Templeton, G. Kalitzin, A framework for near-wall rans/les coupling, in *Center for turbulence research, annual research briefs*, 2005
206. J. Happel, H. Brenner, *Low Reynolds number hydrodynamics* (Kluwer Academic Publishers, 1963)
207. O. Schenk, K. Gärtner, Solving unsymmetric sparse systems of linear equations with PARDISO. Fut. Gener. Comput. Syst. **20**, 475–487 (2004)
208. I. Gustafsson, *On first- and second-order symmetric factorisation methods for the solution of elliptic difference equations* (Chalmers University of Technology, 1978)
209. L.F. Richardson, J.A. Gaunt, The deferred approach to the limit. Trans. R. Soc. London Ser. A Math. Phys. Sci. **226**, 299–361 (1927)
210. A. Ten Cate, C.H. Nieuwstad, J.J. Derksen, H.E.A. Van den Akker, Particle imaging velocimetry experiments and lattice-Boltzmann simulations on a single sphere settling under gravity. Phys. Fluids **14**, 4012–4025 (2002)

211. G. Pianet, A. Ten Cate, J.J. Derksen, E. Arquis, Assessment of the 1-fluid method for DNS of particulate flows: sedimentation of a single sphere at moderate to high Reynolds numbers. Comput. Fluids **36**, 359–375 (2007)
212. S.V. Apte, M. Martin, N.A. Patankar, A numerical method for fully resolved simulation (FRS) of rigid particle-flow interactions in complex flows. J. Comput. Phys. **228**, 2712–2738 (2009)
213. T. Kempe, J. Fröhlich, On Euler-Lagrange coupling and collision modelling for spherical particles, in *8th international ERCOFTAC symposium on engineering turbulence modelling and measurements*, Marseille, France, 2010
214. J.-P. Caltagirone, S. Vincent, C. Caruyer, A multiphase compressible model for the simulation of multiphase flows. Comput. Fluids **50**, 24–34 (2011)
215. V.R. Voller, C.R. Swaminathan, Fixed grids techniques for phase change problems: a review. Int. J. Numer. Methods Fluids **30**, 875–898 (1990)
216. C. Viozat, *Calcul d'écoulements stationnaires et instationnaires à petit nombre de Mach et en maillages étirés*. Ph.D. thesis, Université de Nice-Sophia Antipolis, 1998
217. A. Majda, *Compressible fluid flow and systems of conservation laws in several space variable* (Springer-Verlag, 1984)
218. C. Hirsch, *Numerical computation of internal and external flows. Volume II: computational methods for inviscid and viscous flows* (John Wiley and Sons, Chichester, 1990)
219. P.J. Roache, *Verification and validation in computational science and engineering* (Hermosa, 1998)
220. G. Servant, *Contribution à l'étude et à la modélisation de la propagation des ultrasons de puissance en milieu cavitant*. Ph.D. thesis, Université de Bordeaux 1, 2001
221. G. Servant, J.P. Caltagirone, A. Gerard, J.L. Laborde, A. Hita, Numerical simulation of cavitation bubble dynamics induced by ultrasound waves in a high frequency reactor. Ultrason. Sonochem. **7**, 217–227 (2000)
222. S. Stolz, N.A. Adams, An approximate deconvolution procedure for large-eddy simulation. Phys. Fluids **11**, 1699 (1999)
223. S. Stolz, N.A. Adams, L. Kleiser, An approximate deconvolution model for large-eddy simulation with application to incompressible wall bounded flows. Phys. Fluids **13**, 997–1015 (2001)
224. D. Lakehal, B.L. Smith, M. Milelli, Large-Eddy simulation of bubbly turbulent shear flows. J. Turbulence **3**, 25 (2002)
225. V. Srinivasan, A.J. Salazar, K. Saito, Numerical investigation on the disintegration of round turbulent liquid jets using les/vof techniques. Atom. Spray **18**, 571–617 (2008)
226. J. Chesnel, J. Réveillon, T. Ménard, F.-X. Demoulin, Large eddy simulation of liquid jet atomization. Atom. Spray **21**, 711–736 (2011)
227. P. Sulasalmi, A. Karna, T. Fabritius, J. Savolainen, CFD model for emulsification of slag into the steel. ISIJ Int. **49**, 1661–1667 (2009)
228. L. Li, Z. Liu, M. Cao, B. Li, Large eddy simulation of bubbly flow and slag layer behavior in ladle with discrete phase model (DPM)-volume of fluid (VOF) coupled model. J. Min. Metals Mat. Soc. **67** (2015)
229. D. Lacanette, S. Vincent, E. Arquis, P. Gardin, Numerical simulation of gas-jet wiping in steel strip galvanizing process. Iron Steel Inst. Japan Int. **45**, 214–220 (2005)
230. M. Herrmann, A sub-grid surface dynamics model for sub-filter surface tension induced interface dynamics. Comput. Fluids **87**, 92–101 (2013)
231. P. Sagaut, C. Cambon, *Homogeneous turbulence dynamics* (Cambridge University Press, 2008)
232. S. Fleau, S. Vincent, S. Mimouni, LES modeling with a multifield approach, in *Proceedings of the 4th turbulence and interactions conference TI2015*, Nov 2015, Cargèse, France, Springer, 2016
233. M. Saeedipour, S. Vincent, S. Pirker, Large eddy simulation of turbulent interfacial flows using approximate deconvolution model. Int. J. Multiphase Flow **112**, 286–299 (2019)
234. M. Tavares, S. Vincent, M. Ould-Rouiss, J.-L. Estivalezes, A priori study for the modeling of les subgrid scale terms, in *Proceedings of the 4th turbulence and interactions conference TI2015*, Nov 2015, Cargèse, France, Springer, 2016

235. S. Vincent, M. Tavares, S. Fleau, S. Mimouni, M. Ould-Rouiss, J.-L. Estivalezes, A priori filtering and les modeling of turbulent two-phase flows application to phase separation. Comput. Fluids **176**, 245–259 (2018)
236. W. Aniszewski, A. Boguslawski, M. Marek, A. Tyliszczak, A new approach to sub-grid surface tension for LES of two-phase flows. J. Comput. Phys. **231**, 7368–7397 (2012)
237. M. Klein, S. Ketterl, J. Hasslberger, Large eddy simulation of multiphase flows using the volume of fluid method: part 1–governing equations and a priori analysis. Exp. Comp. Multiphase Flow **1**, 130–144 (2019)
238. S. Ketterl, M. Reiγmann, M. Klein, Large eddy simulation of multiphase flows using the volume of fluid method: part 2—a-posteriori analysis of liquid jet atomization. Exp. Comp. Multiphase Flow **1**, 201–211 (2019)
239. W.J. Layton, L. Rebholtz, *Approximate deconvolution models of turbulence* (Springer, 2012)
240. J. Hasslberger, S. Ketterl, M. Klein, *A-priori assessment of interfacial sub-grid scale closures in the two-phase flow les context* (Flow Turb, Comb, 2020)
241. Various authors, *Test-cases for interface tracking methods*, 2003. http://test.interface.free.fr/
242. S. Vincent, J.-P. Caltagirone, D. Jamet, Test case no 15: phase inversion in a closed box (pn, pe). Multiphase Sci. Technol. **6**, 101–104 (2004)
243. J.R. Bull, A. Jameson, Explicit filtering and exact reconstruction of the sub-filter stresses in large eddy simulation. J. Comput. Phys. **306**, 117–136 (2016)
244. J.-P. Caltagirone, *Discrete mechanics* (Wiley and Sons, 2015)
245. J.-P. Caltagirone, *Discrete mechanics: concepts and applications* (Wiley and Sons, ISTE, 2019)
246. J.-P. Caltagirone, S. Vincent, On primitive formulation in fluid mechanics and fluid-structure interaction with constant piecewise properties in velocity-potentials of acceleration. Acta Mech. Available online, 2020
247. J. Magnaudet. Méthodes de simulation pour les problèmes d'interfaces. *Notes de cours, Ecole de Printemps de Mécanique des Fluides Numérique, La Londe Les Maures, Cambridge*, 1997
248. G. Tryggvason, R. Scardovelli, and S. Zaleski. *Direct numerical simulations of gas-liquid multiphase flows*. Cambridge University Press, 2011
249. S. Vincent, G. Balmigère, J.-P. Caltagirone, E. Meillot, Eulerian-Lagrangian multiscale methods for solving scalar equations - Application to incompressible two-phase flows. J. Comput. Phys. **229**, 73–106 (2010)
250. S. Fleau, S. Mimouni, N. Mérigoux, S. Vincent, Validation of a multifield approach for the simulations of two-phase flows. Comput. Thermal Sci. **7**, 1–17 (2016)
251. S. Mimouni, R. Denèfle, S. Fleau, S. Vincent. Multifield approach and interface locating method for two-phase flows in nuclear power plant, in *SimHydro 2014: modelling of rapid transitory flows*, 11-13 June, Sophia-Antipolis, 2014
252. S. Fleau, S. Mimouni, N. Mérigoux, S. Vincent, Simulations of two-phase flows with a multifield approach, in *Proceedings of computational heat transfer conference CHT-15*, May 2015, New Brunswick, USA, 2015
253. N. Adams, S. Stolz, A subgrid-scale deconvolution approach for schock capturing. J. Comput. Phys. **178**, 391–426 (2002)
254. F. Chow, R. Street, M. Xue, J. Ferziger, Explicit filtering and reconstruction turbulence modeling for large-eddy simulation of neutral boundary layer flow. J. Atmos. Sci. **62**, 2058–2077 (2004)
255. P. Moin, *Fundamentals of engineering numerical analysis* (Cambridge University Press, 2001)
256. P. Sagaut, S. Deck, M. Terracol, *Multiscale and multiresolution approaches in turbulence: LES, DES and hybrid RANS/LES methods: applications and guidelines* (Imperial College Press, London, 2013)
257. F. Raymond, J.-M. Rosant, A numerical and experimental study of the terminal velocity and shape of bubbles in viscous liquids. Chem. Eng. Sci. **55**, 943–955 (2000)
258. M. Germano, U. Piomelli, P. Moin, W. Cabot, A dynamic subgrid-scale eddy viscosity model. Phys. Fluids **3**, 1760–1765 (1991)

259. J. Gullbrand, F. Chow, The effect of numerical errors and turbulence models in large-eddy simulations of channel flow, with and without explicit filtering. J. Fluid Mech. **495**, 323–341 (2003)
260. M. Habisreutinger, R. Bouffanais, E. Leriche, M. Deville, A coupled approximate deconvolution and dynamic mixed scale model for large-eddy simulation. J. Comput. Phys. **224**, 241–266 (2007)
261. H. Jeanmart, G. Winchelmans, Investigation of eddy-viscosity models modified using discrete filters: a simplified "regularized variational multiscale model" and an "enhanced field model." Phys. Fluids **19**, 055110 (2007)
262. J. Mathew, H. Foysi, R. Friedrich, A new approach to les based on explicit filtering. Int. J. Heat Fluid Flow **27**, 594–602 (2006)
263. J.A. Deardoff, A numerical study of three-dimensional turbulent channel flow at large reynolds number. J. Fluid Mech. **41**, 453–465 (1970)
264. M. Milelli, *A numerical analysis of confined turbulent bubble plumes*. Ph.D. thesis, Swiss Federal Institute of Technology Zurich, 2002
265. D. Lakehal, DNS and LES of turbulent multifluid flows, in *3rd international symposium on two-phase flow modelling and experimentation* (Pisa, Italy, 2004)
266. M. Salvetti, F. Beux, The effect of the numerical scheme on the subgrid scale term in large-eddy-simulation. Phys. Fluids, 3020–3022 (1998)
267. B. Vreman, B. Geurts, H. Kuerten, On the formulation of the dynamic mixed subgrid-scale model. Phys. Fluids **6**, 4057–4059 (1994)
268. Y. Zhang, R. Street, J. Koseff, A dynamic mixed subgrid-scale model and its application to turbulent recirculating flows. Phys. Fluids **5**, 3186–3196 (1993)
269. O. San, A. Staples, T. Iliescu, A posteriori analysis of low-pass spatial filters for approximate deconvolution large eddy simulations of homogeneous incompressible flows. Int. J. Numer. Anal. Mod. **1**, 1–18 (2014)
270. F. Najjar, D. Tafti, Study of discrete test filters and finite difference approximations for the dynamic subgrid-scale stress model. Phys. Fluids **8**, 1076–1088 (1996)
271. T. Lund, The use of explicit filters in large eddy simulation. Comput. Math. App. **46**, 603–616 (2003)
272. P. Schlatter, S. Stolz, L. Kleiser, Les of transitional flows using the approximate deconvolution model. Int. J. Heat Fluid Flow **25**, 549–558 (2004)
273. S. Lele, Compact finite difference schemes with spectral-like resolution. J. Comput. Phys. **103**, 16–42 (1992)
274. C. Pruett, N. Adams, A priori analyses of three subgrid-scale models for one-parameter families of filters. Phys. Fluids **12**, 1133–1142 (2000)
275. G.D. Stefano, O. Vasilyev, Sharp cut-off vs. smooth filtering in les. Phys. Fluids **14**, 362 (2002)
276. A. Toutant, M. Chandesris, D. Jamet, O. Lebaigue, Jump conditions for filtered quantities at an under-resolved discontinuous interface. Part 1: theoretical development. Int. J. Mult. Flow **35**, 1100–1118 (2009)
277. M. Herrmann, M. Gorokhovski, A large eddy simulation subgrid model for turbulent phase interface dynamics, in *11th Triennial international annual conference on liquid atomization and spray systems (ICLASS 2009)*, 2009
278. M. Herrmann, *A dual scale volume-of-fluid approach for modeling turbulent phase interface dynamics*, in Technical report (Center for Turbulence Research, 2014)
279. E. Shirani, F. Ghadiri, A. Ahmadi, Modeling and simulation of interfacial turbulent flows. J. Appl. Fluid Mech. **4**, 43–49 (2011)
280. E. Shirani, A. Jafari, N. Ashgriz, Turbulence models for flows with free surfaces and interfaces. AIAA J. **44**, 1454–1462 (2006)
281. S. Stolz, N.A. Adams, L. Kleiser, The approximate deconvolution model for large eddy simulations of compressible flows and its application to shock-turbulent-boundary-layer interaction. Phys. Fluids **13**, 2985–3001 (2001)

282. B. Rembold, *Direct and large-eddy simulation of compressible rectangular jet flow*. Ph.D. thesis, Swiss Federal Institute of Technology Zurich, 2003
283. M. Küng, *Large eddy simulation of turbulent channel and jet flows using the approximate deconvolution model*. Ph.D. thesis, Swiss Federal Institute of Technology Zürich, 2007
284. M. Germano, A new deconvolution method for large eddy simulation. Phys. Fluids **21** (2009)
285. O. San, Analysis of low-pass filters for approximate deconvolution closure modelling in one-dimensional decaying burgers turbulence. Int. J. Comput. Fluid Dyn. **30**, 20–37 (2016)
286. S. Schneiderbauer, M. Saeedipour, Approximate deconvolution model for the simulation of turbulent gas-solid flows: an a-priori analysis. Phys. Fluids **30**, 023301 (2018)
287. W. Aniszewski, A. Boguslawski, M. Marek, A. Tyliszczak, A new approach to sub-grid surface tension for les of two-phase flows. J. Comput. Phys. **231**, 7368–7379 (2012)
288. W. Aniszewski, Improvements, testing and development of the adm-$\tau$ sub-grid surface tension model for two-phase les. J. Comput. Phys. **327**, 389–415 (2016)
289. S. Vincent, J.-P. Caltagirone, D. Jamet, Test-case no 15: phase inversion in a closed box (pc). Multiphase Sci. Technol. **16**, 101–104 (2004)
290. W. Aniszewski, T. Ménard, M. Marek, Volume of fluid (VOF) type advection methods in two-phase flow: A comparative study. Comput. Fluids **97**, 52–73 (2014)
291. J.-L. Estivalezes, W. Aniszewski, F. Auguste, Y. Ling, L. Osmar, J.-P. Caltagirone, L. Chirco, A. Pedrono, S. Popinet, A. Berlemont, J. Magnaudet, T. Ménard, S. Vincent, S. Zaleski, A phase inversion benchmark for multiscale multiphase flows. J. Comput. Phys. **450**, 110810 (2022)
292. R. Lebas, T. Ménard, P.A. Beau, A. Berlemont, F.X. Demoulin, Numerical simulation of primary break-up and atomization: DNS and modelling study. Int. J. Multiphase Flow **35**, 247–260 (2009)
293. G. Tomar, D. Fuster, S. Zaleski, S. Popinet, Multiscale simulations of primary atomization. Comput. Fluids **39**, 1864–1874 (2010)
294. M. Saeedipour, S. Schneiderbauer, A new approach to include surface tension in the subgrid eddy viscosity for the two-phase les. Int. J. Mult. Flow **121** (2019)
295. F. Nicoud, F. Ducros, Subgrid-scale stress modelling based on the square of the velocity gradient tensor. Flow Turb. Comb. **62**, 183–200 (1999)
296. J. Bardina, J.H. Ferziger, W.C. Reynolds, Improved turbulence models based on large eddy simulation of homogeneous, incompressible, turbulent flows. Report TF-19, thermosciences division, det (Mechanical Engineering, Standford University, 1983)
297. M. Saeedipour, S. Puttinger, N. Doppelhammer, S. Pirker, Investigation on turbulence in the vicinity of liquid-liquid interfaces—large eddy simulation and PIV experiment. Chem. Eng. Sci. **198**, 98–107 (2019)
298. P. Minutolo, M.V. Prati, M. Sirignano, A. D'Anna, Emission of particles from practical combustion devices burning methane/natural gas, in *European aerosol conference 2009*, Karlsruhe
299. E.R. Bissell, A.K. Burnham, R.L. Braun, Shale oil cracking kinetics and diagnostics. Ind. Eng. Chem. Process Des. Dev. **24**, 381–386 (1985)
300. F. Sefidvash, Status of the small modular fluidized bed light water nuclear reactor concept. Nucl. Eng. Des. **167**, 203 (1996)
301. S. Shyuer-Ming, P.D. Komar, Sediments, beach morphology and sea cliff erosion within an Oregon coastal littoral cell. J. Coastal Res. **10**, 144–157 (1994)
302. P.M. Holton, D.L. Tackett, K. Willeke, Particle size-dependent leakage and losses of aerosols in respirators. Am. Ind. Hygiene Assoc. J. **48**, 848–854 (1987)
303. A. Hölzer, M. Sommerfeld, New simple correlation formula for the drag coefficient of non-spherical particles. Powder Technol. **184**, 361–365 (2008)
304. M. Moreau, O. Simonin, B. Bédat, Development of gas-particle Euler-Euler LES approach: a priori analysis of particle sub-grid models in homogeneous isotropic turbulence. Flow Turbul. Combustion **84**, 295–324 (2009)
305. O.A. Druzhinina, S. Elghobashi, On the decay rate of isotropic turbulence laden with microparticles. Phys. Fluids **11**, 602–611 (1999)

306. D. Squires, J.K. Eaton, Particle response and turbulence modification in isotropic turbulence. Phys. Fluids **2**, 1191–1204 (1990)
307. A. Ferrante, S. Elghobashi, On the physical mechanisms of two-way coupling in particle-laden isotropic turbulence. Phys. Fluids **15**, 315–329 (2003)
308. A.M. Ahmed, S. Elghobashi, Direct numerical simulation of particle dispersion in homogeneous turbulent shear flows. Phys. Fluids **13**, 3346–3364 (2001)
309. P. Fede, O. Simonin, Numerical study of the subgrid fluid turbulence effects on the statistics of heavy colliding particles. Phys. Fluids **18**, 045103–045120 (2006)
310. H.H. Hu, D.D. Joseph, M.J. Crochet, Direct simulation of fluid particle motions. Theor. Comput. Fluid Mech. **3**, 285–306 (1992)
311. E.A. Fadlun, R. Verzicco, P. Orlandi, J. Mohd-Yusofz, Combined immersed-boundary finite-difference methods for three-dimensional complex flow simulations. J. Comput. Phys. **161**, 35–60 (2000)
312. S. Takagi, H.N. Oguz, Z. Zhang, A. Prosperetti, PHYSALIS: a new method for particle simulation part II: two-dimensional Navier-Stokes flow around cylinders. J. Comput. Phys. **187**, 371–390 (2003)
313. M. Coquerelle, G.H. Cottet, A vortex level set method for the two-way coupling of an incompressible fluid with colliding rigid bodies. J. Comput. Phys. **227**, 9121–9137 (2008)
314. J.A. Simeonov, J. Calantoni, A pressure boundary integral method for direct fluid-particle simulations on Cartesian grids. J. Comput. Phys. **230**, 1749–1765 (2011)
315. K. Höfler, S. Schwarzer, Navier-Stokes simulation with constraint forces: finite-difference method for particle-laden flows and complex geometries. Phys. Rev. E **61**, 7146–7160 (2000)
316. S.H. Cho, H.G. Choi, J.Y. Yoo, Direct numerical simulation of fluid flow laden with many particles. Int. J. Multiphase Flow **31**, 435–451 (2005)
317. H.H. Hu, N.A. Patankar, M.Y. Zhu, Direct numerical simulations of fluid-solid systems using the arbitrary Lagrangian-Eulerian technique. J. Comput. Phys. **169**, 427–462 (2001)
318. R. Glowinski, T.W. Pan, T.I. Hesla, D.D. Joseph, J. Périaux, A fictitious domain approach to the direct numerical simulation of incompressible viscous flow past moving rigid bodies: application to particulate flow. J. Comput. Phys. **169**, 363–426 (2001)
319. Z. Zhang, A. Prosperetti, A second-order method for three-dimensional particle simulation. J. Comput. Phys. **210**, 292–324 (2005)
320. M. Uhlmann, An immersed boundary method with direct forcing for the simulation of particulate flows. J. Comput. Phys. **209**(2), 448–476 (2005)
321. M. Uhlmann, Interface-resolved direct numerical simulation of vertical particulate channel flow in the turbulent regime. Phys. Fluids **20**(5), 053305 (2008)
322. A. Ten Cate, J.J. Derksen, L.M. Portela, E.A. Van Der Akker, Fully resolved simulations of colliding monodisperse spheres in forced isotropic turbulence. J. Fluid Mech. **519**, 233–271 (2004)
323. H. Gao, L.-P. Wang, Lattice Boltzmann simulation of turbulent flow laden with finite-size particles, in *Proceedings of the 7th international conference on multiphase flows ICMF*, 2010
324. F. Lucci, A. Ferrante, S. Elghobashi, Modulation of isotropic turbulence by particles of Taylor length-scale size. J. Fluid Mech. **650**, 5–55 (2010)
325. F. Lucci, A. Ferrante, S. Elghobashi, Is Stokes number an appropriate indicator for turbulence modulation by particles of Taylor-length-scale size? Phys. Fluids **23**(025101), 1–7 (2011)
326. T.W. Pan, D.D. Joseph, R. Bai, R. Glowinski, V. Sarin, Fluidization of 1204 spheres: simulation and experiment. J. Fluid Mech. **451**, 169–191 (2002)
327. J.C. Brändle De Motta, S. Vincent, J.-L. Estivalezes, E. Climent, Fictitious domain methods and penalty techniques for the simulation of turbulent particulate flows, in *ASME fluid summer meeting*, Montreal, August 4–7, 2010
328. A. Benkenida, J. Magnaudet, A method for the simulation of two-phase flows without interface reconstruction. C. R. Acad. Sci. Paris Série IIb **28**, 25–32 (2000)
329. R.M. Cabezón, D. García-Senz, A. Relaño, A one-parameter family of interpolating kernels for smoothed particle hydrodynamics studies. J. Comput. Phys. **227**, 8523–8540 (2008)

330. G.C. Truesdell, S.E. Elghobashi, On the two-way interaction between homogeneous turbulence and dispersed solid particles; Part 2: particle dispersion. Phys. Fluids **6**, 1405–1407 (1994)
331. C. Poelma, J. Westerweel, G. Ooms, Particle-fluid interactions in grid-generated turbulence. J. Fluid Mech. **589**, 315–351 (2007)
332. A. Lefebvre, *Numerical modeling of fluid/particle flows. Accounting of lubrication effects.* Ph.D. thesis, Paris-Sud. Faculté des Sciences d'Orsay, 2007
333. A.M. Ardekani, R.H. Rangel, Numerical investigation of particle-particle and particle-wall collisions in a viscous fluid. J. Fluid Mech. **596**, 437–466 (2008)
334. W.-P. Breugem, A combined soft-sphere collision for immersed boundary method for resolved simulations of particulate flows, in *Proceedings of the ASME 2010 3rd joint US-European fluids engineering summer meeting*, Montréal, 2010
335. B. Maury, A many-body lubrication model. Comptes Rendus de l'Académie des Sci. Ser. I Math. **325**, 1053–1058 (1997)
336. S.L. Dance, M.R. Maxey, Incorporation of lubrication effects into the force-coupling method for particulate two-phase flow. J. Comput. Phys. **189**(1), 212–238 (2003)
337. H. Brenner, The slow motion of a sphere through a viscous fluid towards a plane surface. Chem. Eng. Sci. **16**(3–4), 242–251 (1961)
338. S.E. Elghobashi, On predicting particle-laden turbulent flows. Appl. Sci. Res. **52**, 309–329 (1994)
339. S. Laín, M. Sommerfeld, J. Kussin, Experimental studies and modelling of four-way coupling in particle-laden horizontal channel flow. Int. Heat Fluid Flow **23**, 647–656 (2002)
340. D. Legendre, C. Daniel, P. Guiraud, Experimental study of a drop bouncing on a wall in a liquid. Phys. Fluids **17**, 097105–13 (2005)
341. B.J. Alder, T.E. Wainwright, Studies in molecular dynamics. I. General method. J. Chem. Physics **31**, 459 (1959)
342. B.J. Alder, T.E. Wainwright, Studies in molecular dynamics. II. Behavior of a small number of elastic spheres. J. Chem. Phys. **33**, 1439 (1960)
343. L. Verlet, Computer "experiments" on classical fluids. I. Thermodynamical properties of Lennard-Jones molecules. Phys. Rev. **159**(1), 98 (1967)
344. X. Fang, J. Tang, H. Luo, Granular damping analysis using an improved discrete element approach. J. Sound Vibration **308**(1–2), 112–131 (2007)
345. A. Massol, *Simulations numériques d'écoulements à travers des réseaux fixes de spheres monodisperses et bidisperses, pour des nombres de Reynolds modérés*. Ph.D. thesis, Université de Toulouse Paul Sabatier, 2004
346. S. Tenneti, R. Garg, S. Subramaniam, Drag law for monodisperse gas-solid systems using particle-resolved direct numerical simulation of flow past fixed assemblies of spheres. Int. J. Multiphase Flow **37**, 1072–1092 (2011)
347. R. Beetstra, M.A. van der Hoef, J.A.M. Kuipers, Drag force of intermediate Reynolds number flow past mono- and bidisperse arrays of spheres. AIChE J. **53**, 489–501 (2007)
348. R.J. Hill, D.L. Koch, A.J.C. Ladd, Moderate-Reynolds-number flows in ordered and random arrays of spheres. J. Fluid Mech. **448** (2001)
349. L. Schiller, A.Z. Naumann, Über die grundlegenden Berechnungen bei der Schwerkraftaufbereitung. Ver. Deut. Ing. **77**, 318–320 (1933)
350. A. Gobin, H. Neau, O. Simonin, J.R. Llinas, V. Reiling, J.-L. Sélo, Fluid dynamic numerical simulation of a gas phase polymerization reactor. Int. J. Numer. Meth. Fluids **43**, 1199–1220 (2003)
351. C.-Y. Wen, Y.H. Y, Mechanics of fluidization. Chem. Eng. Prog. Symp. Ser. 100–111 (1966)
352. S. Ergun, Fluid flow through packed columns. Chem. Eng. Prog. 89–94 (1952)
353. T.N. Randrianarivelo, *Numerical study of fluid/solid hydrodynamic interactions: applications to fluidized beds*. Ph.D. thesis, Université Bordeaux 1, 2005
354. C.S. Campbell, C.E. Brennen, Computer simulation of granular shear flows. J. Fluid Mech. **151**, 167–188 (1985)

355. M. Mehrabadi, S. Tenneti, S. Subramaniam, Importance of the fluid-particle drag model in predicting segregation in bidisperse gas-solid flow. Int. J. Multiphase Flow **86**, 99–114 (2016)
356. A. Aguilar Corona, *Agitation of particles in a liquid fluidized bed. Experimental study*. Ph.D. thesis, Toulouse University, 2008
357. J.F. Richardson, W.N. Zaki, The sedimentation of a suspension of uniform spheres under conditions of viscous flow. Chem. Eng. Sci. **3**, 65–73 (1954)
358. R. Di Felice, R. Kehlenbeck, Sedimentation velocity of solids in finite size vessels. Chem. Eng. Technol. **23**, 1123–1126 (2000)
359. C. Rosales, C. Meneveau, Linear forcing in numerical simulations of isotropic turbulence: Physical space implementations and convergence properties. Phys. Fluids **17**, 095106 (2005)
360. N.N. Mansour, A.A. Wray, Decay of isotropic turbulence at low reynolds number. Phys. Fluids **6**, 808 (1994)
361. F. Lucci, A. Ferrante, S. Elghobashi, Modulation of isotropic turbulence by particles of taylor length-scale size. J. Fluid Mech. **650**, 5–55 (2010)
362. C.M. Tchen, *Mean value and correlation problems connected with the motion of small particles suspended in a turbulent fluid*. Ph.D. thesis, University of Delft, 1947
363. J.O. Hinze, *Turbulence* (McGraw-Hill, 1975)
364. E. Deutsch, *Dispersion de particules dans une turbulence homogène isotrope stationnaire calculée par simulation numérique directe des grandes échelles*. Ph.D. thesis, Ecole Centrale de Lyon, 1992
365. J. Lavieville, *Simulations numériques et modélisation des interactions entre l'entraînement par la turbulence et les collisions interparticulaires en écoulements gaz-solide*. Ph.D. thesis, Université de Rouen, 1997
366. P. Fede, *Modélisation et simulation de l'influence de la turbulence sur les collisions dans les écoulements mono- et bi-solides*. Ph.D. thesis, Institut National Polytechnique de Toulouse, 2004
367. G.A. Voth, A. La Porta, A.M. Crawford, J. Alexander, E. Bodenschatz, Measurement of particle accelerations in fully developed turbulence. J. Fluid Mech. **469** (2002)
368. N.M. Qureshi, U. Arrieta, C. Baudet, A. Cartellier, Y. Gagne, M. Bourgoin, Acceleration statistics of inertial particles in turbulent flow. Euro. Phys. J. B **66**, 531–536 (2008)
369. E. Calzavarini, R. Volk, M. Bourgoin, E. LéVêQue, J.-F. Pinton, F. Toschi, Acceleration statistics of finite-sized particles in turbulent flow: the role of faxén forces. J. Fluid Mech. **630**, 179 (2009)
370. J. Bec, L. Biferale, G. Boffetta, A. Celani, M. Cencini, A. Lanotte, S. Musacchio, F. Toschi, Acceleration statistics of heavy particles in turbulence. J. Fluid Mech. **550**, 349 (2006)
371. P.G. Saffman, J.S. Turner, On the collision of drops in turbulent clouds. J. Fluid Mech. **1**, 16–30 (1956)
372. J. Abrahamson, Collision rates of small particles in a vigorously turbulent fluid. Chem. Eng. Sci. **30**, 1371–1379 (1975)
373. J.C. Brändlde de Motta. *Simulation of turbulent flows with finite size particles in dense regime*. Ph.D. thesis, Universitée de Toulouse, 2013
374. A.T. Cate, J. Derksen, L.M. Portela, H.E.A. Van Den Akker, Fully resolved simulations of colliding monodisperse spheres in forced isotropic turbulence. J. Fluid Mech. **519**, 233–271 (2004)
375. B.E. Stapper, W.A. Sowa, G.S. Samuelsen, An experimental study of the effects of liquid properties on the breakup of a two-dimensional liquid sheet. J. Eng. Gas Turbines Power **114**, 32–39 (1992)
376. B.E. Stapper, G.S. Samuelsen, An experimental study of the breakup of a two dimensional liquid sheet in the presence of co-flow air shear, in *AIAA paper 89-0461*, 1990
377. A. Mansour, N. Chigier, Dynamic behavior of liquid sheets. Phys. Fluids A Fluid Dyn. **3**, 2971–2980 (1991)
378. A. Mansour, N. Chigier, Disintegration of liquid sheets. Phys. Fluids A Fluid Dyn. **2**, 706–719 (1990)

379. V. Gutierrez Fernandez, *Experimental study of a liquid sheet disintegration in a high pressure environment*. Ph.D. thesis, ISAE, 2009
380. F. Jaegle, J.-M. Senoner, M. Garcia, F. Bismes, R. Lecourt, B. Cuenot, T. Poinsot, Eulerian and lagrangian spray simulations of an aeronautical multipoint injector, in *Proceedings of the Combustion Institute*, 2011
381. J.M. Senoner, M. Sanjose, T. Lederlin, F. Jaegle, M. Garcia, E. Riber, B. Cuenot, L.Y.M. Gicquel, H. Pitsch, T. Poinsot, Eulerian and Lagrangian large-eddy simulations of an evaporating two-phase flow. Comptes Rendus Mecanique, 2009
382. O. Desjardins, V. Moureau, E. Knudsen, M. Herrmann, H. Pitsch, Conservative level set/ghost fluid method for simulating primary atomization, in *20th annual conference on liquid atomization and spray systems*, vol. 238 (ILASS Americas, 2007), pp. 8395–8416
383. K. Luo, C. Shao, Y. Yang, J. Fan, A mass conserving level set method for detailed numerical simulation of liquid atomization. J. Comput. Phys. **298**, 495–519 (2015)
384. M. Gorokhovski, M. Herrmann, Modeling primary atomization. Ann. Rev. Fluid Mech. **40**, 343–366 (2008)
385. M. Herrmann, A parallel eulerian interface tracking/Lagrangian point particle multi-scale coupling procedure. J. Comput. Phys. **229**, 745–759 (2010)
386. G. Tomar, D. Fuster, S. Zaleski, S. Popinet, Multiscale simulations of primary atomization using gerris. Comput. Fluids **39**(10), 1864–1874 (2010)
387. M. Herrmann, A balanced force refinement level sel grid method for two-phase flows on unstructured flow solver grid. J. Comput. Phys. **227**, 2674–2706 (2008)
388. D. Fuster, A. Bague, T. Boeck, L. LeMoyne, A. Leboissetier, S. Popinet, P. Ray, R. Scardovelli, S. Zaleski, Simulation of primary atomization with an octree adaptive mesh refinement and VOF method. Int. J. Multiphase Flow **35**, 550–565 (2009)
389. S. Popinet, S. Zaleski, Gerris: a tree- based adaptive solver for the incompressible euler equations in complex geometries. J. Comput. Phys. **190**(2), 572–600 (2003)
390. G.S.H. Pau, J.B. Bell, A.S. Almgren, K.M. Fagnan, M.J. Lijewski, An adaptive mesh refinement algorithm for compressible two-phase flow in porous media. Comput. Geosci. **16**(3), 577–592 (2012)
391. M. Herrmann, Detailed numerical simulations of the primary atomization of a turbulent liquid jet in crossflow. J. Eng. Gas Turbines Power **132**(6), 451–466 (2010)
392. X. Li, M. Arienti, M. Soteriou, M. Sussman, Towards an efficient, high-fidelity methodology for liquid jet atomization computations, in *AIAA, 2010-210, 48th AIAA aerospace sciences meeting*, 2010
393. M. Arienti, X. Li, C. Soteriou, C.A. Eckett, M. Sussman, R.J. Jensen, Coupled level-set/volume-of-fluid method for simualtaion of injector atomization. J. Propulsion Power **29**, 147–157 (2013)
394. S. Apte, M. Gorokhovski, P. Moin, Les of atomizing spray with stochastic modeling of secondary break-up. Int. J. Multiphase Flow **29**, 1503–1522 (2003)
395. M.A. Gorokhovski, V.L. Saveliev, Analyses of kolmogorov's model of breakup and its application into lagrangian computation of liquid sprays under air-blast atomization. J. Phys. Fluids **15**(1), 184–192 (2003)
396. J. López, J. Hernàndez, Analytical and geometrical tools for 3D volume of fluid methods in general grids. J. Comput. Phys. **227**(12), 5939–5948 (2008)
397. D.S. Balsara, Divergence-free adaptive mesh refinement for magnetohydrodynamics. J. Comput. Phys. **174**(2), 614–648 (2001)
398. K. Olson, P. Mac Neice, An over of the paramesh AMR software and some of its applications, in *Adaptive mesh refinement-theory and applications, proceedings of the Chicago workshop on adaptive mesh refinement methods, series: lecture notes in computational science and engineering*, eds. by T. Plewa, T. Linde, G. Weirs (Springer, Berlin, 2005), p. 41
399. Y. Ling, S. Zaleski, R. Scardovelli, Multiscale simulation of atomization with small droplets represented by a lagrangian point-particle model. Int. J. Multiphase Flow **76**, 122–143 (2015)
400. I. Franckel, D. Weihs, Shape and stability of a liquid drop moving at low weber numbers. Appl. Sci. Res. 279–294 (1983)

401. B. Déjean, P. Berthoumieu, P. Gajan, *Experimental study on the influence of the liquid and air thicknesses on a planar air-blasted liquid sheet, in 25th European conference—liquid atomization & spray systems–ILASS 2013* (CHANIA, Greece, Sep 2013)
402. A. Lozano, F. Barreras, G. Hauke, C. Dopazo, Longitudinal instabilities in a air-blasted liquid sheet. J. Fluid Mech. **437**, 143–173 (2001)
403. G. Blanchard, *Modélisation et simulation multiéchelle de l'atomisation d'une nappe liquide cisaillée.* Ph.D. thesis, ISAE, 2014
404. D. Zuzio, J-L. Estivalèzes, A parallel adaptive projection method for incompressible two phase flows, in *ICCFD 2010 refereed conference proceedings*. ISSN: 2070-3740, ISSN: 2070-3724, 2010